Lunar Exploration
Human Pioneers and Robotic Surveyors

Springer
London
Berlin
Heidelberg
New York
Hong Kong
Milan
Paris
Tokyo

Paolo Ulivi with David M. Harland

Lunar Exploration

Human Pioneers and Robotic Surveyors

Springer

Published in association with
Praxis Publishing
Chichester, UK

Paolo Ulivi
Cernusco Sul Naviglio
Italy

David M. Harland
Space Historian
Kelvinbridge
Glasgow
UK

SPRINGER–PRAXIS BOOKS IN ASTRONOMY AND SPACE SCIENCES
SUBJECT *ADVISORY EDITOR*: John Mason B.Sc., M.Sc., Ph.D.

ISBN 1-85233-746-X Springer-Verlag Berlin Heidelberg New York

Springer-Verlag is a part of Springer Science + Business Media (*springeronline.com*)

British Library Cataloguing-in-Publication Data
Ulivi, Paolo
   Lunar exploration: human pioneers and robotic surveyors. –
   (Springer-Praxis books in astronomy and space sciences)
   1. Space flight to the moon – History 2. Space race – History
   3. Moon – Exploration – History
   I. Title II. Harland, David M. (David Michael), 1955–
   629.4'353'09

   ISBN 1-85233-746-X

Library of Congress Cataloging-in-Publication Data
A catalogue record for this book is available from the Library of Congress

Project Copy Editor: Bob Marriott
Cover design: Jim Wilkie
Typesetting: BookEns Ltd, Royston, Herts., UK

Printed in the United States of America on acid-free paper

This book is dedicated to all of my relatives;
to my friends Alessia, Andrea, Cristina, Emanuele, Filippo,
Francesco, Giorgio, Giulio, Luigi and Silvia;
and to the memory of Fabio.

Table of contents

Foreword

For millennia the Moon, shining down upon us from above, has served as a lure. Initially, we merely stared at it in wonderment. In 1609 Galileo Galilei turned his first telescope onto it and was astounded to see a rugged terrain. After centuries of peering at it through ever more advanced telescopes, we sent robotic probes – both to map it from lunar orbit and to land on its surface. Then, at long last, we went to explore it in person! Remarkably, this modern assault on the Moon occurred within a period of a decade... and I was fortunate enough to witness it.

Having been born in 1955 I was just old enough to be enthralled a decade later when the first probes – initially a Soviet probe, and a few months later an American one – landed on the lunar surface. Meanwhile, NASA astronauts were flying the Gemini missions in low Earth orbit, rehearsing techniques that would be required for a lunar mission. I was amazed by the picture taken by a Lunar Orbiter peering obliquely over the rim of the crater Copernicus. The newspapers called it the 'Picture of the Century'. Instead of from the astronomer's perspective, looking down from above and from a great distance, it depicted this vast crater, with its terraced walls and complex central peak, from the point of view of a pilot in lunar orbit.

Patrick Moore's television programme *The Sky at Night* – then approaching its tenth anniversary – broadened my horizons. I promptly decided that when I grew up I would be an astronomer. For some reason, most of my friends wanted to become footballers, bank managers or engine mechanics. In 1969 I sat up through the night to watch Neil Armstrong and Buzz Aldrin walk on the Sea of Tranquillity. The next big event for me was Apollo 15, whose mobile camera let us follow Dave Scott and Jim Irwin as they explored a valley in the Apennine Mountains. I think that I was the only child in my school to stay home to watch these moonwalks. I was so happy that, when asked afterwards by one of the teachers if I had been ill, I was truthful: no, I was watching television! I got away with it because I was the 'weird kid' who wanted to be an astronomer. I was the first in the school to sign up for astronomy O-level. Although no-one was teaching this subject, I had the course book written by Patrick Moore, so I knew I could study it on my own; I passed.

As the final Apollo crew returned home in December 1972, the President of the United States, Richard M. Nixon, alarmingly prophesied that it would probably be the last mission for a generation. And it happened that this was a self-fulfilling prophesy, because Nixon had previously instructed NASA to destroy the tooling used for building the mighty Saturn V rocket! By the time that I set off to university to study astronomy, the Skylab project was underway. Sadly, by the time I graduated, the Apollo flights had faded into myth, and some were claiming that they had been faked! Although I expected that the human exploration of the Moon would resume in the 1980s, this was not to be. For a few years after Apollo, the Soviets continued to send automated probes, with limited success – and then they stopped. NASA resisted calls from scientists to mount further robotic missions using a new generation of sensors. In the early 1990s, however, the Department of Defense allowed its Clementine spacecraft to be placed into lunar orbit to undertake multispectral mapping, which it carried out flawlessly, prior to setting off on its primary mission during which, ironically, it was lost. Later in the decade, NASA followed this mission with Lunar Prospector, which verified Clementine's hint of there being water in permanently shadowed craters near the poles. Despite this important discovery – which, if it had been made by an Apollo crew during one of the planned, but cancelled, extended orbital surveys may well have prompted the expansion of that programme – the presence of water, in what one eminent lunar scientist has referred to as 'the most valuable piece of extraterrestrial real estate in the Solar System', seems to have stirred up a wave of indifference. Although Space Shuttle crews are currently assembling an international space station in low orbit, there is no prospect of the Shuttle assembling a vehicle capable of flying to the Moon.

An enormous number of books have been written about the missions to the Moon during the Apollo era. Most have focused on the derring-do of the astronauts, many have described the technology required for reaching the Moon, and some have discussed the politics of the programme. Others, however, considered the science (in my own case I explained how the astronauts served as lunar field geologists). Over the years, many of the documents about the American robotic programmes – relating their development in addition to their achievements – have become so obscure that they are effectively restricted to historians. Ironically, the details of the initially much more secretive Soviet programmes have been released only recently, but again these are not widely available to members of the public. Of course, nowadays, with the Internet, the problem is to sort the wheat from the chaff! I therefore heartily welcome Paolo Ulivi's book, which diligently presents much of this obscure information in a single convenient volume.

David M. Harland
December 2003

Author's preface

Almost forty-five years have passed since that day in September 1959 when a Soviet spacecraft, after a flight of 400,000 km over two days, announced to the whole of mankind that the Moon was ours, and that it would forever remain our territory. Only ten years later, in July 1969, two men walked on its arid and lifeless surface.

This book presents a detailed account of the dozens of spacecraft – American, Soviet, and other nationalities that have reached our natural satellite since the 1950s in order to study it – that have taken symbolic possession of it, even if only to reach more distant and ambitious targets in our Solar System.

In addition to describing the spacecraft and their missions, this book presents the principal scientific results of each mission, and an insight into the politics and management of the various space programmes. Wherever possible, the approach is technical and scientific rather than anecdotal. (The human side of lunar exploration is discussed in other books, and the reader is referred to A. Chaikin's *A Man on the Moon*, listed in the Bibliography.)

This approach has been maintained in the choice of images – in particular, in the third chapter. Instead of a selection of 'astronauts and flags', pictures of instruments, equipment and geological formations were chosen. I have also borrowed several images from NASA's *Exploring Space With a Camera* (published during the 1960s), which provides a colourful pictorial history of the first decade of space exploration.

Although I have made every effort to secure permissions to use contemporary drawings and photographs, in a few cases I have been unable to trace the copyright holders. Nevertheless, because they are so important in illustrating the story, I decided to use them and to assign as complete a credit as possible.

My hope is that, besides finding a place on the bookshelves of space enthusiasts, this book will prompt further serious research that may result in additional books, theses and papers. I also hope (with the best intentions) that it will soon become outdated due to a resurgence in lunar studies.

Paolo Ulivi
December, 2003

Acknowledgements

I would like to thank all of the following people for their help in the writing of this book. My brother Federico for his computer help, and my father, Carlo, for the translations from Russian and for access to his collection of magazines on the early lunar missions of the Apollo programme.

Professor Amalia Ercoli Finzi, my graduation thesis adviser, and Michelle for her help during my studies, when I began to collect much of the material used in this book; Gian Guido, for teaching me some of the secrets of the art of information retrieval; and the staff of the aeronautical engineering department library of the Milan Politecnico, the central Italian library of Florence, and the author's rights section of UNESCO.

James Garry, for the overwhelming quantity of material on Soviet rovers which he provided; Philippe Jung and Jean-Jacques Serra for the details of the French Lunokhod laser reflectors; Peter Pichler and Roman Svihorik for their translations from Czech; Dennis Laurie for the information on the TrailBlazer probe; Heather Franz, for Wind orbital figures; and Raffaele Bonacchi, Martin Brown, Michael L. Ciancone, Phillip Clark, Luca Coren, Paolo D'Angelo, Michael K. Heney, Andrew Lawler, Danilo Marangio, Dmitry Pieson, and everyone who provided suggestions, criticism and corrections.

Paul Blase, Richard Blomquist, Bonnie Cooper, Dwayne A. Day, Eric M. Galimov, David Gump, Thomas Gold, Brian Harvey, Sergei Matrossov, Takahashi Nakajima, Stewart Nozette, Patrick Roger, Dave Schrunk, Vladislav Shevchenko, Nikolay Vvedensky, Graham Woan, Dave R. Woods, and everyone who supplied images and permissions for their use.

Lastly, I would like to thank the referees who evaluated this book for Springer–Praxis, David M. Harland for his help in fields as diverse as refereeing, image acquisition and copyright, Bob Marriott, who had the titanic task of turning my wobbly English and Italian sentence structure into something readable, and the Chairman of Praxis, Clive Horwood, for his enthusiasm for this project, and for his boundless patience.

List of illustrations

1

Pioneers and first exploits

1.1 THE BEGINNING OF THE SPACE RACE

It was the summer of 1957. The fame of Elvis Presley was rising rapidly, the shooting of the 'greatest movie of all time', *Ben Hur*, was close to completion, and a new, terrible weapon debuted on the world stage when the Soviet 8K71 InterContinental Ballistic Missile (ICBM) made its first successful flight from a secret base in Kazakhstan. Behind this Soviet feat was a group of engineers of several disciplines, led by the Chief Designer, Sergei Pavlovich Korolyov.

Korolyov was born in 1907 in Jitovir, Ukraine. From his boyhood he was an aviation enthusiast, and he designed and built a glider when he was eighteen. After being refused by the military aviation school, he graduated at Moscow Bauman University in the early 1930s. A few years later, after designing and building some more gliders and airplanes, he became interested in the infant discipline of rocketry, and joined the Moscow section of the paramilitary group GIRD (Gruppa na Isucheniyu Reaktivnogo Dvisheniya – Group for the Study of Rocket Propulsion), of which he became one of the leading personalities. Here his colleagues included the pioneer Fridrikh Arturovich Tsander, who had previously built the OR-1 (the first liquid-fuel Soviet rocket engine, of just 1.4 N thrust), and Valentin Petrovich Glushko, of the Leningrad section, who later also became a leading rocket designer and builder.

The scope of the GIRD was to build small experimental rockets in pursuit of the dream of one day launching a satellite around our planet, following the path indicated by Konstantin Tsiolkovsky, the Soviet 'Father of Astronautics'. The first rocket built by the Soviet group flew, with little success, on 17 May 1933; but later devices, such as the beautiful GIRD-X, which was 220 cm long and had a 0.68-kN-thrust engine, flew at up to 3,000 km/h and reached a height of a little more than 1 km.

In 1937 the promising future of Soviet rocketry began to be jeopardised when Joseph Stalin ordered the arrest and execution of Marshal Mikhail Tukhachevskii, under whose protection the rocketeers had been working. Some months later Glushko was put under house arrest, and in June 1938 Korolyov (probably denounced by Glushko himself) was also arrested during the night. Sentenced to forced labour, he was first sent to the Magadan gulag, then to the infamous mines of Kolyma, and finally, upon being rescued by Andrei Nikolaevich Tupolev (a famed aeronautical engineer who was also under arrest), was transferred to an aeronautical design bureau consisting entirely of convicts. Korolyov regained his freedom in 1944, when the allied armies began collecting information about the fearsome German V2 (Vergeltungswaffe – 'revenge weapon') missile.

Other Soviet rocketeers fared much better during the Second World War. In particular, Glushko supervised the flights of the first Soviet rocket-propelled airplane – a modification of a Korolyov glider called RP-318 (Raketnyi Planer, rocket powered glider) – and built several rocket engines, including the RD-1 (Reaktivny Dvigatel, rocket engine) and the RD-1KhZ of 2.9 kN thrust that was tested on some mixed propeller- and rocket-powered fighters built by the design bureaux of Yakovlev, Sukhoi and Lavochkin.

Fuelled by kerosene and nitric acid, Glushko's water-cooled rocket engines never entered production because the acid fumes could damage the wooden structure of the airplanes, and the self-igniting hypergolic fuels caused several accidents during which the engine blew up in flight.

The Soviet Union came second in the race to the V2 secrets, succeeding only in occupying the testing site at Peenemünde and the Nordhausen concentration camp where the V2s were built by Hitler's slaves, and in capturing some 'second-class' engineers. The leading engineers fled to the West. Some months after the end of the war in Europe, both Glushko and Korolyov participated as observers in Operation Backfire, during which the British fired some captured V2s.

The first Soviet V2 (known as R-1 or 8A11 to the Soviets, and as SS-1a (Scunner) in the West) – assembled from original components by captured German engineers under the direction of Korolyov – flew in October 1947, and was followed by two medium-range missiles: the infamous R-11 or 8K11 (known in the West as the SS-1b, Scud) and the 8K51 (SS-3, Shyster).

For the next extremely ambitious project, Korolyov's design bureau teamed with the research division of the Defence Ministry and with the mathematics department of the Academy of Sciences. It was decided to build a gargantuan missile to carry an atomic warhead over a range of 12,000 km. As in all of the previous projects, Glushko's team worked on the rocket engines.

Several years later, during the spring of 1957, the first 8K71 missile was ready for launch from Tyuratam, in Kazakhstan.* The first two tests on 15 May and 12 July failed, but the third test, on 21 August, was almost successful, the dummy warhead disintegrating in sight of its target in Kamchatka. The existence of this new terrible weapon had become known to the CIA just a few days earlier, as on 5 August an 'invulnerable' U-2 spy-plane had taken the first pictures of the launch pad. The 8K71

---

*   The Tyuratam missile range is better known as the Baikonur Cosmodrome, after a small town several
    hundred kilometres away.

– later renamed, during its very brief military service, R-7 (Semiorka – 'Little Seven') by the Soviets, and SS-6 (Sapwood) by Western intelligence – was quite different from the contemporary American missiles. It was much heavier, not only because of its very long range, but also because the Soviet nuclear weapons of the time were very heavy and cumbersome.

The missile, which implemented much of the design work of German engineers under Hermann Gröttrup for the unbuilt G-4 (3,000-kg warhead, 3,000-km range) and G-5 (a cluster of five G-4) missiles, was built around a core stage using kerosene and liquid oxygen, to which were attached four booster stages of conical shape and using slightly different engines. The total height was more than 34 metres, and the first stages were powered by no less than twenty primary combustion chambers, augmented by a dozen small vernier engines. Korolyov, of course, was well aware of the potential of the 8K71 as a satellite launcher, and as early as 29 August 1955 he had sent a memorandum to the Supreme Soviet suggesting the use of the missile to place a satellite in orbit around the Earth and to send probes to the Moon.

Korolyov's plan was approved on 30 January 1956 and a small number of engineers and scientists were put to work on a large scientific satellite, designated Object D, being the fifth payload designed for the 8K71. (Objects A, B, V and G were different versions of the nuclear warhead.) By the end of 1956, however, the development of the satellite was running behind the development of the missile, and it was decided to begin the parallel development of the much simpler PS (Prosteishii Sputnik, simplified satellite) programme.

Thus after a second successful test of the ICBM on 7 September 1957, the time came for the PS-1 – a small metal sphere of 83 kg and 56 cm diameter, equipped with two transmitters and four whip antennae. To carry the satellite into space, the launcher was slightly modified – in particular in the event-timing and in the systems of guidance, engine feeding and tank pressurisation. PS-1 reached Earth orbit with the name of Sputnik (Satellite) late in the evening on Friday, 4 October 1957. But this was not enough. To properly celebrate the fortieth anniversary of the Russian Revolution, Nikita Khrushchev asked Korolyov for a new exploit, and on 3 November 1957, Sputnik 2 was launched: it was a 510-kg satellite carrying a dog, Laika.

In the United States, the first government-funded studies on the establishment of an artificial satellite were undertaken in 1945, and the RAND (Research and Development) Corporation report was published the following year. Nothing happened, however, until 1954, when a secret RAND report was published detailing the advantages of a military space-based reconnaissance system. The project, called WS-117L (Weapon System), was immediately approved, and led to several military satellite programmes, of which the most famous products are the CORONA spy satellites. It was immediately recognised that the establishment of a spy satellite network could clash with a major point of international law: the right of one country to overfly another country above the atmosphere. For this reason, in parallel with project WS-117L a scientific satellite project was begun. The scientific satellite was to fly first, as the US contribution to the International Geophysical Year (July 1957 to December 1958), and its real purpose was to provide a basis for space law so that the

Soviet Union could not complain when the US began to send spy satellites overhead.

Three projects were presented, from which the winner, destined to be the first US artificial satellite to fly (or, as was then thought, the world's first) would be drawn. The *first* project, sponsored by the Army, was simply called Orbiter, and reused as much existing technology as possible, including the medium-range Redstone missile, designed by Wernher von Braun's German team. The *second* project was called Vanguard, and was sponsored by the Naval Research Laboratory. Vanguard, too, was to reuse existing technology, for its launcher would be based on the Viking sounding rocket. A *third* project, sponsored by the Air Force, would use the Atlas ICBM, which was then under development.

On 3 August 1955, the choice was made: the first US satellite would be Vanguard. The primary reason for this choice was political, as Vanguard was mainly a civilian project, reusing civilian rocket technology and unlikely to impede military missile development. A year later the Army tried to overthrow that decision, but to no avail. As a result, on 20 September 1956 the first Jupiter C – a refinement of the Redstone that was also to be used for project Orbiter – was launched. The missile reached a range of 5,390 km and a height of 1,097 km. Had not the Department of Defense, fearing the launch of a clandestine Army satellite, ordered that the fourth stage be replaced with an inert mock-up, the US could have beaten the Soviets into orbit by a whole year!

Instead, the news of the launch of Sputnik left the whole world astonished. While the United States was preparing to launch the tiny 1.8-kg Vanguard satellite, the Soviets had launched, without any apparent problem, an 83-kg satellite. (Someone even considered this to be a typing error: the Soviets surely meant 8.3 kg!) The Soviet Union had to be grateful for the ability of both Korolyov and Glushko, and, paradoxically, had benefitted from the fact that its heavy nuclear weapons required such a powerful missile.

The US Army reiterated its case, stating that it was ready to use one of the remaining Jupiter C rockets to launch a satellite within three months – and this time it received approval to take the launcher out of storage and prepare it in the event of the failure of Vanguard. On 6 December the first Vanguard missile fell back on the pad at Cape Canaveral after rising just a few metres. The launch was derided in the Press as a 'flopnik', or 'kaputnik'! On 31 January 1958 (1 February, GMT) the Army fared much better and, using the Jupiter C, now renamed Juno 1, succeeded in carrying into orbit the small 15-kg Explorer 1 satellite, which provided physicist James Van Allen with the first great scientific discovery of the space age: the innermost of several radiation belts encircling our planet. Explorer 1 was also the first satellite equipped with a simple stabilisation system, as it was designed to maintain its attitude by spinning on its longitudinal axis. This experiment proved unsuccessful, as the damping effect of the whip antennae was neglected – so much so in fact that from the first orbit the satellite spun around an axis perpendicular to the expected axis, thus introducing an unforeseen Doppler effect on its telemetric radio signals.

The first stage of the Juno I was a stretched Redstone – a medium-range missile, based on the V2, and designed by von Braun's team – and the upper stages were clusters of solid-fuelled 'Baby Sergeant' experimental rockets, each of which was 120 cm long, with a diameter of 15 cm, a mass of 26 kg, and a burning time of just six seconds! The satellite itself was integrated into another such rocket, which sat atop the cluster and provided the final impetus to achieve orbit.

On 5 February came the second failure of Vanguard. On 17 March, however, Vanguard 1 at last reached orbit ... and there it remains – the oldest artificial object still in orbit.* In the following weeks, Explorer 2 failed to reach orbit, but Explorer 3 made it, and the Soviets made two attempts to launch the huge Object D. The first time, the launcher exploded 88 seconds into the flight, and the launch of the replacement, later renamed Sputnik 3, had to be hurried in order to support the Italian Communist Party in the general election of 25 May 1958. Unfortunately, because of this urgency the on-board tape recorder could not be tested properly, and it failed, impeding the Soviets' first attempt to study the Van Allen radiation belts.

By this time, both the Soviets and the Americans considered themselves ready to aim for the Moon.

1.2 THE PIONEERS

There has been a constant interest and fascination in the Moon throughout the history of mankind. The first example of a crude depiction of lunar maria, dating from the third millennium BC, can be seen in the Irish neolithic temple of Knowth, while the Chinese *Huai Nan Zi* fairy tale of the fourth/third century BC is one of the first examples of a description of a trip to the Moon.

The Greeks made the first known lunar studies. Anaxagoras – a philosopher of the fifth century BC – was one of the first to state that the Moon's nature is 'terrestrial'; Aristarchus of Samos – better known for being the first to theorise heliocentrism – wrote *On the Sizes and Distances of the Sun and Moon*; and Plutarch wrote *On the Face in the Lunar Disk*. From Greek literature also come descriptions of fantastic travels to the Moon. The first known example of such a voyage is in Antonius Diogenes' *Wonders beyond Thule*, written in the first century, and of which we possess only a short abstract. In contrast, we possess both works by Lucianus of Samosata, describing such a trip: *Icaromenippus* and *True History*, the latter being a simple parody of Antonius Diogenes' work. However, one of the most intriguing archeological finds from ancient Greece is the Antikythera mechanism – a complex clockwork device dating from the first century BC designed to simulate the lunar phases.

During the Middle Age the nature of the dark spots in the lunar disk was discussed by Ibn-Ahmad Ibn-Rushid (Averroës) and by Albertus Magnus, but most of all by Dante Alighieri in the *Convivio* and in some length in the *Comedy* (*Paradise, II*).

Interest in our natural satellite was revived during the Renaissance, when Dutch

---

* Its orbital lifetime is estimated to be about 1,000 years, during which time the weak atmospheric braking will cause its orbit to decay.

painters – in particular, Jan Van Eyck – painted the first accurate depictions of the lunar surface, when Ludovico Ariosto described a trip to the Moon in *Roland Enraged*, and when Leonardo da Vinci sketched the first naked-eye map of the Moon and correctly explained the phenomenon of 'earthshine'. Leonardo also indicated the possibility of using two lenses to magnify the spots on the face of the Moon; the same possibility was stated some years later by Gerolamo Fracastoro. However, the telescope was not 'officially' invented until 1608.

The first telescopic observations of the Moon were carried out by Thomas Harriot in England during the summer of 1609, and by Galileo Galilei in Italy a few months later. These observations renewed both the interest of novelists – including Johannes Kepler, Francis Goodwin, Cyrano de Bergerac and Rudolf Erich Raspe – and the curiosity of scientists, and a lunar cartographic system was soon put into place. After the early attempts of Flemish astronomer Michel Florent van Langren (better known by the Latinised form of his name, Langrenus) in 1645 and of Polish astronomer Johannes Hevelius in 1647, the Italian Jesuit priests Giovanni Battista Riccioli and Francesco Grimaldi published a lunar map in 1651, the nomenclature of which is mostly still in use. Telescopic study came of age in 1837, when Wilhelm Beer and J.H. Mädler published their extremely accurate map. At the same time, there was a revival in fantastic literature and the descriptions of trips to the Moon began to acquire modern characteristics. In 1835 Edgar Allan Poe published *The Unparalleled Adventure of One Hans Pfaall*; in 1857 Ernesto Capocci, the Director of Naples Observatory, published his novel *On the First Voyage to the Moon Made by a Woman in the Year 2057*; in 1865 Jules Verne published *From the Earth to the Moon*, followed five years later by its sequel *Around the Moon*, describing the trip to the Moon by three men on board a shell launched by a 274-metre-long cannon; and in 1901, H.G. Wells published *The First Men in the Moon*.

The first examples of gunpowder-powered rockets date to the end of the first millennium AD, being later used by the Chinese against the Mongols, and by the Mongols against Japan and to conquer Baghdad, from where they came to Europe. Rockets had a brief period of glory during the European wars of the fourteenth and fifteenth centuries, before being replaced by more accurate firearms. After a long period of oblivion, rockets were rediscovered during the eighteenth century, at the time of the colonial expansion toward the East.

The first Western engineer to have his name connected to rocketry was the Englishman William Congreve, who developed kilometre-range rockets to be carried by infantry units. Congreve's work was later developed by another Englishman, William Hale, who discarded the long wooden guidestick used in previous rockets and introduced fins inside the nozzle to set the projectile into a rotating motion, which provided stabilisation. Congreve and Hale rockets – which were psychological rather than offensive weapons – were used by the armies of several countries.

At the beginning of the twentieth century, rocket science began the long process leading to its eventual maturation. The incendiary rockets of Le Prieur were mounted on French and British aircraft during the First World War, and were used against observation balloons.

Robert H. Goddard – a physics professor from Massachusetts – was the first man to conduct a serious scientific study into solid-fuelled rockets. In 1919, to justify some funds he was receiving, Goddard published the booklet *A Method of Reaching Extreme Altitudes* in which, besides describing the results of his experiments, he presented the results of the first scientific study on the possibility of reaching the Moon. Goddard investigated the possibility of reaching the Moon with a rocket that carried photographer's magnesium powder, in order to record the explosion made on impact. Instead of carrying out a simple theoretical study, in October 1916 Goddard set out an experiment to establish the minimum powder mass to be carried by the rocket. By observing at night, from his Worcester (Massachusetts) home, the magnesium flashes produced in air-evacuated glass ampules located some 3,600 metres away, he determined that the flash produced by the ignition of 0.0029 grammes of magnesium was barely visible, but that the flash produced by 0.015 grammes was plainly visible. From these data he computed that, using a 30-cm telescope to observe the impact, the rocket should carry 1.2 kg of magnesium for the flash to be barely visible and 6.27 kg to be plainly visible. Goddard estimated that to carry this mass to the Moon, it would be necessary to use a rocket of some 15 tonnes launch mass, able to accelerate the payload to escape speed. He also noted that 'the plan of sending a mass of flash powder to the surface of the Moon, although a matter of much general interest, is not of obvious scientific importance'. This was the first concept of an interplanetary mission without humans onboard the spacecraft: the first space probe.

The publication of Goddard's booklet caused quite a stir in the United States, and the front pages of some newspapers carried sarcastic comments about it. Goddard therefore continued his research in complete secrecy, and in 1926 launched the world's first liquid-fuel rocket, which made a flight of a few tens of metres. Three years later he received a visit from Charles Lindbergh – the first man to fly solo across the Atlantic Ocean – to whom he exposed his projects, hoping to receive new funding to build his Moon rocket. After asking for $1 million to reach to Moon, Goddard received from the Guggenheim foundation, on behalf of Lindbergh, $25,000, which was at that time a considerable amount of money. However, by avoiding confrontation with other scientists and by refusing to publish his results in technical papers, Goddard's place in the history of astronautics was destined to remain marginal; and by the time he died in 1945, other groups in Germany, in the Soviet Union and in the United States had achieved significantly better results.

Some copies of Goddard's booklet crossed the ocean to Europe, where his work was received with far less scepticism – particularly in Germany. Here it reached the hands of Hermann Oberth, who in 1923 published *The Rocket in Interplanetary Space*, in which he described for the first time a mission to circumnavigate the Moon and to take pictures of the otherwise perpetually hidden far-side.

During the 1920s, two important associations studying rocket propulsion were formed in Germany. The first, led by Friedrich Wilhelm Sander and Max Valier, and financed by automotive tycoon Fritz von Opel, led to the first flight of a rocket-powered aircraft: the Ente (Duck) glider which utilised two solid-fuelled rockets. Beside this futuristic type of research, the same association also took a look back in time, and designed a cannon capable of firing a 7-metre-long shell to the Moon. To improve the gun's performance, it was to be emplaced atop a 4,900-metre high mountain. Of the second association, VfR (Verein für Raumschiffahrt – Spaceflight Society) were part, and among the others were Wernher von Braun, celestial mechanics pioneer Walter Hohmann, who wrote a book entitled *The Attainability of Other Celestial Bodies*, and the popular science writer Willy Ley who, after fleeing to the United States upon Adolf Hitler's rise to power in 1933, further perfected Goddard's idea. To counter the problem of inclement weather preventing a terrestrial observer from seeing the momentary flash as a magnesium-laden spacecraft hit the Moon, Ley proposed the impact on the Moon of 0.5 kg of high explosive and 4.5 kg of white powder – possibly powdered glass – that once dispersed on the surface would form a patch more brilliant than the surroundings.

Besides building some rocket prototypes with little or no success, the members of VfR acted as scientific consultants on Fritz Lang's film *The Woman in the Moon*, for which they built a rocket. This film is usually remembered because Lang, in order to make the moment of launch even more dramatic, decided to have it preceded by the first 'countdown'.

After Hitler seized power, the contacts between VfR and the Wermacht (the army of Nazi Germany) led to the transferral of all rocket research to the secret base at Peenemünde on the coast of the Baltic Sea, where the rockets of the 'A' (Aggregat) series were perfected, leading to the infamous A4, *alias* V2. Although little is known of this type of research, it seems that the German researchers never abandoned the dream of building rockets for space and for lunar exploration. Thus, the A4's successor would be the huge A10, designed to reach the United States, to be followed by A11 and A12, which would be capable of placing up to 27 tonnes in Earth orbit.

Besides Germany, other countries carried out studies on rocket propulsion – in particular, France, due to Robert Esnault-Pelterie, and Italy, where in 1923 Luigi Gussalli published a book with the prophetic title *Is it Already Possible to Attempt a Trip from the Earth to the Moon?*, in which he proposed the use of a multi-staged rocket and analysed the logistical requirements of such a mission, and where Ettore Cattaneo built a rocket-powered glider following the German lesson. In Britain, where the law banned private rocket research, the British Interplanetary Society was formed in 1933, and in 1939 it published the first serious study of human lunar flight. This envisaged a 30-metre-tall rocket with a diameter of 6 metres and made of more than two thousand smaller solid-fuelled rockets. The lunar lander, carrying three astronauts, was bell-shaped, and included systems for course correction during the flight and for attitude control. The navigational system comprised an extremely advanced inertial platform and a clever optical system for star observation called Coelostat. Despite having some extremely modern aspects, the project showed the lack of physical knowledge of the day, and carried nothing for protection during re-entry. In 1947 this project was updated, this time envisaging a liquid-fuel launcher.

Meanwhile, in the United States, other groups had been carrying out research into rocket propulsion. One of the most important of these groups was the then-obscure laboratory that went on to build Explorer 1 and most of the American lunar and planetary spacecraft: the Jet Propulsion Laboratory. The laboratory was created at the end of the 1930s by a group of doctoral students under Professor Theodore von

Kármán of Caltech, and was led by its first director, Frank J. Malina, and by the Chinese born Tsien Hsue-shen (or, according to a more modern transliteration, Qian Xuesen).* During the Second World War, JPL perfected the first modern solid-fuelled rocket, which was used by heavy bombers for rocket-assisted take off, and after the war it continued to work for the Army, designing and building a series of liquid-fuel missiles that had military designations. The first engine was called Private, the next Corporal, and the last Sergeant. The most famous JPL product of that time was a small-scale prototype of Corporal called WAC Corporal.† In 1950 the Air Force's Eastern Test Range at Cape Canaveral on Florida's Atlantic coast was inaugurated by fitting WAC Corporals as second stages to V2s and firing them to altitudes of 400 km. While working on this historic project, some laboratory engineers computed (just for fun) that by using the full-scale Corporal missile and a cluster of anti-aircraft Loki solid-fuelled rockets, it would be possible to hurl to the Moon ... an empty beer can!

1.3   THE MILITARY AIM FOR THE MOON

At the end of the Second World War it was still unclear whether radio waves of a short enough wavelength could pass through the barrier of the ionosphere. To prove this, British scientist Arthur C. Clarke – in the famous paper, published in October 1945, in which he first proposed the concept of geostationary telecommunication satellites – suggested trying to send radio waves to the Moon in the hope of recording their echo after they were reflected by its surface.

The first to succeed was the US Army Signal Corps with Project Diana, using a war-surplus radar installation. The experiment was made even more complicated because the radar antenna could not be steered in elevation, and could thus send radio waves to the Moon only at moonrise and moonset. After several attempts, on 10 January 1946 a faint pulse reflected from the rising Moon was recorded 2.5 seconds after transmission. This experiment caused quite a stir, as for the first time, everyone realised that the moment of man's first physical contact with our natural satellite was not very far into the future. The second group to succeed was the Hungarian Tungsram company, which had begun working on the experiment as early as March 1944, but had had to restart work from scratch several times because of the war. Using a war-surplus transmitter/receiver unit and a custom-built antenna (incorporating a small telescope for aiming at the Moon), the team obtained its first

---

*   The Hungarian-born Malina is a unique case in the debate of scientific morality applied to rocket science. While other 'dreamers' such as Goddard, von Braun and Korolyov never objected to the idea of using their knowledge for designing and producing weapons, Malina, after working for the US Army during the war and realising that military research would dominate the activities of the laboratory for many years to come (and in fact until 1958), silently resigned and went to work for UNESCO (United Nations Educational, Scientific and Cultural Organisation).

†   In today's politically correct world, WAC stands for Without Attitude Control. In a better story, WAC Corporal is so named as a joke. WAC was the Women's Army Corps. As this was *a short slim rocket* to test Corporal, it was named the WAC Corporal.

radio echo on 6 February 1946, less than one month after the US team, which had used a less sensitive instrument.*

Based on these results, a US classified programme called PAMOR (Passive Moon Relay) was set up to map the position of the Soviet missile early warning system radar sites. Aided by the Earth's rotation, antennae in the continental US were able to collect the echo of the radio waves reflected by every single radar off the lunar surface.

As a spin-off of this work, between 1954 and 1962 the US Navy used the same technique of reflecting radio waves off the lunar surface to establish a radio link between the continental US and Hawaii during ionospheric storms. It can thus be said that our natural satellite was also man's first communication satellite!

During the same months of Project Diana, US astronomer H.H. Nininger had published a theory advocating that the entry into the Earth's atmosphere of ejecta from either the formation of the larger craters or the eruption of the lunar volcanoes had created a class of natural glasses called tektites. To prove his theory he proposed the use of two new technologies developed during the Second World War – guided missiles and atomic weapons – to create, on the lunar surface, a disturbance large enough to 'make possible a contribution of lunar material to our planet'. Nininger also estimated that the entry speed of the fragments into the Earth's atmosphere would be several times lower than that of a typical meteorite, and thus the chances of surviving the entry were deemed large. (Tektites were later shown *not* to have come from the Moon.)

In the following years, projects involving lunar probes and lunar launchers began to proliferate. In 1956 the RAND Corporation proposed a probe to be launched on an Atlas ICBM derivative to dig into the lunar surface at a speed of 550 km/h; the Massachusetts Institute of Technology began to study a recoverable lunar probe; the US Air Force proposed to launch a lunar probe using a four-stage rocket consisting of the booster of the Navaho vertically launched intercontinental-range cruise missile as the first stage, a Redstone as the second stage, and clusters of solid-fuelled rockets as the third and fourth stages; Ford proposed a five-stage launcher based on a combination of the X-17 experimental rocket and the Vanguard; Aerojet Corporation proposed a five-stage derivative of its scientific Aerobee rocket; and the Martin company studied a launcher combining the Titan ICBM and the Vanguard missile. The Army, during the long debate on the relative merits of the Orbiter and Vanguard projects, proposed using an improved Jupiter C to send 45 kg to the Moon.

In 1957 Krafft Ehricke, an Atlas missile designer, and Nobel prize-winning physicist George Gamow, proposed a small probe called Cow (after a nursery rhyme) that would fly by the Moon before returning to Earth one week after launch. A follow-on version would be preceded by an atomic bomb designed to raise a cloud of vaporised rock. The second probe would then fly through the cloud and return lunar surface samples to Earth. Crazy, but true!

---

* There is also evidence that US Naval Research Laboratory researchers had unsuccessfully tried to retrieve radio echoes from the Moon as early as 1928.

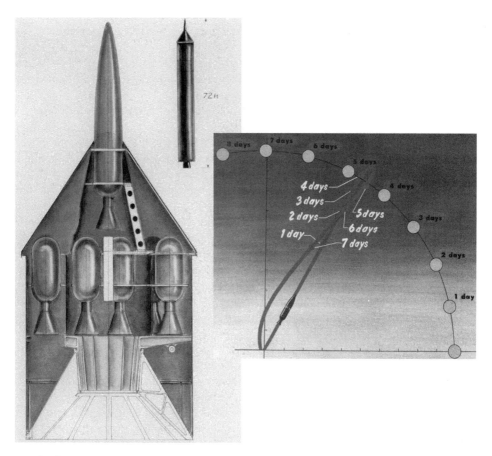

The first American lunar probe, Project Red Socks – a response to the Soviet launch of Sputnik. (Left) high-speed stages; (right) trajectory to the Moon, and return. All of the planned achievements for this project were eventually accomplished by Soviet spacecraft: the first far-side photography (Luna 3, 1959), the first lunar satellite (Luna 10, 1966), and the first circumlunar flight and Earth return (Zond 5, 1968). (Courtesy JPL/NASA/Caltech.)

At the end of October 1957, in response to the Soviet launch of Sputnik, JPL proposed its own project of lunar exploration called Project Red Socks. This project had three main objectives: to take pictures of the Moon's hidden hemisphere, to improve spacecraft guidance techniques, and to impress the world. It was proposed to launch no less than nine probes of two different types. The first would use a technology very similar to JPL's Explorer, would have a mass of 6.75 kg, and would orbit the Moon as early as June 1958. The second would take pictures of the hidden side of the Moon, and would return them to Earth by a re-entry vehicle exploiting technologies developed using the Jupiter missile. Seven other missions were supposed to put 54-kg satellites in orbit around the Moon between January 1959 and the end of 1960. The objectives of these following missions were unclear, but one suggestion by the laboratory's director, W.H. Pickering, was to detonate a nuclear weapon on the surface of the Moon in order to collect, as Nininger had proposed, any lunar rock that would have been hurled to our planet by the explosion, and to produce 'beneficial psychological results'. Not bad for a nation that had yet to launch an artificial satellite of its own!

President Dwight Eisenhower approved the first American lunar programme on 27 March 1958. The project was to be managed by the Advanced Research Projects Agency – a military agency created to coordinate the US military space programme. Based in part on Project Red Socks, the plan called for five probes, three to be built by the USAF and two by the Army. (It must be remembered that to begin with, all American space projects – including scientific projects – were managed by the military.)

In parallel with the lunar exploration project, the Pentagon signed a contract with the Yerkes Observatory near Chicago to prepare a photographic atlas of the visible side of the Moon. This was to be used to choose the landing sites of future missions.

The ARPA plan was clearly overly ambitious from the beginning. Despite the lack of deep-space navigation experience and the very low success rate of their launchers, the United States, with the USAF probes, aimed at putting a satellite in orbit around the Moon at the first attempt! This was a development strategy that would later be called 'all up' testing.

The design of the USAF probe resembled a pair of squat cones joined at their bases by a cylindrical strip. The fibreglass structure was 45 cm tall and 72.5 cm in diameter. On one apex was a cluster of eight small rockets for trajectory control. This pack could be jettisoned once used. On the other apex was a 13.3-kN-thrust solid fuelled TX-8-6 motor derived from the Falcon air-to-air missile, to brake the probe into lunar orbit. The initial plan was to carry a single scientific instrument: a simple camera yielding 1.6-km ground resolution from an altitude of 1,600 km, using a tilting mirror which scanned the lunar surface, and sending its image to an infrared lead sulphide photodiode. But after the discovery of the Van Allen radiation belts in June, more instruments were added to a total mass of 18 kg. These additions included a radiation counter, a simple search-coil magnetometer to measure any lunar magnetic field, and a microphone to detect the impact of micrometeorites – a danger to spaceflight that would prove much less fearsome than expected. The small probe was covered with a removable black-and-white stripe pattern to maintain a stable 28° C internal temperature, and was sterilised by exposure to ultraviolet rays to avoid, in case of impact, any future search for lunar microorganism being compromised by Earth bacteria. The radio communication system consisted of two different antennae: a dipole antenna for receiving commands from Earth, and a loop antenna coated on the outside of the central cylinder for telemetry transmission. The probe incorporated two transmission systems using the same frequency and quite different powers. The low-power system was to be used for 'routine' telemetry, while the high-power system would be used exclusively for camera image transmission, for which a second battery package was provided.

The launcher would be the Thor–Able, using the intermediate range Thor missile as the first stage, the second stage of the Vanguard, and a solid third stage. The mass of the probe was 38 kg – almost the maximum payload that the rocket could carry to the Moon.* Sixty hours after launch, a retro-rocket would place the probe into an orbit that would take it within 29,000 miles of the Moon. Five hours later a timer would switch on the camera, the batteries of which were to last just a few hours. The total mission duration was estimated at two weeks.

On the Soviet side, Korolyov's aforementioned memorandum of 1955 suggested the start of development of a series of lunar exploration probes. In April 1957 a design group dedicated to lunar spacecraft was formed, and four different probes, collectively known as Object E, were studied. The first model, E-1, had a mass of 170 kg, and its mission was to impact the Moon while collecting some simple scientific data during its approach. E-2 and E-3 – with a mass of 280 kg and differing in details of both the scientific payload and the attitude control system – were to take pictures of the far-side of the Moon and relay them to Earth. The final model, E-4, with a mass of 400 kg, would detonate an atom bomb on the visible hemisphere to provide a dramatic visual confirmation of the impact and to perform a remote chemical analysis of the soil vaporised in the explosion. This probe had the shape of a naval mine, with detonator rods over all its body. The yield of the weapon is not known, but in consideration of the state of the art of nuclear technology in the Soviet Union in the 1950s, and the very limited mass of the probe, it must have been quite low. (For example, the tactical RDS-4 Natasha atom bomb, which entered service in 1953, yielded 30 kT, and weighed, in its 'free fall' version, 1,200 kg.)

The E-1 probe, based on the design of the PS-1 satellite, was a sphere of 80 cm diameter on which were mounted four whip antennae, two wire aerials and a long magnetometer boom. In addition to the search-coil magnetometer, the scientific payload included a number of radiation counters, cosmic ray counters, and some piezoelectric micrometeoroid detectors having a characteristic 'window' shape. For temperature control, the outside of the probe was subjected to a particular treatment which made it extremely reflective, while the interior was pressurised with 1,300 hPa nitrogen, the circulation of which in weightlessness was ensured by a small fan. The interior of the probe was divided into four different sections: radio and command systems, scientific payload, silver-mercury batteries, and USSR emblems.

For lunar launches, two uprated versions of the 8K71 using an upper stage were under scrutiny: 8K72 and 8K73. The first model used an upper stage (stage E) with a Kosberg RO-5 kerosene/liquid oxygen engine. To assure the safe ignition of the upper stage prior to the separation of the lower stage, it was decided to mount it on a lattice truss structure so that the exhaust gases could not build up. This technique was to become a characteristic of Soviet rockets. The second model – designed to launch the heavier E-2, E-3 and E-4 probes – used a more powerful upper stage with a Glushko RD-109 hydrazine/liquid oxygen engine.

---

*   All of the American and Russian lunar missions of the 1950s used the 'direct ascent' technique, without insertion in parking orbit. The launches all took place from the northern hemisphere, which constrained the launch window to the periods when the Moon was south of the celestial equator.

A Soviet E-1 probe on its handling cart. Note the window-shaped micrometeoroid detectors and the cup-shaped ion detectors.

The development of the E-4 culminated with the construction of a full scale mock-up, before being abandoned for two main reasons. First of all, there was the risk of a launch abort, with the nuclear weapon falling back to Earth.*

---

*   On 25 July 1962, during an operation Dominic test, a Thor missile was launched from the US base on Johnston Island, in the Pacific Ocean, carrying a 100-kT warhead that was to explode in the high atmosphere, but it experienced an engine failure and was blown up on the pad, completely destroying it and contaminating it with plutonium. In three other tests of the same series something went wrong during the flight, and the warhead's 'safe' (non-nuclear) self-destruction system was activated, resulting in contamination of islands in the Pacific Ocean.

Technicians in radiation suits inspect the engine of the nuclear-tipped Thor rocket that exploded on the Johnston Island launch pad on 25 July 1962. Parts of the pad had radiation counts as high as 1,000,000 CPM (counts per minute) – 100,000 times higher than ambient radiation. One of the reasons for the cancellation of lunar nuclear probe projects was the possibility of a similar catastrophic launch failure. (Courtesy US Department of Defense.)

Second, it was discovered that, there being no lunar atmosphere to form the characteristic mushroom-shaped cloud, a nuclear or conventional explosion on the Moon would be visible only as a very brief flash of light and possibly a small dust cloud.*

Soon after the cancellation of E-4, the E-3, 8K73 and E-2 were also cancelled, the latter being replaced by the E-2A probe, having an improved photographic system and using the 8K72 as a launcher. It was also decided, as suggested by the famed Soviet astronomer Iosef Shklovski, to release 2 kg of sodium from the E-1's escape stage after the probe was on its way, to enable visual observation of the sunlight reflected by the small comet-like cloud to confirm its trajectory.

---

*   Again, consider the case of the US Dominic nuclear test of 20 October 1962, when a 60-kT bomb was detonated in the tenuous upper atmosphere, at a height of 147 km. According to eye witnesses, no fireball was visible, but instead a circular blue region inside a blood-red ring was seen, and it disappeared in less than a minute.

1.4   A DEPRESSING START

Between April and July 1958 the United States carried out three qualification flights of the Thor–Able launcher, of which two were successful. During the summer, the fourth Thor–Able was mounted on pad 17A at Cape Canaveral, and a small group of pressmen witnessed history's first lunar launch attempt, four minutes late, at 12.18 GMT on 17 August 1958. After 77 seconds of flight, the Thor's turbo-pump seized, and the missile exploded at a height of 15,200 metres. The payload's telemetry was erratically received for another 123 seconds before the vehicle impacted in the ocean.* The launch, however, attracted the interest of the public, and was the first to be broadcast, with some time delay, on television.

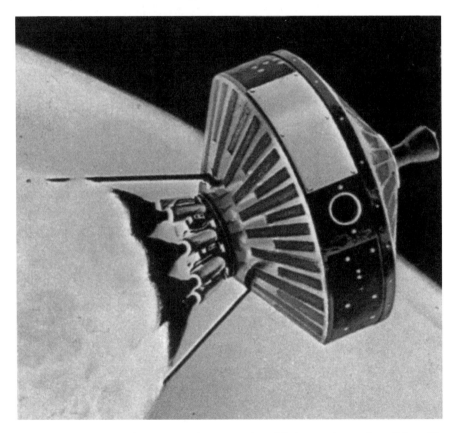

An artist's depiction of a USAF Pioneer probe with its vernier engines firing. The camera's infrared port is visible on the waist, and to its left is the polished-metal micrometeoroid detector. The Falcon engine nozzle can also be seen (right).

---

*   The lunar launch, today called Pioneer 0, or Able 1, proved so popular that a toy company produced a scale model kit of the Thor–Able and the lunar probe. This kit is now one of the most highly sought by collectors.

Upon the creation of the National Aeronautics and Space Administration (NASA) on 1 October 1958, the entire lunar programme was transferred to the new organisation. Its first launch, on 11 October, a mere ten days after foundation, was that of the second USAF lunar probe, Pioneer 1 (or Able 2), whose launcher, two days before lift-off, was missed by a few tens of metres by the debris of an exploding military missile. In keeping with NASA's desire for openness, it was to be the first launch to be shown live on television.

The flight of Pioneer 1 went without a hitch up to third-stage burn, when it was discovered that because the Able stage had cut off prematurely – possibly due to an accelerometer programming error – the probe would not reach the Moon, even after the small manoeuvre rockets were fired; it would soon fall back to the ground after a parabolic flight. The only possibility of at least partially saving the mission was to try to fire the Falcon engine to reach a very high Earth orbit (130,000 × 30,000 km); and then if a way could be found to turn the spacecraft through 180°, a fly-by of the Moon could be attempted. Unfortunately there was no way of turning the probe, and eight attempts to start the engine – four commanded from Hawaii, and four from Florida – failed because the onboard batteries had become so cold in space that their temperature had fallen to below 2° C, and they could not provide the necessary power. The spacecraft reached a peak of 115,000 km from Earth – less than one third of the distance to the Moon – and fell back, disintegrating in the atmosphere after a flight of 43 hrs 17 min. The lunar mission had been a failure, but the probe relayed interesting data on the Van Allen belts, measuring their extent despite a radiation counter malfunction, and on the micrometeoroids, of which only 0.0052 impacts per second per square metre of sensor were recorded. Simple uncoded signals were reportedly transmitted by using the spacecraft as a relay between the tracking stations at Jodrell Bank in England and Mileston Hill in Massachusetts, in a demonstration of transoceanic space communications which are today commonplace.

The second NASA lunar probe was launched one month later, and just one day late, on 8 November 1958. Pioneer 2, also known as Able 3, was quite similar to its predecessors, but incorporated some modifications. The Naval Ordnance Test Station (NOTS) camera was replaced by a lighter one with better resolution and functioning in the visible part of the spectrum. It was built by Space Technology Laboratories (STL), and its location was changed from the waist to the lower cone.* A cosmic ray detector was added, and the batteries were replaced after the failure of Pioneer 1. For this probe too, a transatlantic radio relay experiment was planned. The Thor–Able launcher also was modified, with a different staging system and a complex system of integrating accelerometers designed to shut off the engines at the desired speed. However, all of these improvements were rendered useless when the third stage did not fire at all. Pioneer 2 burned up in the atmosphere over Africa after a 45-minute flight that peaked at an altitude of 1,600 km.

---

* The NOTS camera was later mounted on some Transit satellites, and recorded some images of Earth in 1960. In August 1959 an STL camera similar to that on Pioneer 2 was mounted on Explorer 6 and recorded the first very crude TV image of Earth ever taken from a satellite.

The first image of the Earth, taken by the American Explorer 6 satellite in August 1959. The darker area at upper left is said to be clouds over the Hawaiian Islands. The same camera was mounted on Pioneer 2, and so this image was representative of what could be expected from lunar pictures taken by early Pioneer probes.

Like the Americans, the Soviets initiated their lunar programme with the flight testing of the launcher. The first 8K72 arrived in Tyuratam in June 1958, but as the test of stage E was not contemplated it was replaced by a mock-up having the same geometry, mass and inertia moments. The missile was fired on 10 July, but what happened next is still not clear. According to one account of events, one of the boosters lost its pressurisation and the whole launcher was removed from the launch pad; and according to a second version the same problem happened some seconds into the flight and the launcher fell back onto the pad. But whatever the truth, the result was a fiasco.

The Soviet team then prepared a new rocket and the first E-1 lunar probe in the hope of beating the American Pioneer that was to be launched on 17 August. The launch was scheduled for 18 August, but there was a series of problems with the engines and some delay in the preparation of the probe. Thus, when news of the American failure arrived, and the pressure to proceed was relaxed, the launch was postponed. If the Soviet probe had been launched as planned, it would have beaten its American rival to the Moon because the more powerful booster would have sent it on a faster trajectory. On 23 September the first Soviet lunar probe left Tyuratam's Pad 1, but after just 93 seconds the rocket exploded. An identical probe was therefore prepared for the next launch window, but on 11 October, exactly fifteen hours after the failure of Pioneer 1, the 8K72 exploded 104 seconds into the flight.

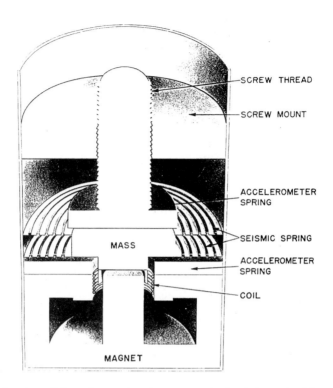

A (low-quality) drawing of the seismometer that was to be landed on the Moon by project A119 in order to exploit the scientific potential of a lunar nuclear explosion. The seismometer also doubled as an accelerometer which could provide data during the landing. (Courtesy Armor Research Foundation, US Department of Defense.)

All development activities were then stopped, and an investigation was carried out. The analysis of the two failures showed that the addition of stage E had aggravated a well-known problem of vibration in the 8K71, which moved its centre of mass so that after 100 seconds the rocket would suffer a longitudinal resonance phenomenon, leading to its destruction. The problem was eliminated by modifying the piping feeding the engines of the boosters. A modified launcher with the third E-1 was launched on 4 December – but this time a lubrication problem in the extreme conditions of the high atmosphere caused the central stage to lose 70% of its thrust 245 seconds into the flight, after which the engine completely stopped.

In parallel with the Pioneer probes, the US Air Force embarked on a top-secret project called A119 – described euphemistically as a 'study of lunar research flights', and only revealed 42 years after its conception. It was probably based on a still-secret RAND Corporation study, begun in 1956, aimed at putting a nuclear warhead on the Moon. The same idea was shared by Edward Teller, the 'father of the hydrogen bomb', who in February 1957 proposed exploding an atomic bomb at some distance from the lunar surface to observe the fluorescence induced in it, or even directly on the surface to observe what kind of disturbance it might cause. Moreover, after being mentioned in Project Red Socks, the idea of the emphatically-named InterSpatial Ballistic Missile (ISBM) was analysed in some detail by engineers of Lockheed Space and Missiles Division, who determined that an 11-kT bomb carried by an Agena rocket would have enough time to explode before being crushed in the impact with the Moon, despite travelling at a relative speed of some 2.5 km/s, close to the 'speed of information' (the speed of sound) in the metal.

Ironically, such a project would probably have fallen by the wayside had there not been moves by both superpowers to introduce a ban on nuclear tests. The Soviets unilaterally ceased tests on 31 March 1958, then resumed on 30 September. After America unilaterally ceased on 31 October, the Soviets resumed their ban in December. With both sides abstaining, talks were initiated to establish a total test ban but due to lack of agreement the Soviet Union resumed testing on 1 September 1961, followed by the United States fourteen days later. Nevertheless, for almost three years the two superpowers did not explode a single nuclear weapon. In this climate of uncertainty it is not surprising that the US military considered moving their own tests into space, giving them an aura of scientific respectability. During this period, Project A119 was begun by the US Air Force Special (nuclear) Weapons Center – its main aim being to send to the Moon, without any warning, a fission atomic bomb to impress the Soviets and their allies.

Very few details of the project have been revealed, and those were mostly concerned with the scientific aspects. From the spring of 1958 there was participation by a small group of scientists of the Armor Research Foundation of the Illinois Institute of Technology, who acted as scientific consultants. This group included many well-known scientists such as Leonard Reiffel, who was project Chief Scientist and was later to be the Apollo lunar missions scientific instrumentation manager; Gerard P. Kuiper, a Dutch-born planetologist; and Kuiper's doctoral student Carl Sagan, the future famous planetary astronomer, scientific populariser, and author of the science fiction novel *Contact*. Counting on the accuracy of the launcher – far too optimistically estimated as 'a couple of miles' at the Moon's distance – it was decided to detonate the bomb on the night side close to the terminator, in order to maximise visibility.

In contrast to the similar Soviet E-4 project, the American plan never reached the hardware stage. We thus ignore the intended launcher – probably an uprated version of the Atlas or Titan ICBM – and the characteristics of the bomb. The final scientific report – signed by Reiffel and recently made public through the Freedom Of Information Act – envisaged the possibility of using a weapon with a yield as great as 1 MT (one million tonnes of TNT). Contrary to Kuiper's calculations, the crater created by the explosion would not have been visible. A 1-MT bomb would have made a crater smaller than 400 metres. The smallest object that can be resolved telescopically, even in excellent observing conditions, is about 1 km in diameter. The explosion would have facilitated studies of the thermal characteristics of the surface exposed to the explosion's heat; the internal structure of the Moon, and if one to three seismometers had been placed on the surface in advance of the explosion, constraints on the size of the metal core (if it has one); the composition of the rocks; and the presence of a magnetic field. Other scientific investigations related to the project were envisaged by Sagan, but we know only the names of these: 'Possible Contribution of Lunar Nuclear Weapons Detonations to the Solution of Some Problems in Planetary Astronomy', and 'Radiological Contamination of the Moon by Nuclear Weapons Detonations'. This second paper, in particular, dealt with the effects of radioactive fall-out on the surface of the Moon, which could have altered lunar geology research for centuries to come; the military, of course, were not concerned by such pollution. However, in his report Reiffel was sceptical of the opportunity of staging such a mission, noting that the reaction of the 'unprepared' public to the explosion of an atom bomb on the Moon would probably have been negative. As the mission was mainly designed as a public relations exercise, it is not surprising that the project was terminated in January 1959. Alas, many documents on project A119 were destroyed during the 1980s by the Illinois Institute of Technology, and it is therefore improbable that other information will surface in the future.*

Fortunately, the *Partial Test Ban Treaty*, in force since October 1963, and the *Treaty on the Principles Governing the Activities of the States in the Exploration and on the Use of Outer Space, including the Moon and the other Celestial Bodies*, signed four years later, prohibited the stationing of nuclear weapons in space, thus ending all nuclear-tipped lunar probe projects; but alas, not before the United States exploded six weapons in space (three during operation Argus in 1958, using the small W-25 warheads weighing 98.9 kg and yielding 'only' 1.7 kT, a very good candidate for the A119 mission, and three during the aforementioned operation Dominic) and the Soviet Union four more.

An interesting piece of trivia is that in 1957 the US press published the story of a Soviet space launch that would take a nuclear bomb to explode on the Moon during the eclipse of 7 November, to commemorate the anniversary of the Soviet revolution. Although one major television channel aired the lunar eclipse live, nothing, of course, happened. However, as we have seen, both the Americans and the Soviets were really planning to be the first to achieve this, and it is quite curious that in the prevailing political climate the Soviet scientists could hint at the existence of such a project, while the American scientists could not do so.

Finally, it must be noted that before the fall of the USSR, Western intelligence reported at least three more failed Soviet lunar shots in 1958 – on 1 May, 25 June and 15 November – all of which have been denied by the Russians.

1.5   A SORT OF SUCCESS

After the three USAF Pioneer failures, it was time for the Army's two lunar probes. These were completely different spacecraft: a gold-plated fibreglass cone covered with black and white stripes, 51 cm long and with a base diameter of 23 cm, on which

---

*   At the same time, the US was working on project Orion, aimed at developing a spaceship propelled by a sequence of atomic detonations. The less powerful version of the ship was designed to be able to land 1,200 tonnes on the lunar surface!

was mounted a short antenna. The stabilisation system was new and very 'cute': the probe would separate from the launcher spinning 415 times every minute, two small 28-gramme masses, each connected to a 1.5-metre-long wire would be deployed, and the spin would be slowed to a dozen revolutions per minute before the wires were cut. The probes were built by JPL, working under contract with NASA. According to the original plan, the first probe would carry two Geiger–Mueller tubes to record high-energy particles, and a photoelectric sensor designed to detect the Moon from a distance of 32,000 km. The spacecraft would fly by the Moon and possibly, with some luck, hit it 33 hrs 45 min after lift-off. The second probe would carry a 1.5-kg camera which would be activated by a photoelectric sensor to take a single picture of the lunar far-side from a distance of 24,000 km with a surface resolution of 32 km. For this mission, the spacecraft would be placed in a very elliptical Earth orbit which would take it around behind the Moon. Following the discovery of the Van Allen belts, it was realised that this radiation would have ruined the photographic film. A design for a small electronic camera was therefore studied. Each spacecraft was equipped with a 180-mW transmitter, had a mass of just 6 kg, and was to be launched by the new Juno II missile, a distant relative of the Juno I that launched Explorer 1. It was a four-stage vehicle combining the intermediate-range Jupiter missile (with a single S-3D engine of 666 kN thrust; the Juno I first stage had a thrust of just 370 kN) and a slightly improved version of the three solid-fuel upper stages of the Juno I launcher. The total time from second-stage firing to fourth-stage burn was just 27 seconds. To protect the payload from the thermal load of the launch, both the probe and the upper stages were covered by an aerodynamic fairing which also contained the guidance system platform and a system of cold gas thrusters for attitude control. The payload capabilities of the new launcher were even more ridiculous than that of Thor–Able: a mere 7 kg to the Moon.

Deeming the radio tracking systems used up to that day to be inadequate for lunar missions, and even more inadequate for the planetary missions that were to follow, JPL received approval to construct a facility specifically for tracking deep space missions, and this was completed within the tight schedule of six months. The 26-metre antenna – located at Goldstone, California – was the first of NASA's Deep Space Network. The antenna was used in November 1960 to test the techniques to be used in connection with the Echo passive relay communication satellite (an aluminised mylar balloon of 30 m diameter) by bouncing radio waves off the Moon.

Pioneer 3 reached Cape Canaveral after travelling from California to Florida inside a small container designed to fit an airline seat, and was launched from Pad 5 on 6 December 1958. A series of minor and major problems followed: the first-stage shut off 176.3 seconds into the flight instead of 180 seconds; it was out of the optimal trajectory; the upper stages impressed a speed 316 m/s less than required; and when the probe separated, the stabilisation system failed. The thermal control system was, however, completely successful, and the internal temperature remained constant at 43° C. Pioneer 3 reached a height of some 102,000 km before falling back to Earth and disintegrating over Chad in Africa, which was then still a French colony, after a total time of flight of 38 hrs 6 min. Nevertheless, it was not without result, because it discovered the second, outer, Van Allen belt.

It had been a bad year for space exploration, with even distribution of failures for both the Americans and the Soviets. However, activity in 1959 was immediate. On 1 January, Khrushchev announced that during that year, Soviet probes were to reach to Moon; and the following day, 2 January, Korolyov's team tried again.

This time the fourth E-1 probe, variously referred to as Luna 1, Mechta (Dream), and First Cosmic Rocket, was placed on a hyperbolic trajectory towards the Moon. It was just the fourth successful Soviet space launch, after the three Sputniks! (In the West, the first Soviet lunar probes were initially called Lunik – a contraction of Luna and Sputnik – but in 1963 were renamed by the Soviets, Luna (with stress on the 'a').)

Stage E was also put in an Earth-escape trajectory, carrying a payload of cosmic ray detectors, which increased the total mass of the scientific payload (probe plus sensors) to 361.3 kg. The 'sodium comet' experiment was released 113,000 km from Earth, and was seen from the Indian Ocean as a small nebulosity of fifth or sixth magnitude. The observations of the cloud not only enabled Soviet scientists to pinpoint the position of the probe, but also to study, for the first time, the behaviour of gases in the environment of space. Being an E-1 probe, the objective of Luna 1 was to impact on the Moon, but due to a launcher programming error it flew by at a distance of some 6,400 km – less than twice the Moon's diameter – 34 hours after launch and became the first man-made object to enter solar orbit (like a planet) – an

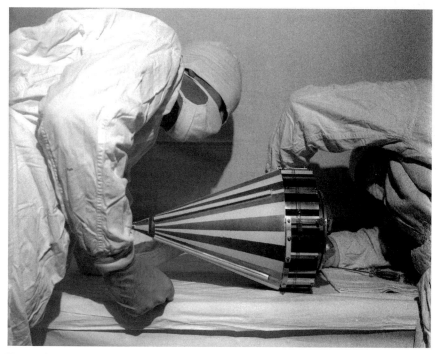

JPL technicians in surgical clothes prepare the tiny Pioneer 3 probe before launch. (Courtesy JPL/NASA/Caltech.)

orbit where it still resides, with aphelion at 199 million km, perihelion at 147 million km, and a period of 443 days. The batteries lasted for 62 hours, and when they were finally exhausted the spacecraft was 600,000 km from Earth. The major scientific merit of the mission was the discovery of the 'solar wind': a flux of plasma that originates from our star and fills the whole Solar System.

Despite the planned impact not taking place, the Soviet press were instructed to depict the mission as a great success, which indeed it was. It was then time for the Americans to retaliate – which they did with Pioneer 4, the final ARPA probe. After the discovery of the second Van Allen radiation belt, the payload was once again redesigned to include two Geiger–Mueller tubes, one of which had a thin lead cover to provide calibration for the other.

The launch of Pioneer 4 was initially scheduled for 2 January 1959 (the same day as Luna 1), but was delayed until 3 March because of the necessity of modifications to the launcher and to the probe itself. But once again there were problems. The fourth stage shut off 1 second late and followed a slightly different trajectory than planned, so that the probe crossed the orbit of the Moon some 59,500 km from its target – so far that the photoelectric sensor did not activate. The probe, however, collected some interesting data on the deep space environment and it was possible to communicate with it for 82 hrs 4 min, up to a distance of 655,000 km, which set a new distance record. Like Luna 1, it eventually entered solar orbit, in this case with a perihelion at 147 million km, aphelion at 171 million km, and a period of 394.75 days. Pioneer 4's partial success went a long way towards soothing injured American pride, but the Soviets were now in a very strong position to reassert their lead.

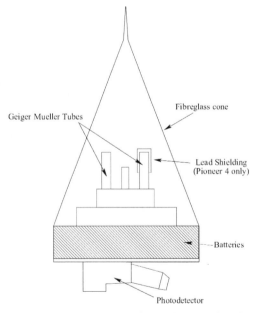

Pioneer 4 differed from Pioneer 3 because of the addition of lead shielding on one of the Geiger tubes in order to provide calibration for the unshielded tube.

1.6   A DATE WITH THE UNKNOWN

After the partial failure of Luna 1, a new version of the E-1 probe was prepared, having a different arrangement of the magnetometer, six instead of four charged particle detectors, and improved micrometeoroid detectors. It was slightly more massive, at 192 kg. The launch of the first probe of the new E-1A version was scheduled for 16 June, but on that day it was discovered that the third stage had been filled with normal kerosene instead of the required high-density kerosene. The tanks were therefore unloaded and again filled. The probe lifted off two days late but one of the gyros controlling the attitude of the rocket broke down 153 seconds into the flight, and it was necessary to activate the self-destruct system.

A new 8K72 was then prepared, but this time, on 9 September, the core-stage engine did not reach lift-off thrust and was shut down. Twenty seconds later the boosters were also shut down, and the launcher was hurriedly removed from the pad to make room for a replacement, which lifted off on schedule on 12 September.

The launch proceeded without any problems, and some hours later Soviet ships on station in the Pacific Ocean picked up telemetry from the probe indicating that it was some 31,000 km from Earth and travelling at a relative speed of 4.9 km/s. These data and positional fixes were fed into the computers, and the engineers then knew for certain that an impact was imminent. When the probe was 77,000 km from Earth, Radio Moscow's top speaker, Yuri Levitan, announced the launch of the second Soviet lunar probe: the Second Cosmic Rocket, or Luna 2. The stage E once again carried sodium. The artificial comet was seen from the Soviet Union as a cloud of magnitude 4 or 5 with an angular diameter of 2 arcmin located 35 arcmin from ε Aquarii. The Czechoslovak Skalnate Pleso Observatory estimated its magnitude as 7.5. While Luna 1 missed its target, Luna 2 did not fail. Tracked by the Jodrell Bank radio telescope in England, it hit the Moon two minutes late at 21 hrs 2 min 24 sec GMT on 13 September, at a relative speed of 3.3 km/s and an incidence angle of around 60°. (At that time, the mass and the distance of the Moon were only approximately known.) The probe disintegrated on impact, and released onto the surface the non-scientific part of its payload: two spheres, one of 9 cm diameter, the other of 15 cm, covered with Lenin's portraits and hammer-and-sickle emblems. Stage E, carrying two similar but larger spheres, also hit the Moon half an hour later. The impact point – determined with some uncertainty as 31° N, 1° E, or 29°.1 N, 0° E, close to the large crater Archimedes – was assigned, during the 1970s, the official name Sinus Lunicus (Lunik Gulf).

Some 'trustworthy' astronomers at the Budapest–Szabadsághegy Observatory stated that they had observed and drawn the cloud of dust raised by the impact. It appeared like a rapidly expanding small dark ring, which then became more and more faint. The maximum diameter of the cloud was said to have been between the diameter of the nearby craters Archimedes and Aristillus, and thus of some 60 km, and it was said to have been visible for between 10 and 58 minutes. Moreover, the cloud apparently had a clearly defined shadow from which its height could be computed as between 500 and 900 km. These observations were deemed dubious from the start, and difficult to explain only by the kinetic energy of the probe or the upper stage of its launcher. Furthermore, no-one has ever successfully taken pictures of the impact of a man-made object in the visual range of the spectrum – not even when the huge Saturn S-IVB stages were impacted on the Moon.

Luna 2 detected no measurable lunar magnetic field within 55 km of the surface, nor any trace of lunar Van Allen radiation belts. Moreover, the probe observed the solar wind for the second time, recording a flux of more than 200,000,000 ions per sec per cm^2 of detector (four Faraday cups).

Three days after the impact, Khrushchev arrived in the US for the first official visit of a Soviet Secretary. On meeting Eisenhower, he handed him two replicas of the emblem spheres, and then proceeded to recall the success of the Soviet spacecraft, adding: 'We have no doubt that the excellent scientists, engineers and workers of the United States of America who are engaged in the field of conquering the cosmos will also carry their pennant over to the Moon. The Soviet pennant, as an old resident, will then welcome your pennant, and they will live there together in peace and friendship.'

The next lunar launch dispatched a much more sophisticated probe: the E-2A Automatic Interplanetary Station, or Luna 3. After lift-off on 4 October 1959, the probe was sent not into a hyperbolic Earth escape trajectory, but on a very eccentric Earth orbit with an inclination of 80° to the equator which would take it behind the Moon.

Luna 3 – the only one of its kind – was different in both the payload and in the mission. It was a small spin-stabilised cylinder topped by two hemispheres, 1.3 metres long and with a mean diameter of 95 cm and a maximum width of 1.2 metres. Six aerials protruded out of the cylinder, while its interior was pressurised at 230 hPa for thermal control. Power was generated by several solar panels, by which the body was almost completely covered. One of the hemispheres carried a Sun sensor for attitude determination, and the other a Yenissei-2 camera with two objective lenses: a 200-mm f/5.6 and a 500-mm f/9.5. The camera was loaded with several metres of special 35-mm film (which, according to a Russian source, was a portion of the film carried by an American Project Genetrix spy balloon) that would be developed on board, scanned with a resolution of 1,000 × 1,000 pixels, and relayed to Earth for reconstruction of the images. At 278.5 kg, Luna 3 was heavier than E-1 probes; but it was not required to reach escape speed, which freed up to 150 kg on stage E for scientific instruments. The aim of the mission was to take the first pictures of the far side of the Moon. Because of constraints on the launch window, the E-2A could photograph the eastern portion of the far side only in October, or the western portion only in April.

The flight was everything but normal. The radio system worked at a power lower than its design specifications, and the internal temperature was too high; but finally, late on 6 October, Luna 3 flew over the south pole of the Moon, was deflected to the north, and, on reaching a distance of 65,200 km over the far-side on the morning of 7 October, slowed its rotation almost to a stop by using gas jets, then began to take pictures with exposures of 1/200, 1/400, 1/600 and 1/800 of a second. The session began at 03.30 GMT, at a distance of 65,200 km and at selenocentric coordinates 16°.9 N, 117°.6 E, and ended at 04.10 GMT – when the mechanical shutter of the

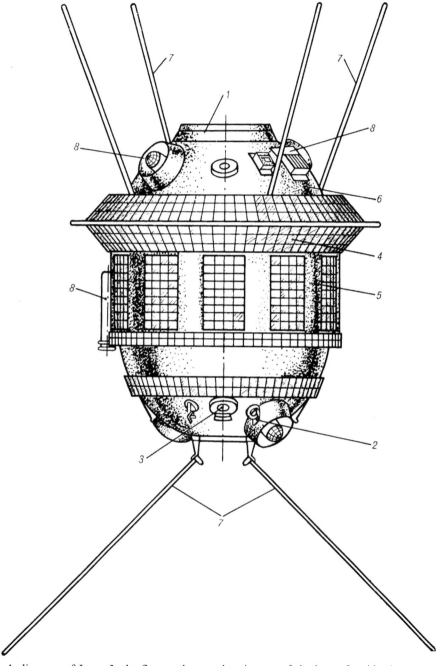

A diagram of Luna 3, the first probe to take pictures of the lunar far-side. 1, camera port (closed); 2, attitude control jets; 3, Sun sensor; 4, solar panels; 5, solar panels; 6, hermetic body; 7, antennae; 8, scientific instruments.

500-mm lens jammed – at 68,400 km and at 17°.3 N, 117°.1 E. On that day the Moon was five days 'old', the far-side was almost completely illuminated (70%), and the Sun shone almost directly over it.

Luna 3 then began spinning again. The Moon had perturbed its trajectory, taking it back over the northern hemisphere (without the Moon's intervention, the probe would have approached perigee from the south), now in an 18-day orbit with apogee at 480,000 km and perigee at 48,000 km. As viewed from the Soviet Union, the elevation of Luna 3 over the horizon was steadily increasing, and the probe's point soon became circumpolar, being thus easily tracked by the Kochka antenna, near Simeis in the Crimea, and the Petropavlovsk antenna in Kamchatka. To improve the signal-to-noise ratio, the authorities closed the streets to motor vehicles for several kilometres around Simeis, and ordered radio silence on every ship in the Black Sea near the Crimea. The probe could use two different data transmission rates: a slow rate able to return a single picture in 30 minutes, for long distances, and a faster rate able to return a picture in 15 seconds, for small distances. A first attempt to receive the images using a slow transmission rate was made as early as 8 October. However, only noise was received. On the fifth attempt, however, two images were received from a distance of 470,000 km, and later, with the probe much closer to Earth, fifteen more of a total of twenty-nine. The quality of the images was quite poor due to transmission 'noise', and the contrast was low due to the imaged hemisphere being completely lit, but they clearly showed that the far-side is devoid of the large dark maria that dominate the near-side, although there were several small dark spots, one of which was promptly named by the Soviets Mare Moscoviense (Sea of Moscow) and another was assigned to Konstantin Tsiolkovsky. As the American geologist Nathaniel Southgate Shaler had inferred in 1903 from the bright 'rays' that poke over the limb, the far-side is heavily cratered. One particularly prominent such crater was named Giordano Bruno.* A bright streak running across the far-side was named 'Soviet Mountains', but it turned out to be one of Bruno's rays. A large 1,500-km dark area near the southern limb of the disk was named Mare Ingenii (Sea of Ingenuity). It was later discovered that Luna 3 had for the first time imaged the Aitken basin – the largest lunar crater – and the name Mare Ingenii was given to a smaller formation, just 270 km across. The images also showed, for the first time, the true shape and extent of several limb features: Mare Humboltianum, Mare Smythii, Mare Australe and Mare Marginis.† The spacecraft returned some data on the deep space environment, and also recorded ion fluxes in the solar wind.

Before completing its first orbit, Luna 3 failed, and remained dead during a second fly-by of the Moon at a distance of about 50,500 km on 24 January 1960, and until re-entry into the Earth's atmosphere during its eleventh orbit. The stage E that had taken it into space also ended up in a similar orbit, but it cannot be found in any catalogue of artificial space objects, since it was never detected by any Western observatory.

---

*   The Italian philosopher (1548–1600) from Nola, who was burned by the Inquisition.
†   Two of these were named after the German explorer Alexander von Humbolt (1769–1859), and the English astronomer William Henry Smyth (1788–1865).

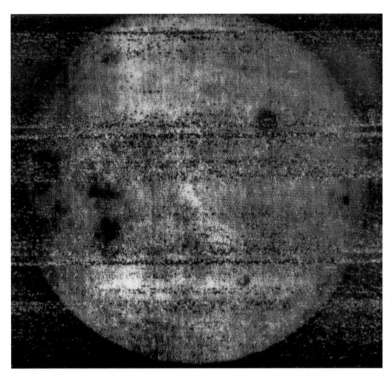

The second wide-angle image of the lunar far-side taken by Luna 3. The left-most quarter of the image shows near-side features, the dark spot at upper right is Mare Moscoviense. and the small dark spot to the right and below the centre of image is the crater Tsiolkovsky with its bright central peak. (Courtesy NASA/NSSDC.)

1.7   PROJECT ATLAS–ABLE

NASA, meanwhile, had begun its own project of planetary exploration incorporating four STL-built probes to the Moon and to Venus. After the apparently successful flight of Luna 1, however, the agency decided to restore its honour by converting the project to a series of three lunar orbiters with masses close to 175 kg. These were fibreglass spheres of 1 metre diameter, inside which was an instrument platform, a large spherical hydrazine tank and two smaller nitrogen tanks to pressurise the fuel tank. The probe had two different engines, with a total firing time of 1,700 seconds: the first for trajectory correction during the flight from the Earth to the Moon, and the second for orbit insertion ($6,400 \times 4,800$ km for the first probe, $4,000 \times 2,250$ km for the second – although according to another source, $4,800 \times 3,200$ km – and $4,300 \times 2,400$ km for the third). On the outside of the sphere were attached four solar panel wings and fifty-two small metallic propellers that, on deforming with the heat, would expose black or white areas, thus providing some passive control of the internal temperature.

The launcher of these new probes would be Atlas–Able, a combination of the Atlas ICBM and the two Thor–Able upper stages. While Atlas was about to become (after a long series of failures) a very reliable missile, Able was troublesome from the start, as proved by the many failures of the Vanguard satellites and of Pioneer lunar probes.

The first NASA probe – called Atlas–Able 4, or Pioneer P-3 – was prepared for launch at the end of summer 1959. Its scientific payload was overwhelming: a triple-coincidence telescope for high-energy radiation, an ion chamber and a Geiger–Mueller tube for medium energy and a scintillation counter for low energy radiation, a search-coil magnetometer and a flux-gate magnetometer (much more accurate), a simple 1-kg camera very similar to that on Pioneer 2, and a VLF radio receiver to study interplanetary particles and magnetic fields. During an engine test prior to launch, the Atlas booster exploded; although fortunately without the spacecraft being mounted on it. Once a new Atlas was prepared, the launch on 26 November did not fare much better. The fibreglass fairing protecting the probe collapsed under the aerodynamic pressure after 45 seconds, and the third stage and its payload were ripped away from the missile. The launcher itself exploded when the second stage ignited prior to staging.

The second probe of the series was launched almost a year later, on 26 September 1960. Atlas–Able 5A, or Pioneer P-30, had a different scientific payload: a plasma detector, a cosmic ray telescope, an ion chamber and a Geiger–Mueller tube, a search-coil and a flux-gate magnetometer, a high-energy proton spectrometer, a scintillation counter for low energy particles, and a micrometeoroid sensor for determining the flux and momentum of interplanetary dust. The camera was no longer part of the payload. Unfortunately, the rocket shut off early, and the probe re-entered the atmosphere over South Africa after a flight of 17 min. For the first time, however, the trajectory control engine of a spacecraft was fired in space, which was a useful test.

The third and last probe of the series was Atlas–Able 5B, or Pioneer P-31, launched on 15 December 1960. It was destroyed after only 68 seconds when the launcher exploded following the premature ignition of its second stage.

Intermixed with America's attempts, the Soviet Union launched two probes similar to Luna 3, but with better radio systems, on 15 and 19 April. The mission of these two probes – called E-3, reusing an earlier designation – was to take pictures of the western part of the far-side of the Moon from a closer distance, and thus with increased resolution. On the first launch, stage E shut off too early, and the probe reached only 200,000 km into space before looping back to Earth. It was thus decided to launch an identical probe before the 'window' closed. It was one of the most spectacular failures of the whole Soviet space programme. One of the boosters did not ignite correctly, its thrust reaching only 75%, and separated. Two more boosters separated at a height of some 20 metres, and the four components of the launcher continued flying on their own. Three of them then fell back on the pad while the fourth crashed on a railway line, missing a group of engineers by only 40 metres. It is worth noting that by the time these two probes were to reach their destination, the far-side would be only 50% illuminated, although this included the

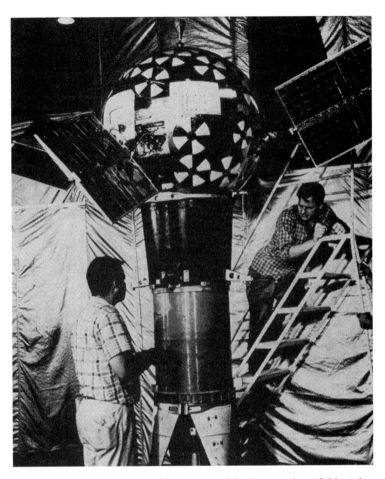

The American Pioneer P-30 lunar orbiter. None of the three probes of this series reached space.

30% that had been in darkness for Luna 3. A new E-3 probe could have been launched to the Moon in October 1960, but at that time the launch base was busy working on the 1M probes designed to reach Mars (both of which failed because of launcher problems). The E-2 and E-3-type lunar probes went on to become the centrepiece of a very successful intelligence operation. After it was learned that the model on show at a Mexican fair was, in fact, a fully functional probe, the CIA 'kidnapped' for one night the truck carrying it, disassembled and photographed the probe, taking note of every detail, including the serial numbers of several of its components.

After the two April lunar failures, Korolyov's team worked on completing the prototype of the piloted Vostok spacecraft to be launched in May. The CIA learned of an impending launch, and ordered a new overflight of the Soviet Union by one of its U-2 spyplanes. Such aircraft had already flown over Tyuratam four times – on 5 and 26 August 1957, 9 July 1959, and 9 April 1960 – without ever taking a picture of a Semiorka on the launch pad, and this was the main target of the mission on 1 May. Francis Gary Powers and his U-2 were shot down over Sverdlovsk. The spaceship prototype, called Korabl-Sputnik (Cabin-Satellite, or Spaceship), was placed in orbit on 15 May.

The two April failures marked the end of the career of the 8K72 as a deep-space launcher, but it remained in service and launched the Vostok spaceships and many satellites for different uses. Its retirement, after more than 1,500 launches, came only in 1991.

Korolyov meanwhile, started thinking about three new families of lunar probes: E-5, E-6 and E-7. The E-5 would enter lunar orbit. The probe and its payload were based on the E-1, and the only objective of the mission (cancelled at the end of 1959) was to beat the announced NASA Pioneer orbiters. The E-6 would land on the lunar surface, with the first landing quite optimistically scheduled for 1960, while the E-7 would conduct an orbital photographic mission. In 1961, with the programme in trouble, the E-7 was cancelled in favour of a simpler specialised version of the E-6 lander.

At the end of the first round of lunar exploration, the situation could not have been worse for the Americans, who had had seven failures and a partial success (Pioneer 4); while the Soviets, after many failures, had some Lenin emblems on the lunar surface, and were beginning to name the features of the far-side, using names such as Tsiolkovsky – after the man who had inspired their space programme.

2

The race: 1960–69

2.1 JPL TAKES THE LEAD

After having set America's strategy in the first round of lunar exploration missions, inspired by its Project Red Socks, JPL wanted to be involved in the next round too. To this end, in October 1958 it began studying a plan which included both lunar and planetary unmanned exploration. The plan was called Vega, and was approved in January 1959 after the reportedly successful Soviet Luna 1 mission. As initially designed, the plan included the development of a new mid-weight launch vehicle called Atlas–Vega. This would consist of the Atlas ICBM as the first stage, a modified Vanguard first stage as the second stage, and a new hydrazine/nitrogen tetroxide third stage which would be developed by JPL.

The new launcher was designed with several missions in mind: an experimental meteorological satellite, an experimental geosynchronous telecommunication satellite, a two-crew space capsule, a 600-kg Mars probe to be launched in October 1960, a similar lunar probe to be launched in December 1960, and a Venus probe to be launched in February 1961, all of which were intended to provide America with a lead over the Soviets.

However, many factors conspired against Vega, including the scarcity of funds (NASA had just approved the expensive Mercury project, the objective of which was to send a human into space), and NASA's naivety (the agency believed that it could attain a launch rate of one still-untested Atlas–Vega every two months, even though the launch pad for it was still to be built). The main problem, however, was not technical but managerial. The military were neither involved nor interested in the launcher, and were, moreover, secretly developing the Atlas–Agena B with broadly similar characteristics.

Meanwhile, the inaugural Atlas–Vega launch was postponed to 1961, which effectively ruled out the launch of the Mars probe. As no-one wanted to launch the Venus probe on the very first Atlas–Vega, it was delayed until 1962. By moving the timing of the project, the development of the rocket began to conflict with the development of the powerful Atlas–Centaur, intended to become the standard NASA launcher, so it was decided to cancel the last two Atlas–Vega vehicles and completely overhaul the project, which by then consisted of two meteorological satellites and four lunar probes. The mission of the lunar probes was never decided in detail, but interesting proposals were to take twenty high-resolution pictures of the lunar surface from an 800-km-high orbit using a photographic read-out system similar to that which the Soviets had used on Luna 3, a 200-kg soft lander and a 300-kg hard lander.

In December 1959 – eleven months after the start of the project – NASA cancelled Atlas–Vega and adopted Atlas–Agena B. Just one month later, JPL revised its lunar exploration project, named Ranger, to fit Atlas–Agena. The initial plan consisted of two versions of a standardised spacecraft, referred to as Block I and Block II.

Block I probes were to carry out the first two technology demonstration missions. They would be put into a very high Earth orbit, with a perigee at 60,300 km and an apogee at 1.1 million km, to test innovative features of the design. The most important innovation was the three-axis stabilisation system, employing small jets of compressed nitrogen to keep the main axes of the probe in a fixed attitude. A set of sensors kept the solar panels face-on to the Sun for power generation, and another sensor kept the high-gain antenna pointing towards the Earth for communication. The main advantage of the three-axis stabilisation system was the high pointing precision for such scientific instruments as a camera. Another important innovation was the possibility of correcting the trajectory during flight, to remove any launcher guidance error.

The Ranger probes used a common 152-cm-wide hexagonal base (inherited from the Vega project) housing the electronics, batteries, gyroscopes, attitude sensors and two transmitters – one of 0.25 W and the other of 3 W. Mounted on this base were two trapezoidal solar panels, producing 150–210 W, which increased the total span of the spacecraft to 5.2 metres. Under the base were three tanks for 1.1 kg of nitrogen and the 1.22-metre-diameter parabolic high-gain antenna.

In the case of Block I, on top of the base was mounted a pyramidal truss carrying the scientific instruments and a bulky omnidirectional antenna, which increased the total height to 4 metres.

On the Block II spacecraft – which were to go to the Moon – the instrument truss's place was taken by a small hard lander and its solid-fuel retro-rocket. It had a total mass of some 330 kg, was 3.2 metres tall, and used a hydrazine monopropellant engine for course corrections. The instruments carried on the base included a 6-kg Los Alamos-built gamma-ray spectrometer mounted on the tip of a 180-cm boom, a radar altimeter, and a vidicon camera built by RCA yielding 200×200-pixel resolution.

The lander, built by Ford Aeronutronics, consisted of a 43-kg 64-cm-diameter balsa sphere, inside which, floating in a cavity filled with fluid, was the 31-cm diameter 25-kg probe. The probe was bottom heavy to maintain the proper attitude once on the Moon. Only two scientific instruments were carried on the lander: a thermometer, and a single-axis seismometer consisting of a metallic coil inside which a spring-mounted magnet was able to move; a seismic displacement was computed by measuring the current induced in the coil by the magnet's movement. The

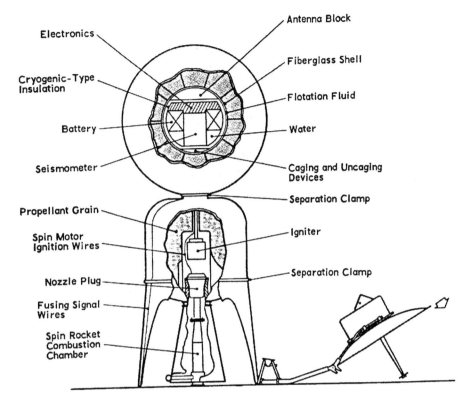

Electronics

Antenna Block

Cryogenic-Type
Insulation

Fiberglass Shell

Battery

Flotation Fluid

Seismometer

Water

Caging and Uncaging
Devices

Separation Clamp

Propellant Grain

Spin Motor
Ignition Wires

Igniter

Nozzle Plug

Separation Clamp

Fusing Signal
Wires

Spin Rocket
Combustion
Chamber

A drawing of the mission module of the Ranger Block II spacecraft. It included the spherical hard-lander, the braking retro-rocket, and the radar altimeter, which also doubled as a scientific instrument of its own.

experiment – which was possibly based on a similar instrument under development for the A119 project – had a mass of 3.3 kg, and was sensitive enough to detect any displacement of the magnet greater than one millionth of a millimetre. The seismometer was expected to survive an impact at up to 250 km/h, being immersed in n-heptane. This fluid was then to be released on the lunar surface after the balsa shell was punctured by a pyrotechnic system. The instruments shared a small battery-powered 0.05-W transmitter. The thermal control system used 1.7 litres of water, which was enough to keep it running for between two and six weeks on the Moon. As an alternative to the lander, a penetrator designed to bury itself into the lunar surface at very high speed was proposed, but was deemed too futuristic.

Once again, for fear of bacterial contamination of the Moon, the probe was sterilised by heating all its components to 125° C and then kept in an ethylene oxide-saturated atmosphere before and during the launch.

It can easily be seen that the Ranger probes were over-engineered, since, for example, the short flight to the Moon did not require solar panels; batteries would have sufficed. But JPL wanted to accumulate some experience for future interplanetary missions for which solar panels were a necessity. In parallel with Ranger, in fact, JPL

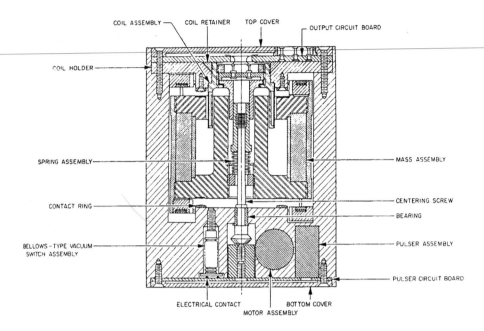

A cutaway of the Ranger Block II lunar seismometer. This was the most advanced instrument carried by US lunar probes during the 1960s. (Courtesy JPL/NASA/Caltech.)

began to design a family of Mariner interplanetary probes using the same basic architecture.

In July 1960, Lockheed – the builder of the Agena – presented its computations of the Atlas–Agena B's payload on a trajectory towards the Moon. Realising that the Ranger Block II was at least 34 kg overweight, JPL modified it by manufacturing several structural components in magnesium, drilling holes in panels in order to save mass, and removing some redundant components such as a low-power transmitter and back-up attitude determination and control systems.

2.2   TWO FAILED TESTS

Ranger 1 carried no less than nine scientific instruments to study the terrestrial magnetosphere, cosmic rays, solar wind and micrometeoroids, including a wide-field telescope that would take images of the Earth in the Lyman-α (atomic hydrogen) emission region of the electromagnetic spectrum. It was delivered by JPL in June 1961, travelling across the country in a purpose-built climate-controlled van. The countdown began three days late on 28 July 1961, and was stopped twice. The next day, the whole Cape Canaveral area experienced a power shortage. On 30 July, while still sitting on the pad, the spacecraft decided that it was now in space, powered up

A mock-up of a Surveyor Block II spacecraft. The instruments shown include a neutron activator (extended on the left), a 1.8-metres-long drill (left of centre), and a geophysical probe (extended on the right). The spacecraft was intended to carry four cameras.

the instruments, and attempted to deploy the solar panels inside the aerodynamic launcher shroud. The cause of this unusual malfunction was never determined, but the probe had to be removed from the pad, disassembled, checked piece by piece, and reassembled.

The fifth launch attempt, on 23 August, was successful. Unlike earlier probes, Ranger was not to use 'direct ascent', the Atlas was to place the Agena into a low 'parking orbit', prior to it firing its engine to head for its objective. Unfortunately, the Agena, instead of firing for 90 seconds to reach the final orbit, shut off after just a few seconds due to a faulty fuel valve, leaving the spacecraft in a very low Earth orbit. Nevertheless, Ranger 1 proved what it could do by deploying its solar panels and aligning them with the Sun. However, because the probe was not designed to cope with an orbital cycle of sunrise and sunset every ninety minutes, it used its attitude-control gas in just one day, and re-entered the atmosphere one week later.

A full investigation into the Agena's systems was carried out before Ranger 2 was declared ready for launch. The Atlas performed flawlessly on 17 November 1961, but again the Agena did not fire correctly, leaving the spacecraft in an even lower orbit (153×234 km). The probe, carrying a scientific payload similar to that of its twin, re-entered a few hours after launch. The only powered instrument – Vela Hotel – was funded by the military. This instrument, which JPL had reluctantly included on its spacecraft, was later used on twelve Vela satellites to monitor atmospheric nuclear tests by observing gamma-rays.*

The cause of the failure of the Ranger 2 launch was found in the Agena attitude control system, which allowed the rocket to spin very fast on its longitudinal axis, thus pushing the propellants to the tanks' walls instead of towards the bottom, where the engine pumps were located.

2.3   RANGER, SURVEYOR, AND KENNEDY'S CHALLENGE

In the launch schedule, it was now time for the three Ranger Block II spacecraft. As before, the Atlas would put the Agena into parking orbit. At a scheduled time, the engine would start again to push the spacecraft to its rendezvous with the Moon. After shutting off, the Agena would separate and manoeuvre to enter solar orbit. Five minutes after separation, Ranger would deploy and orient the high-gain antenna and solar panels. During the short trip, it would be possible to execute a single course correction, after which the gamma-ray spectrometer boom would be deployed.

The flight plan dictated that one hour before intersecting the Moon's orbit, some 7,250 km from its surface, the probe would aim its camera at the Moon, and it was possible that after this manoeuvre the solar panels would no longer generate power, leaving power generation to the cleverly-designed batteries, which used potassium hydroxide as an electrolite, which also provided for sterilisation.

In this new attitude, the camera – a vidicon sensor mated to a 20-cm-diameter JPL Cassegrain telescope – would take one image of the lunar surface every 13–15 seconds. The camera was expected to return a total of one hundred images. Later still, some 24 km from the surface, the radar altimeter – which meanwhile had been collecting data on the surficial material and its radio reflectivity – would command the lander to separate, fire a small rocket to set it spinning, and then fire its retro-rocket. Once the lander had been brought to a halt, some 300 metres above the ground, the spherical payload would be ejected to fall freely and strike the surface at a speed of 200 km/h. Consideration was given to mounting two grenades with 0.45 kg of explosive on the retro-rocket to calibrate the seismometer one hour after landing, but this idea was never implemented. The main spacecraft would smash into the ground, and be destroyed.

Beginning in May 1961, Ford carried out the first lander fall tests by releasing it from a helicopter flying over the Mojave desert. However, five tests revealed that something was wrong. The payload, and in particular the seismometer, had serious trouble surviving a 3,000-g deceleration impact. A second series of test in October showed that the probability of payload survival increased if sterilisation was not carried out.

Later that year, the payload performance of the Atlas–Agena was recomputed, and it was discovered that, because of a computational error, this was greater than

---

*   These satellites discovered one of the greatest mysteries of modern astrophysics: gamma-ray bursts, on which light began to be shed only in the late 1990s.

was previously thought. Unfortunately, some 50 kg of redundancies had already been deleted from the probes – in the power distribution, radio, and attitude control systems.

Meanwhile, before the Ranger probes were launched, NASA had approved a new and more ambitious family of JPL lunar probes: Surveyor, of which two versions – Block I and Block II – were being studied. In contrast with its predecessors, Surveyor would soft-land on the Moon, carrying only engineering instruments (Block I) or some 157 kg of scientific instruments (Block II): cameras, robotic manipulators to move rocks and dig trenches, seismometers, magnetometers, and so on. The total mass of the probe was to reach 1,100 kg, which was far beyond Atlas–Agena's capabilities. The launcher would be the Atlas carrying the new Centaur as its upper stage, the exceptional performance of which derived from the use of liquid oxygen and hydrogen. This combination of propellants – first tested on the Centaur itself – delivered a 40% greater thrust with respect to traditional fuels, but posed many new technical problems. The first Atlas–Centaur, launched in May 1962, exploded well before reaching orbit.

By the time the contract for Surveyor Block I hardware was awarded to Hughes Aircraft, a third version – the Surveyor B orbiter – was also being studied.

A drawing of the Prospector rover, designed by Marshall Space Flight Center. The two parabolae are the high-gain antenna and the mercury solar concentrator for power.

At the same time, on 25 May 1961 President Kennedy presented NASA with the difficult task of resurrecting the United States' public image. This was the beginning of the Apollo programme, the objective of which was to land a man on the Moon before 1970. Suddenly, funds began raining on the space agency. To supplement Ranger and Surveyor, NASA began to design new unmanned probes including Ranger Block III (which would carry out a simple photoreconnaissance mission prior to hitting the surface), Lunar Orbiter (of which more will be told later) and an ambitious project called Prospector.

The Prospector family derived from a June 1960 Marshall Space Flight Center study of lunar missions that could be launched by the new Saturn I rocket. The first was a 1-tonne lander using a complex system of rockets and air-bags to land a 315-kg scientific payload on the surface of the Moon. The second and most interesting project was a large wheeled rover of 660 kg, able to travel at least 80 km on the surface. The design of the rover was quite unusual, comprising two large inflatable 4.9-metre-diameter wheels and a third, far smaller, wheel on the back. The probe would use an electrical generator using mercury heated on a solar concentrator as a working fluid. The project was adopted by NASA during the same year, and was to be managed by JPL. The first missions were to be carried out during 1965–66, and two versions of the probes were initially under study. The first was a large 1-tonne rover carrying up to 130 kg of scientific instruments: a magnetometer, a gravimeter, a camera, some surface composition spectroscopic instruments, a mass spectrometer to study a possibly thin atmosphere, a gas cromatographer to search for improbable traces of microorganisms, and other instruments such as thermometers, thermal flux sensors, and so on. Some consideration was also given to loading small explosive charges on the rover, to be deployed on the surface for seismological studies in which seismic waves from the explosions would be detected by other landers such as the Block II Surveyors. In NASA's plans, the rover would evolve into a vehicle able to explore the far-side of the Moon. The second version was a large lander, of relatively simple technology, designed to collect and return to Earth up to 11 kg of samples from any point of the surface. A multitude of such missions would provide a comprehensive characterisation of lunar geology which a few human landings would never provide (and, in fact, did *not* provide). It was also expected to be able to collect samples, using a special drill, from a maximum depth of 1.5 metres. With time, more versions of the Prospector spacecraft were proposed: an organic sample carrier with which to take animals to the Moon and back to Earth to study the effects of radiations on living beings, a beacon for Apollo landings, and a logistic version to carry fuel and consumables to the astronauts. In the end, however, Prospector increased in weight without control, swinging between the powerful Saturn I and the gargantuan Saturn V launcher, and the project – designed to act as a pathfinder for human landings at a time when these were still far into the future – essentially became a sort of substitute. Thus, one year after approval, in 1962, Congress withdrew funding.

2.4   THREE FAILURES AND AN UNEXPECTED SUCCESS

The first Ranger Block II, Ranger 3, took off from Cape Canaveral on 26 January 1962, targeted to an area near the eastern border of Oceanus Procellarum just south of the lunar equator, this site being dictated by the requirement for the Earth to be close to the zenith for seismometer transmissions.

The launcher experienced no problem, and the Agena–Ranger complex entered a perfect parking orbit. Although the Agena was re-ignited at the correct time to put Ranger 3 on an Earth escape trajectory, a guidance error resulted in a 32,000-km Moon-miss distance, which was far too large for the small hydrazine engine to correct. The spacecraft would cross the Moon's orbit some fourteen hours before its target reached the same spot. It was, however, decided to attempt a course correction in order to bring the spacecraft as close as possible to its intended target and to take some pictures of it. Unfortunately it was not possible to test the lander separation manoeuvre, as this had to be commanded by the radar altimeter. The course correction produced the expected change in velocity, but because of another guidance error it was given in the incorrect direction, so that the miss distance was increased to 36,785 km!

Some 50,000 km from the Moon, the camera was powered on and commanded to take some pictures. But at this very moment, disaster struck. The high-gain antenna lost track of the Earth, and the onboard computer jammed. Only a few very 'noisy' images of the sky were received, with fiducial marks added. The only positive aspect of the mission was that Ranger 3 collected some 30 hours of gamma-ray spectrometer data, providing the first characterisation of the radiation background in a range of energies between 0.1 and 2.6 MeV. The disturbance caused by the metallic body of the spacecraft was also clearly detected, as gamma-ray counts before the deployment of the instrument's boom were double those after deployment. Concluding its unlucky mission, Ranger 3 entered solar orbit, with a perihelion of 147 million km, an aphelion of 174 million km, and a period of 406 days.

Ranger 4 was launched just three months later, on 23 April. This time the Atlas–Agena worked as designed, and placed the probe on a collision course for the Moon. When the time came to establish communications, however, there was no answer except for the transponder system. The spacecraft's clock had stopped, preventing the deployment of the antenna and solar panels and the normal working of the telemetric system. The craft was followed, using the lander's transmitter, until it disappeared behind the lunar limb. Calculations indicated that it crashed a few minutes later on the far-side at 15°.5 S, 130°.7 W, near the crater Paschen.* There was some celebration, because for the first time an American probe had reached the Moon. Nevertheless, the Soviets had succeeded in doing so three years earlier, with a perfectly functioning probe instead of a 'radio-equipped idiot'! (It has been reported that the crater produced by the impact has been identified in Lunar Orbiter pictures, but this is probably untrue.)

---

* Friedrich Paschen (1865–1947), German physicist

Ranger 4 was the first American-built probe to hit the Moon, but its mission was otherwise a complete failure. It is shown here without its solar panels, during launch preparations. The hard lander is painted in a black-and-white pattern for thermal control. The cylindrical object between the two white avionics bays is the camera objective, and the radar altimeter antenna is barely visible behind the retro-rocket. (Courtesy JPL/NASA/Caltech.)

After this string of failures, America gained some relief from an unusual Ranger probe. NASA planned to launch, in 1962, two large 500-kg Mariner A spacecraft to Venus, but as the day of the launch approached, it became clear that the rocket that was to take Mariner A into space – the aforementioned Atlas–Centaur – would not be ready in time. (The official NASA policy on the naming of deep-space probes required that lunar probes be given names connected with geological exploration – Ranger, Surveyor and Prospector – and interplanetary probes names connected with sea exploration – Mariner, Voyager and Navigator, the latter being a proposed probe to explore the Sun, comets, Mercury and Jupiter.) With less than a year to go to the 'launch window', NASA accepted a bold JPL proposal to launch two stripped Ranger Block Is (renamed Mariner–Ranger, or Mariner R) to Venus with a 9-kg scientific payload to collect some basic scientific data on the cloud-enshrouded planet. Based on the experience of previous Ranger flights, the two missions would quite probably fail. The first spacecraft – Mariner 1 – was launched by an Atlas–Agena on 22 July 1962, but suffered one of the most bizarre mishaps: a missing dash in the transcription of the launcher's guidance algorithm meant that the Atlas overcompensated for minor velocity errors. The flight of the first US probe to Venus lasted less than five minutes before the self-destruct system was activated.

Although the cause of the guidance fault had not yet been found, it was decided to proceed with Mariner 2 before the launch window closed. The Atlas performed almost flawlessly, the Agena restart was flawless, and the probe kept running until 14 December, when it became the first man-made object to fly by, and return data from another planet. Mariner 2 banished the myth of Venus as a twin planet to Earth, a clouded planet with tropical vegetation, and a climate similar to that of our own Carboniferous period; it was an inferno with a surface hot enough to melt lead.

With Mariner 2 on its way to Venus, it was time for Ranger to prove its worth. The Agena payload performance computation error had been finally corrected, and JPL had added redundancies, so Ranger 5, the last of the Block II family, was launched on 18 October 1962 after some extensive modifications to the Agena, to the Ranger onboard computer – which failed on the pad some 50 minutes before lift-off – and to other spacecraft components. The flight was flawless for about 75 minutes, after which the solar panels shorted out and power had to be drawn from the batteries. Some five hours of gamma-ray spectrometric data were received, in the undeployed configuration only, before the batteries were exhausted during a course correction manoeuvre that was carried out in the ignorance of the actual spacecraft attitude. Ranger 5 flew by the Moon at a distance of some 724 km and, despite the main spacecraft being inert, provided the first reliable measurement of the lunar mass, based on the Doppler effect on the radio signal of its lander. It was tracked up to a distance of 1,271,110 km, before it entered solar orbit at 157 million km aphelion, 142 million km perihelion, and $0°.39$ inclination to the ecliptic.

With the end of the Ranger flights, an investigation was carried out in order to understand why all the spacecraft had failed. It was concluded that the probes had insufficient redundant systems, and had been degraded by the sterilisation procedures. In particular it was discovered that the ethylene oxide atmosphere used to kill bugs could damage electrical wire insulation, making it very brittle and therefore susceptible to damage by the vibration at launch.

In the case of Ranger 5, a possible connection was found between the spacecraft failure and one of the space nuclear tests of America's Dominic series, which was carried out a few days before and which had already caused other satellites to fail.*

2.5   MOSCOW CRIES TOO

After the Luna 3 mission of 1959, the Soviet Union did not return to the Moon for

---

*   In particular, the explosion had knocked out the first British scientific satellite, Ariel, and the Telstar 1 telecommunication satellite owned by AT&T.

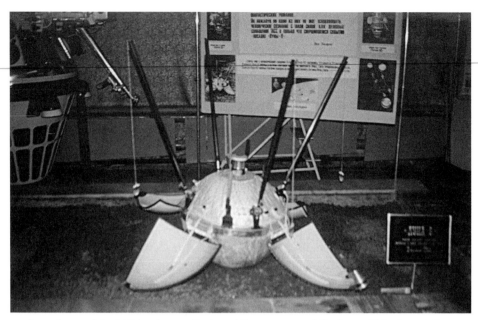

A mock-up of the E-6 lander on display at the museum of the Lavochkin Association. (Photograph copyright Wolfgang Hausmann.)

some years, during which time it tried unsuccessfully to launch several spacecraft to Mars and Venus, and succeeded in putting a man in space for the first time.

The Tyuratam cosmodrome, meanwhile, had suffered the most gruesome catastrophe in the history of rocketry. On 24 October 1960 an R-16 ICBM developed by Mikhail Yangel (to become known in the West as SS-7) exploded on the launch pad, killing between 50 and 150 engineers and technicians. Among the victims was Marshal Mitrofan Nedelin, chief of the Red Army's Strategic Rocket Force, on whose death the TASS press agency invented the fake story of an air accident.

In 1963, with a new family of probes (the E-6), a new launcher (the 8K78 – a development of Semiorka which substituted stage E with stages I and L) and using the parking orbit technique, it was time for the Soviet Union to resume its lunar programme.

Stage I was derived from the new R-9 or 8K74 ICBM developed by Korolyov. It was powered by an engine developed by Semyin Kosberg, fuelled by liquid oxygen and kerosene stored in two spherical tanks, and was used to take the payload and stage L to the Earth parking orbit. Once in space, stage L would fire to achieve escape speed. This stage was also fuelled by liquid oxygen and kerosene, and was provided with compressed gas jets for attitude control, and four small solid-fuel ullage engines to settle the fuel prior to main engine ignition in weightlessness. Using the new launcher – first tested in 1960 and named Molniya (Lightning) after its typical payload, a family of communication satellites – the Soviet Union could hurl some 1,600 kg to the Moon or 1,200 kg to nearby planets. Its introduction was traumatic, as testified by eight failed planetary launches in ten attempts between 1960 and 1962.

The E-6 probe comprised an engine module, a cruise module and a mini-lander, and was 2.7 m tall. The propulsion module was built around a single KTDU-5A (Korrektiruyushaya Tormoznaya Dvigatelnaya Ustanovka – course correction and braking engine) rocket engine developed by Alexei Isayev having a thrust of 45.5 kN. It was fuelled by a hypergolic combination of amine and nitric acid, with amine carried in a toroidal tank and nitric acid in a spherical tank. Every probe carried enough fuel to slow the spacecraft's speed, with respect to the Moon, from 2.6 km/s to 10–30 m/s. On top of the propulsion module was the pressurised cylindrical cruise module, which contained batteries, computer, and radio communication systems. (In contrast with the US Rangers, the Soviet E-6 relied on battery power only for the flight from the Earth to the Moon.) Outside the cruise module were mounted two compartments carrying the attitude control system, nitrogen gas tanks and radar altimeter. These compartments were jettisoned a few kilometres above the Moon, to reduce the probe mass for retro-rocket firing.

On top of the cruise module was the hard lander, protected by a thermal cover and by two hemispherical air-bags which, upon being inflated to 1,013 hPa, enabled

A Soviet E-6 probe in its cruise configuration, showing the cruise stage that was standard for all of the E-6 family probes. Nearest to the camera is the radar altimeter block with its parabolic antenna. The lander is enclosed in the thermal shroud at top. (Courtesy Jean-Jacques Serra.)

it to survive impact at speeds of 4–24 m/s. The lander – also known as the 'egg' – was a 100-kg, 58-cm-diameter metallic sphere, using batteries for power generation and water for thermal control. This sphere was bottom heavy, and incorporated four spring-loaded petals on the top hemisphere in order to maintain an upright position. The petals also acted as an ingenious antenna system, and during the flight protected the main scientific instrument, a simple 3.6-kg Volga camera which was aimed toward the lander's zenith where, mounted on a 2-degree-freedom moveable support, was a tiny mirror. Moving the mirror, the camera could record a full panorama of the landing site in about an hour. The maximum camera resolution was 5.5 mm at a distance of about 1.5 metres. In the camera's field of view were mounted three dihedral mirrors to produce three-dimensional views of short strips of the surface, and dangling from the four whip antennae used for communications were small calibration targets – which were also used to measure the tilt of the lander. Korolyov proposed to load the lander with a 5-kg instrument package consisting of a seismometer, a magnetometer and a cosmic ray detector. Although none of these instruments ever found a place in an E-6 probe, others eventually did.

The E-6 mission profile was quite similar to the Ranger mission profile. The probe was first to be placed in a parking orbit, and be hurled to the Moon at the correct time by stage L. During the flight, a single course correction manoeuvre would be carried out, after which, rapidly closing to the Moon, the probe would align its engine thrust line with the velocity vector and inflate the air-bags to protect the lander. Some 70 km above the lunar surface, the radar altimeter would command the jettisoning of the external compartments, and then ignite the braking engine. The retro-rocket would shut off a few hundred metres over the surface, after which its role would be taken over by four 245-N-thrust engines. Five metres above the Moon, a boom-mounted contact sensor would sense the surface and release the lander, which would be pushed away from the crashing cruise stage by the pressure in the air-bags. After bouncing for some time, the two air-bag hemispheres would separate to leave the metal sphere free to come to rest. It would then open its petals and take its picture.

In contrast with the Rangers, the Soviet E-6 probes had a serious operational limitation. They could perform velocity changes in a direction perpendicular to the lunar surface only, and could thus land successfully only where celestial mechanics produced an approach trajectory that was vertical relative to the surface: namely, in Oceanus Procellarum very close to the western limb of the lunar near-side.

The first of the new probes took off from Tyuratam on 4 January 1963, and was stranded in parking orbit after an electrical system fault prevented ullage rocket ignition. The second attempt, on 3 February, did not even achieve parking orbit because a malfunction at upper stage separation incapacitated the whole attitude control system; the launcher and its payload fell into the Pacific Ocean. The third launch, on 2 April, was successful and upon setting off for the Moon it was named Luna 4. However, it experienced an attitude determination system malfunction, did not correct its trajectory and, after flying by the Moon at a distance of 8,336 km some 88 hours after lift off, entered an unstable terrestrial orbit of 89,800 × 698,000 km, from which it has since slipped into solar orbit. The only scientific data collected were solar wind and cosmic ray measurements. The cause of the failure was never identified, but an investigation found dozens of potential defects, some of which were in the attitude determination system.

It is worth noting that two Italian radio hams – the Judica-Cordiglia brothers, who were well known for claiming to have intercepted transmissions from 'phantom' Soviet cosmonauts – stated that they had recorded six images of the Moon's surface taken by Luna 4. No other Western intelligence station achieved anything similar, and no other E-6 probe is known to have taken images on its way to the Moon, so this claim is certainly spurious, if only because the National Security Agency, the 'blackest' of American intelligence agencies, intercepted the spacecraft's telemetry during the fly-by, noted its extreme complexity, but did not find any trace of an image.

The next Soviet launch took place on 11 November. This time, the payload of the Molniya rocket was not an E-6; it was an engineering test of a generic spacecraft designated 3MV-1A, which was designed to explore Mars and Venus. After a long string of failed missions to the two nearest planets, the Soviets had decided to gain some deep-space navigation experience by performing a simpler flight to the Moon. The scientists were also interested in continuing the far-side mapping, taking new images with better resolution. This time, however, stage L lost its attitude shortly before ignition, and fired in an incorrect orientation. After being stranded in a low parking orbit, it was named Cosmos 21.

Starting in 1961, Korolyov's team had also begun to develop an entire family of extremely powerful new space launchers, with a maximum payload in excess of 80 tonnes. In parallel, Vladimir Chelomei's design team was developing its own heavy launcher, named UR-500, or Proton, which was able to lift up to 21 tonnes to Earth orbit. This was the beginning of the Soviet human lunar programme.

Up to this time, the race to the Moon had involved two competitors only; but a third contender appeared briefly in 1961 when a ten-year plan was proposed by Europe. This included the development of a launcher based on the British Blue Streak ballistic missile, and of a liquid oxygen and hydrogen fuelled launcher for a small lunar lander or orbiter, possibly as soon as 1967. The proposal led to the formation of two European consortia – the European Launcher Development Organisation (ELDO) and the European Space Research Organisation (ESRO) – from which the European Space Agency (ESA) was created in 1975. However, the only part of the plan to actually be developed was the Blue Streak-based launcher, named Europa, which failed four orbital launch tests out of four. The first successful European launcher (using a liquid hydrogen/liquid oxygen upper stage) was Ariane, first launched in 1979. The first European lunar probe did not get off the ground until 2003, but it is interesting to consider how the race for the Moon might have turned out had Europe been involved.

2.6 A STRING OF SUCCESSES

As already mentioned, after the beginning of the Apollo project Ranger Block III was approved, but with much less ambitious objectives than Block II. The mini-

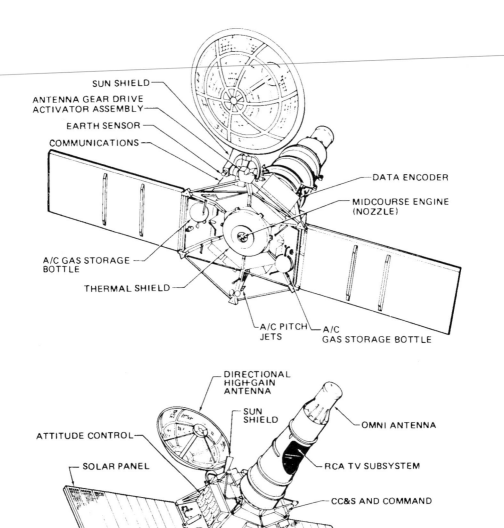

SUN SHIELD

ANTENNA GEAR DRIVE
ACTIVATOR ASSEMBLY

EARTH SENSOR

COMMUNICATIONS

DATA ENCODER

MIDCOURSE ENGINE
(NOZZLE)

A/C GAS STORAGE
BOTTLE

THERMAL SHIELD

A/C PITCH
JETS

A/C
GAS STORAGE BOTTLE

DIRECTIONAL
HIGH-GAIN
ANTENNA

SUN
SHIELD

OMNI ANTENNA

ATTITUDE CONTROL

SOLAR PANEL

RCA TV SUBSYSTEM

CC&S AND COMMAND

A/C GAS
STORAGE
BOTTLE

POWER

ATTITUDE CONTROL ROLL JET

BATTERY (2)

ATTITUDE CONTROL YAW JET

ATTITUDE CONTROL PITCH JET

BACKUP TIMER

PRIMARY SUN SENSOR PAD (4)

The Ranger Block III spacecraft.

lander had been discarded, and the power generation system had been improved by the addition of a spare battery and the use of the rectangular solar panels of Mariner R instead of the trapezoidal Block I and Block II panels. Each panel was insulated in order to prevent the accident that incapacitated Ranger 5. A new, more powerful and more precise course correction system had been added, the computer was modified, redundancies were increased, and the sterilisation procedure, having been found to be useless and a potential source of trouble, was discontinued.

In place of the lander, a conical 1.5-m-tall aluminium tower housed six vidicon cameras, of two different groups using different computers and transmission lines. The first group was called the F (Full scan) channel, and comprised two cameras – one with a 25-degree field of view and a 25-mm focal-length objective, and another with an 8.4-degree field of view and a 75-mm focal-length objective. This channel produced a 1,152-scan-line image every 2.56 seconds. The second group was the P (Partial scan) channel, comprising two 25-mm and two 75-mm focal-length cameras. Only 7% of each picture taken by each camera was relayed to Earth, providing up to five frames per second until a fraction of a second before impact. The P channel produced 300-scan-line images, and the last frame was expected to yield a ground resolution as high as 30 cm. The cameras were to be switched on from Earth, but if the probe did not receive any command it was instructed to turn them on automatically 67 hrs 45 min after lift-off.

Originally, the probes were to carry 22.5 kg of additional scientific instruments – in particular, to record the radiation dosage to which the astronauts would be exposed; but none of these were ever actually carried, which led to criticism that was to accompany the rest of the American lunar programme. Moreover, the targets for the Block IIIs were mostly in the great equatorial maria, with the main objective being to locate sufficiently smooth landing sites for the Apollo flights. Once this had been achieved, JPL's understanding was that Block IV and Block V would be committed to scientific missions. Block IV would have been quite similar to Block III, with the addition of a gamma-ray spectrometer and a surface scanning radar, and Block V would have reintroduced a mini-lander, carrying either a camera that would break the balsa shell to take pictures from the lunar surface, or an improved version of the Block II seismometer. In addition, some of these probes were to carry SURface Mechanics Capsules (SURMEC) consisting of small tennis-ball-sized penetrators to measure the bearing strength of the surface. But the design of the Apollo Lunar Module landing gear was decided even before Block III spacecraft began flying, and the more advanced probes were cancelled, thus depriving the project of even more scientific and technical meaning.

The principal scientific investigator of the last Ranger flights was Gerard P. Kuiper, the scientific consultant to the US lunar atomic bomb project who had at first derided the Luna 3 images of the far-side as 'gross extrapolations from marginal data or even outright forgeries'.

The first of the Block IIIs – Ranger 6 – lifted off from Cape Canaveral's Pad 12 on 30 January 1964, more than two months late, after the discovery of several defective parts in its electric circuits. At 364.7 kg against 340 kg, it was significantly heavier than its predecessors. Its target was located slightly north of the lunar equator, in

The only scientific payload of the Ranger Block III spacecraft consisted of these six cameras. The two centre cameras form the full-scan channel (at top is the wide angle camera, and at bottom is the narrow angle camera), while the other four cameras form the partial-scan channel.

Mare Tranquillitatis, near the crater Julius Caesar.* The launch was perfect, but for the fact that for some reason the camera telemetry was activated for a minute or so immediately following the Atlas booster's separation. The initial trajectory determination revealed that the spacecraft would miss the Moon by about 100 km, but the impact was assured once the course correction manoeuvre had been carried out. Some 2,000 km above the surface of the Moon, approaching 2.5 km closer every second, the command to turn on the camera systems was sent. Everyone hoped for the best ... but only static streamed out of space. A string of emergency commands was sent, but to no avail. On 2 February, Ranger 6 crashed in Mare Tranquillitatis, at 9°.39 N, 21°.51 E, without returning a single image. NASA was hit hard by the outcome of the mission. From the public's point of view it was a failure, despite the success of the course correction and the accuracy of the impact. The launch of Ranger 7 was postponed while no less than three investigations were carried out in parallel. Six weeks later, the failure report was released. It concluded that a spurious electrical signal of unknown origin had turned the cameras on during launch where, because of the still quite dense atmosphere, the power systems of both camera channels had arced and shorted out their power supplies. It was later discovered that during booster separation some 180 kg of hot, conductive gases escaped from the Atlas's tanks, and found its way into the spacecraft's shroud through the small hatch which, up until the moment of launch, had provided access for the cables which had enabled the ground to monitor the spacecraft's systems, and it was this that had caused the short circuit. To prove this, a recording of the launch was examined, and it clearly showed sparks flying out of the shroud at the exact time of camera telemetry turn-on. The report of another investigation board, however,

* Julius Caesar, Roman general 100–44 BC.

This Ranger 7 image was the first image of the Moon ever taken by an American spacecraft. It clearly shows the crater Alphonsus (the target of the Ranger 9 mission) at centre right, with Ptolemeus (153 km) above and Arzachel (97 km) below. Other features include the Davy Catena (just below the second mark from the right of the first line), and the shallow crater Guericke (58 km; the large crater on the left), to the west of which the probe was eventually to impact. (Courtesy JPL/NASA/Caltech.)

bitterly criticised the probe's design. Not enough redundancy was included, mainly in the camera power system. It was proved that in such a breakdown both camera channels would be rendered useless. As the only objective of the mission was photography, the whole mission would have then been a total loss. The access hatch in the shroud was modified to form a hermetic seal and, after modifications, Ranger 7 was readied for launch in the summer of 1964.

Meanwhile, on 21 March and 20 April the Soviets had unsuccessfully launched two more E-6 probes. The first – launched on a modified Molniya, designated 8K78M – was lost because of a problem in the third stage fuel valves, and the second did not reach Earth orbit because the launcher shut off too early after an electrical system failure.

Ranger 7 was launched on 28 July, and headed for Mare Nubium. Once again, the probe's initial trajectory would have resulted in a fly-by of the Moon, and a course correction manoeuvre was carried out. Some 2,277 km above the lunar surface, on 31 July, the camera turn-on command was sent and, after 90 seconds of camera heating, the F channel images began streaming back to Earth, followed some 3½ minutes later by the P channel images. The probe impacted the Moon 17 minutes later at a speed of 2.62 km/s, at 10°.63 S, 20°.6 W, some 13 km from the planned site, after taking 4,316 pictures, of which the last – taken at a distance of around 300 metres, and only partially relayed before the probe crashed – showed metre-sized details. This was the first success of the US lunar programme after no less than twelve failed or inconclusive missions. The crater produced by the impact was later identified in an Apollo 16 picture. It was a small white crater with a diameter of just 14 metres. The area of the impact was later called Mare Cognitum (Sea of

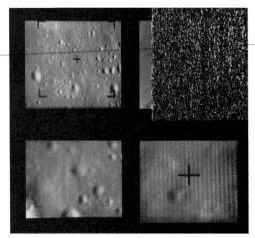

The last images captured by the Ranger 8 partial-scan camera chain. The images are (clockwise from upper left) P3, P4, P2, P1. P3 was taken from a height of 1.09 km, P4 is incomplete (as the spacecraft was destroyed during the transmission), P2 was taken from a height of 640 metres, and P1 was taken from a height of 1.52 km. (Courtesy NASA/NSSDC.)

Knowledge), and a small, 6,800-metre-diameter crater, some 50 km from the impact site, was dedicated to Kuiper after his death in 1972.

The Ranger images caused a controversy in the scientific world. Based on early science fictional depictions, it had been expected that the lunar surface would be very rough, so its smooth appearance was a surprise. This was interpreted to mean that there was a blanket of fragmented rock – a 'regolith' – but there was no agreement regarding its thickness. Kuiper's estimate (1.5 metres) and the estimate of the geologist Eugene M. Shoemaker (15 metres) differed by a whole order of magnitude. The most correct scientific opinion was surely that of Harold Urey (winner of the Nobel Prize for Chemistry, in 1934) who described the images as 'inconclusive'. The critics of the project had little trouble in proclaiming the results of the mission to be second-rate science.

The next Ranger probe – Ranger 8 – was launched more than six months later, on 17 February 1965. Its target in Mare Tranquillitatis was close to the remains of Ranger 6. As on the previous missions, a course correction was required. During the 27-minute manoeuvre, however, the telemetry was repeatedly interrupted, which hinted at some problems in the high-gain antenna pointing system. Thus, when control over the probe was regained it was decided not to risk losing contact during the terminal manoeuvre to point the cameras along the velocity vector, so they were left pointing down obliquely. Although the images would be slightly blurred by the lateral motion, the changing perspective, as the spacecraft slid 'sideways' across many tens of degrees, provided a bonus, particularly because the part of the track already imaged by Ranger 7 would facilitate stereoscopic views.

Twenty-three minutes before impact, the cameras were turned on, and a total of 7,137 pictures were taken up to the impact. The probe crashed at 2°.71 N, 24°.63 E –

A close-up view of the Ranger 8 impact area taken by Lunar Orbiter 2 in November 1966. Both craters C1 and C2 are good candidates for being the crater made by the impact. In particular, C1 has a bright asymmetrical halo of material aligned with the direction of travel of the probe, and C2 has a tiny central peak – a feature observed in other artificial lunar craters. The area is less than 400 metres across.

but its story did not end there. In a three-metre-resolution image taken by Lunar Orbiter 2, two small craters were found aligned with the direction of motion of Ranger 8, either of which could be the crater produced by the impact. The first has a hint of a recent asymmetrical spray of ejecta, and the second, some 14 metres in diameter, shows an unusual central peak. It may be that at some time in the future a small roving vehicle will visit the two craters to make a detailed inspection.

The next two launches were again Soviet. On 12 March 1965, stage L of the Molniya rocket did not ignite because of an electrical problem, and the E-6 probe – named Cosmos 60 as a cover – remained in low Earth orbit for five days. During this time the satellite collected data using a gamma-ray spectrometer. This was the last time this instrument was mounted on a Soviet landing probe, but a similar one with a narrower measurement field from 0.6 to 2 MeV was carried by the Luna 10 and Luna 12 orbiters.

Up to this time, the Soviets had launched six E-6 probes, of which no less than four had failed because of a fault in the electrical system, itself connected to the attitude control system designed by the Piliugin office. For the E-6 probes, the Soviets elected not to use the attitude control system of the Molniya upper stage and instead use the probe's system to control the escape manoeuvre – which contributed to the string of failures.

After the flight of Cosmos 60, it was finally decided to separate the two attitude control systems, as was the case for planetary probes. At the same time, the E-6's system was modified by substituting some of its components – in particular, the converter from the batteries' 30-V direct current to 40-V three-phase current. This component was suspected of being responsible for the failure of at least three missions. Unfortunately, the next Soviet probe, launched on 10 April, failed to achieve parking orbit after the third-stage oxygen tank lost pressure.

It was now the time for the last US Ranger probe, which was targeted – after a suggestion by Kuiper – to investigate one of the then-most interesting scientific sites on the Moon: the large crater Alphonsus* near the centre of the near-side.

The interest in this crater, derived from a long series of observations of strange phenomena like fogging of its floor or dark, short-lived spots. On 26 October 1956, astronomer Dinsmore Alter at Mount Wilson Observatory, California, used a 152-cm telescope to obtain images of the crater through coloured filters and noted that those which he took using a blue filter showed some details as slightly blurred with respect to similar details in nearby craters. This observation compelled Soviet astronomer Nikolai A. Kozyrev to begin monitoring Alphonsus with the 152-cm Zeiss telescope of the Crimean Observatory. On 3 November 1958, when he was recording a spectrogram of the crater near (lunar) local sunset, he noticed that the characteristic central peak appeared 'more brilliant and more white than usual', only to return to its previous status less than a minute later. Perplexed, he stopped the exposure of the spectrogram, and then began a second, 'control' exposure. The analysis of the first spectrum showed an anomalous emission line centred at 4.737 nm. Kozyrev interpreted this as a volcanic eruption that had released molecular carbon. He interpreted a second such observation on 23 October 1959 as a high-temperature lava flow. Some years later, the spectrogram was examined in detail by the scientific community, and was better interpreted (by Soviet scientists!) as a cold-gas venting event rather than a volcanic eruption. NASA hoped to shed some light on the mystery with Ranger 9. In case the lift-off had to be delayed, however, other sites of scientific interest were chosen – namely the craters Copernicus and Kepler, and Schröter's Valley.†

Ranger 9's lift-off on 21 March 1965 was perfectly timed, and after a slight course correction the probe headed to its target. The cameras were turned on 20 minutes

---

*   Alphonsus the Wise, King of Castille, who in the thirteenth century published the most important mediaeval astronomical almanac, the *Alphonsine Tables*.

†   Nicolaus Copernicus (1473–1543) formulated the heliocentric theory; Johannes Kepler (1571–1630) formulated the laws of planetary motion; and Johann H. Schröter (1745–1816) was a German selenographer.

out, and the impact was aired live on television for the first time. Ranger 9 took 5,814 pictures – the last of which showed details only 25 cm across – before crashing on the floor of Alphonsus at 12°.83 S, 2°.37 W. The crater it produced was later identified in Apollo 16 images.

None of the images taken by Ranger 9 showed anything that could be described as the result of a recent volcanic eruption, and Kozyrev's observation was therefore considered at best as a cold-gas release event. But after being re-examined in 1999, even this possibility was discarded. The modifications in the visual appearance were probably due to the Sun setting on the crater; and as for the spectrogram, Kuiper had long ago pointed out that a gas cloud in front of the illuminated peak should have produced absorption lines rather than emission lines! What Kozyrev recorded was probably just a trivial and frequent telescopic tracking error which occurred during the 30-minute exposure.

The whole Ranger programme cost $267 million in 1965 – equivalent to more than $1 billion nowadays – and provided some interesting data on our natural satellite, including a hint of the irregularity of its gravitational field. Unfortunately, now that the technology was at last mature, there was no political interest in funding six more probes equipped to provide high-quality data from scientifically interesting sites at the paltry additional cost of $72 million.

When the Ranger programme was still unfolding, NASA cancelled JPL's proposed Surveyor B and ordered the development of a much smaller orbital mapper that could be launched by an Atlas–Agena rather than the Atlas–Centaur, the development of which was running late. The primary objective of this Lunar Orbiter project was to provide very-high-resolution (i.e. objects 3 metres across) images of the proposed Apollo landing sites. As JPL was seriously overburdened with work on the Ranger, Surveyor and Mariner probes, NASA's Langley Research Center was tasked with developing the lightweight orbiter. As with Surveyor, the

Alphonsus (118 km) taken by Ranger 9, clearly showing some of the dark spots on the floor of the crater. The spacecraft was to impact between the central peak and the rima at the eastern (right) rim. (Courtesy NASA/NSSDC.)

building of the Lunar Orbiters would be undertaken by the private industry. Five firms responded to the tender for the new probes. Hughes and TRW (previously known as STL) proposed spin-stabilised spacecraft, Martin Marietta proposed a three-axis stabilised spacecraft, Lockheed proposed a version of its Samos and CORONA spy satellites which would place an entire Agena stage in orbit around the Moon, and Boeing proposed a small three-axis stabilised orbiter reusing many of the technologies being developed for the Apollo spaceships. On 10 May 1964, Boeing was awarded a contract calling for the construction of five spacecraft.

Meanwhile, the development of the Centaur upper stage was surmounting the problems of using cryogenic propellants. The second launch – which was simply to put a Centaur stage into space without any further manoeuvre – was successful; the third was a complete failure; the fourth took a Surveyor dynamical simulator into orbit, but the re-ignition of the Centaur, which was to place the payload in a very eccentric orbit similar to a lunar transfer orbit, did not succeed; the fifth, on 2 March 1965, was a failure; and the sixth achieved the task assigned to the fourth.

The development of the Surveyor Block II probes was also continuing. The mass of the lander had, moreover, experienced such a growth that an *ad hoc* version of the Atlas–Centaur using liquid hydrogen and fluorine – an extremely powerful fuel, but corrosive and highly toxic – would be required.

In particular, to gain a greater appreciation of a landing site, NASA had decided to include small ground-controlled roving vehicles. Several proposals were scrutinised, of which the most unusual came from Space General Corporation, which proposed a 60-kg rover with a small camera and a tiny manipulator, powered by a moveable solar panel and travelling on three pairs of legs at a top speed of about 3 km/h. The finalists were Bendix Corporation, with a proposal for a robot

The six-wheeled General Motors Surveyor rover. (Courtesy JPL/NASA/Caltech.)

mounted on four tracks, and General Motors, with a proposal for a rover made of three hinged sections, each of which had a pair of large wheels. As the exact mechanical characteristics of the lunar surface were unknown, the two robots, each with a mass of 50 kg, were subjected to a series of extreme tests to demonstrate their ability to travel over terrain ranging from very fine dust to solid lava. The General Motors design proved to be generally superior.

2.7   THE SOVIETS ARE BACK

With the Ranger series over, a fourteen-month hiatus followed, during which the United States did not launch a single probe to the Moon. This left the initiative to the Soviets, who scored with two new feats: the first soft landing and the first lunar satellite.

As the Korolyov office was occupied with other development work, on 2 March 1965 all work on unmanned lunar and planetary probes was reluctantly assigned to the Lavochkin design bureau, led by Georgi Babakin.

The Lavochkin design bureau was formed at the end of the 1930s, and produced some of the finest Soviet fighter aircraft of the Second World War. After the war its prestige began to decline, and it was partly eclipsed by Mikoyan and Gurevich (the famous MiG), Yakovlev, and later Sukhoi. During the 1950s Lavochkin quit aircraft design and began working in other fields. Among the products of these years which must be remembered is Buryia (Storm) – a winged ramjet-propelled cruise missile developed as an alternative to Korolyov's ICBMs. During the 1960s the bureau finally found its place as a designer and builder of robotic spacecraft.

On 9 May 1965, two months after Ranger 9, the Soviets launched Luna 5, which was very similar to Luna 4 but had a mass of 1,476 kg. Once again, the probe's attitude control system failed, orienting it in the wrong direction during the course correction manoeuvre. It began to spin immediately after the engine ignited, but was eventually stabilised. A second, back-up course correction manoeuvre failed, and the probe lost its attitude again just 40 km from the Moon, at the time when the braking rocket should have fired. Nevertheless, it successfully demonstrated a few of the programmed operations, up to the deployment of the lander's thermal cover, before it crashed on 12 May, more than 700 km from its target, at 31° S, 8° W (but according to another source, at 2° S, 25° W). The Rodewish Observatory in East Germany declared that it had taken a sequence of pictures of the impact. These images – taken every 15 seconds – were said to show a dust cloud which expanded to a maximum dimension of 230 × 80 km (!) before settling down in 20 minutes; once again, this information was probably false.

Luna 5 inaugurated the new US space reconnaissance station, cleverly built in Asmara, Ethiopia, on the same longitude as Yevpatoria (the main Soviet deep space tracking station) in order to have above the horizon the same spacecraft that were above the horizon for the Soviets.

Luna 6 was launched one month later, on 8 June. This time, during the course correction manoeuvre an incorrect ignition command was uploaded to the probe,

A Soviet 3MV-class probe similar to Zond 3. This is a Venus lander probe, which differs
from Zond 3 in respect of the spherical lander capsule (bottom) and the position of the
solar panels.

which resulted in it burning its fuel to depletion. It entered solar orbit after passing
some 160,000 km from its intended target. To collect the maximum amount of data
from the otherwise futile mission, the complex lander separation manoeuvre was
attempted for the first time.

The next Soviet lunar probe did not belong to the Luna family, and its original
purpose was not even a lunar mission. This spacecraft – Zond 3 – was launched on
18 July 1965, on a solar orbit that would have taken it very close to the Moon. It was
a spacecraft of the 3MV class – of the same kind as the probe which the Soviets had
attempted to test on 11 November 1963, then sent one each to Venus (2 April 1964)
and to Mars (30 November 1964) as the first two flights in this Zond (Probe) series.
A typical 3MV was comprised of two different sections: the orbital section and the
planetary section. The orbital section was a pressurised cylinder of 1 metre diameter,
housing the computer, batteries, telemetry and probe housekeeping systems. This
cylinder had two wings of solar panels spanning 4 metres, carrying on their tips two
hemispherical radiators for thermal control. The orbital section also had an Isayev
KDU-414 course correction engine, a 2-metre-diameter parabolic high-gain antenna,
and micrometeoroid, magnetic field and cosmic rays sensors. Under the orbital

A mosaic of the lunar far-side, taken by Zond 3. The large dark area is Mare Orientale, at the western lunar limb, and the small dark area at the edge of the lunar disk is the near-side crater Grimaldi. The object at lower left is a camera calibration target.

section was mounted the planetary section, which, in contrast to the generic orbital section, was adapted to the mission and could be an atmospheric entry probe – as was the case with Zond 1, launched toward Venus – or a scientific and photographic module – as was the case with Zond 2, launched toward Mars – although in neither case did the planetary section have the opportunity to operate because the communications systems of the orbital sections fell silent *en route*. In fact, Zond 3 was to have flown to Mars with Zond 2, but technical problems meant that it missed its launch window. When contact was lost with Zond 2 in April 1965, it was decided to send Zond 3 on a trajectory that would enable the instruments in its planetary section to examine the Moon, as a stand-in for Mars, and then wait until it was far from the Earth to transmit the data as a realistic test of the communications system.

Zond 3 carried an experimental attitude control system using magnetohydrodynamic plasma engines and a photographic payload – a 90-cm-diameter pressurised sphere housing a camera for taking ultraviolet spectra, an imaging camera, and an ultraviolet and infrared spectrophotometer. The two cameras used a common film to record three ultraviolet spectra and twenty-five images. This film strip was then to be developed on board, scanned, and relayed to Earth. The image size would be comparable to those of Rangers Block III, at $1,100 \times 860$ pixels. Because of the small mass of the probe (950 kg), the trip to the Moon took just 33 hours. Then, some 11,570 km over the far-side, Zond 3 began to take pictures – one every 135 minutes – until the film was completely used. After this imaging session the probe continued its flight on a solar orbit, simulating a flight to Mars. The images were relayed when the probe was more than 2 million km from Earth. Their quality was clearly superior to the Luna 3 images. The ground resolution was 5 km, and the imaged area included some portions of the Moon's far-side never before seen. The new images confirmed that the far-side is devoid of large maria, and that Mare Parvum (Little Sea) – which some Earth-bound observers claimed to have seen under exceptional circumstances – did not exist. Zond 3 showed, for the first time, the scale and shape of Mare Orientale (Eastern Sea) – a vast 'bull's-eye' formation at (ironically) the *western* limb of the lunar near-side, discovered during the 1930s by the English amateur astronomer Hugh Percival Wilkins, and rediscovered in the course of the preparation of the USAF-sponsored lunar atlas. The ultraviolet spectra revealed that the Moon is less bright in the ultraviolet portion of the spectrum than in the infrared. However, this revelation was not of particular interest, as the instrument had been designed to study the Martian atmosphere. Zond 3 also provided some interesting data on 200-kHz low-frequency radio emissions from Jupiter.

Moving further from Earth, the probe then executed a course correction manoeuvre, and twice relayed images of the Moon: once when 31 million km away, and then when further still. Contact was lost in March 1966, when the probe was 153.5 million km from its mother planet. Frustratingly, after this perfect test, Venera 2, a virtual twin of Zond 3 launched on 12 November 1965, was lost shortly before it reached Venus.

With the next probes, the Soviets were taken back to the hard reality of lunar flights. Luna 7 was launched on 4 October, after a month's delay due to a launcher control system problem. The next day it executed a course correction manoeuvre, which was successful despite some attitude control system problems. Closing with the Moon, the probe was to orient itself with the lunar perpendicular direction, using Sun, Earth and Moon sensors to determine its attitude. Unfortunately, one of these sensors had been mounted incorrectly, and lost sight of Earth. The programming required the spacecraft to abort retro-rocket ignition in this eventuality, and so on 7 October it slammed into the Moon at 9° N, 49° W, near the crater Kepler.

At this point the Kremlin, upset by the long string of failures, began to threaten to cancel the whole lunar probe programme, but it was saved through the intervention of Korolyov on Babakin's behalf.

By the time of the next launch, on 3 December, it had been decided to switch to a lower-inclination parking orbit at 51°.9 rather than 65°, which provided for an increased probe mass; as a result, Luna 8 had a launch mass of 1,552 kg. Like its predecessor, it successfully completed its course correction and continued with its approach to the Moon. Some minutes before the retro-rocket was to ignite, the two air-bags protecting the lander were inflated, but one of them was ripped open by a sharp plastic bracket mounted on a lander petal, and began to vent gas into the vacuum. This had the effect of a small rocket, and put the probe into an uncontrolled spin. The engine started, but shut off after just 9 seconds, as the centrifugal effect of

A full panorama of the Luna 9 landing site (here divided into three parts), prepared chiefly from segments of the third (5 February) panorama. Several spacecraft structures are visible, including aerials, petals and the radiation detector (at top right). At the extreme top left is one of the calibration targets, which also doubled as tilt-meters. At the extreme bottom right, one of the dihedral mirrors can be seen. (From 'An appreciation of the Luna 9 Pictures', *Astronautics and Aeronautics*, May 1966, 40–50 (a publication of AIAA).)

the rotation caused fuel starvation. Luna 8 crashed at 9°.13 N, 63°.3 W, in Oceanus Procellarum.

2.8   TWO SOVIET TRIUMPHS

The first lunar mission of 1966 was a Soviet probe – Luna 9 – which left Tyuratam on 31 January. It was the first E-6M (a slightly modified E-6). Because of the failure of Luna 8, air-bag inflation was delayed until after the braking rocket had fired. For this reason, the nitrogen tank containing gas for the air bags was moved from one of the side modules to the cruise stage. The instrumentation of the probe was modified, too. A radiation counter was mounted on the probe, and the Volga camera was replaced by a new camera – much lighter, at just 1.5 kg – with a power consumption of one sixth the previous camera and yielding a maximum ground resolution of 2 mm at a distance of about 1.5 metres. Every 6,000-line panorama could be relayed to Earth in about 100 minutes, or less if the scan line number was reduced.

After being placed in a lunar transfer orbit, Luna 9 executed its course correction manoeuvre. Some 8,300 km from the Moon it stopped its spin of once every 90 seconds, and readied itself for the retro-rocket firing. At 75 km the radar altimeter sensed the approaching Moon, jettisoned the side modules, fired the retro-rocket, and inflated the air bags. At 250 metres above the surface, the main engine turned off and the four outrigger engines continued to brake the descent. At 5 metres above the surface the contact sensor was triggered, and the egg was projected to a distance of several metres as the rest of the probe impacted at very low speed (22 km/h). The Soviets had taken every possible precaution to ensure that residual fuel did not explode upon impact. The sphere bounced several times, and finally stopped at 18.45.30 GMT on 3 February 1966, in Oceanus Procellarum, at 7°.13 N, 64°.37 W. It began transmitting 250 seconds later, and 15 minutes afterwards it took the first test picture of the lunar surface, on which the Sun had just risen. This image was a disappointment for the Soviets, for it had very little contrast, no doubt due to the low light level of the lunar dawn – the Sun was just 3°.5 above the horizon. Twenty-four hours later, the probe was commanded to take the first complete panorama of the landing site. This unencoded image was intercepted by the British radio telescope at Jodrell Bank, where it was recognised to be a simple teletyped facsimile image and then released to the press some time before the Soviets did the same. The Soviets were very upset by the 'theft', and by the fact that the British image used an incorrect horizontal-to-vertical size ratio which made the surface look more rugged than it actually was. The US Asmara intelligence station also intercepted the images, but these were, of course, never made public. Luna 9 landed inside a 25-metre-diameter crater, near the rim, and was inclined at some 15° with respect to vertical. Close to the spacecraft, no less than fifteen rocks and a dozen craters could be seen, their sizes ranging from half a metre to several metres. This first panorama finally put to rest the theory by Tommy Gold that a vehicle landing on the Moon would sink into fine lunar dust. A second panorama, taken some hours after the first, showed that the egg had moved slightly as the disturbed soil settled, and was now inclined at no less than 22°.5. This fortuitous movement provided the opportunity to create stereoscopic views of small parts of the surface. The probe took a total of four panoramas, the last of which was taken when the Sun was about 40° above the horizon and so was extremely 'noisy':

Image	Date (1966)	GMT (approx.)	Solar elevation (°)
Partial scan	February 3	19.00–19.10	3.5
First panorama	February 4	01.50–03.30	7–8
Second panorama	February 4	15.30–17.10	14–15
Third panorama	February 5	16.00–17.40	26.5–27.5
Fourth panorama	February 6	∼20.00–21.00	∼40.5–41

Some radiation data was also returned until, on the evening of 6 February, after a total of 485 minutes of transmission from the surface of the Moon, the batteries finally expired. Although the landing method was not as elegant as that to be used by NASA's Surveyor, the Soviets had once again beaten the United States. Luna 9's landing site was later named Planitia Descensus (Plain of Landing).

Paradoxically, however, Soviet engineers were not rejoicing about their success, as the man who had led their space programme and achieved many successes, Sergei Korolyov, not witnessed it – he had died two weeks earlier, on 16 January, after a comparatively simple surgical operation which, however, proved lethal, and which was not helped by his having spent six years in a gulag. Only upon his death was it deemed possible for the West to be told the name of the man behind the early space firsts. After a period of public mourning he was interred solemnly in the wall of the Kremlin. The death of Korolyov also marked the end of the 'heroic' period of the Soviet space programme, and the US lead was only a matter of time.

At the beginning of February, Babakin convinced the Academy of Sciences to approve the building of a simplified lunar orbiter, in order to beat the Americans, whose Lunar Orbiter was scheduled to fly during the summer. Just one month later

A mock-up of the Soviet Luna 10 spacecraft. The main spacecraft (far right) is mounted on a standard E-6 cruise stage. (Courtesy Jean-Jacques Serra.)

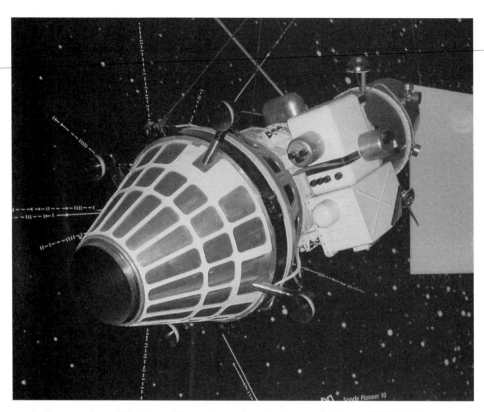

A close-up view of the Luna 10 mock-up, showing the single Isayev KTDU-5A engine and the four smaller outrigger engines. (Courtesy Jean-Jacques Serra.)

on 1 March, the Soviet Union launched its first lunar orbiter, designated E-6S (Sputnik). Instead of Luna 9's egg there was a 1.5-m-long cylinder with a diameter of 75 cm. This was based on the structure of some scientific satellites of the Cosmos family. However, despite the string of recent successes, stage L of the launcher lost roll control and did not reignite, and the probe, stranded in parking orbit, was named Cosmos 111. On 31 March its twin was launched as Luna 10. This time events proceeded smoothly, and at 18.44 GMT on 3 April the spacecraft was braked into a lunar orbit of $350 \times 1,017$ km, with an inclination to the lunar equator of $71°.9$ and a period of 178 minutes. During the first orbit, the engine and cruise stage – identical to those of Luna 9 – were discarded. Inside the probe were mounted micrometeoroid and radiation sensors, an infrared sensor to measure the lunar heat budget, and a gamma-ray spectrometer. A three-axis magnetometer was mounted at the tip of a 1.5-m-long boom. The main apparatus, however, had no scientific purpose: the spacecraft was used to relay to the twenty-third Congress of the Communist Party of the Soviet Union a simple electronic version of the *Internationale*. It has since been revealed that this instrument had partially malfunctioned on the day of the Congress 'performance', and that a recorded rehearsal was played.

The Luna 10 magnetometer confirmed the Luna 2 readings, providing only an upper limit to the lunar magnetic field as 10^{-5} of that of the Earth. The gamma-ray spectrometer detected the difference in composition between the maria and the highlands. The results were crude, but at least suggested the absence of any large bodies of granite or silicic (acidic) ashflow tuff (ignimbrite) from explosive volcanism in the highlands; the dark maria appeared to be basaltic lava flows. In another experiment, the occultation of the probe by the lunar disk was observed by using the radio carrier to investigate the presence of plasma near the surface, but the results were inconclusive. The micrometeoroid sensor detected 198 impacts in 40 days, but the most important discovery made by the spacecraft was the first hint at the extreme irregularity of the lunar gravitational field.

Luna 10's batteries eventually became exhausted on 30 May, by which time the probe had completed 460 orbits in 56 days.

2.9   REVENGE

Just four month after the Soviets, the United States was finally ready to launch its elegant and definitely superior soft-lander, Surveyor 1. During the five-year development, it had been decided to use Surveyor Block I to 'path find' for the Apollo lunar missions. The Block I family was therefore to be an engineering-only spacecraft. After seven of this type, it was planned to launch three instrumented Surveyor Block II. In the end, just like Ranger Block IV and Block V, no funds would be forthcoming for these advanced missions.

Surveyor Block I was 3 metres tall, and was built on an extremely light tetrahedral metallic truss, the mass of which was just 27 kg. Three extremities of the truss carried 4.27-m-span circular landing pads, while the fourth carried a boom on which was mounted a planar high-gain antenna and the solar panels in a characteristic 'book' configuration. Inside the truss was mounted a spherical solid-fuelled retro-rocket with a diameter of 94 cm and a thrust of 45 kN, which would be used to brake the probe prior to landing and then be jettisoned.

Also on the truss were mounted the electronics, sealed inside two boxes wrapped in seventy-five layers of mylar insulation and painted with a mirror finish for thermal control. Two different booms carried the low-gain antennae. A second propulsion system was mounted at the bottom of the truss. This included three vernier engines using monomethylhydrazine and a mixture of nitric acid and nitrogen tetroxide for course correction, attitude control and the final phase of the descent to the lunar surface. The instrumentation had a mass of 30 kg, and included several engineering sensors such as strain gauges, accelerometers, thermometers, and two cameras. The first camera was mounted only on the first two probes, and was intended to image the lunar surface during the descent; but it was never used. The second camera was mounted on a 1.65-m-long boom, and was equipped with a carousel of colour filters and polarisers, and with a zoom lens with focal lengths ranging from 25 mm to 100 mm. The 600×600-pixel images taken with this camera could be relayed to Earth in 3.6 seconds by using the high-gain antenna, or 62 seconds, with 200-line resolution, using the low-gain antennae.

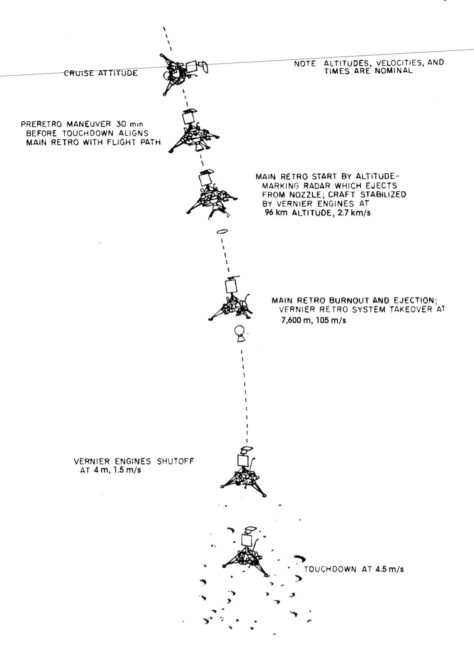

CRUISE ATTITUDE

NOTE ALTITUDES, VELOCITIES, AND
TIMES ARE NOMINAL

PRERETRO MANEUVER 30 min
BEFORE TOUCHDOWN ALIGNS
MAIN RETRO WITH FLIGHT PATH

MAIN RETRO START BY ALTITUDE-
MARKING RADAR WHICH EJECTS
FROM NOZZLE; CRAFT STABILIZED
BY VERNIER ENGINES AT
96 km ALTITUDE, 2.7 km/s

MAIN RETRO BURNOUT AND EJECTION;
VERNIER RETRO SYSTEM TAKEOVER AT
7,600 m, 105 m/s

VERNIER ENGINES SHUTOFF
AT 4 m, 1.5 m/s

TOUCHDOWN AT 4.5 m/s

The descent and landing sequence of the Surveyor probes. (Image adapted from Parks, R. J., 'Surveyor I Design and Performance'; paper presented at the XVII International Astronautical Congress, Madrid, 1966.)

Surveyor 1 – which had a mass of 995 kg, most of which was the 655-kg retro-rocket – was launched by the first operational Atlas–Centaur on 30 May 1966. The powerful Centaur stage put the probe on a direct ascent trajectory to the Moon. The target was the southern part of Oceanus Procellarum, several hundred kilometres from Luna 9. During the flight, a trajectory correction manoeuvre was carried out and with the exception of the low-gain antennae, which failed to deploy, everything went smoothly. At 96 km above the lunar surface the radar altimeter – the antenna of which was cleverly mounted inside the braking rocket's nozzle – sensed the presence of the Moon, and commanded retro-rocket ignition. Forty-two seconds later the braking rocket burned out and separated, and a few seconds later the probe, slowed to a speed of 3 m/s by the three vernier rockets, touched the lunar surface, the undeployed low-gain antennae snapped, the craft bounced once, and finally settled in place. It was 2 June 1966, and Surveyor 1 had successfully landed at 2°.46 S, 43°.23 W, about 14 km from its target. The engineers and scientists at JPL were astonished – they had landed on their first attempt!

After relaying about 36 minutes of telemetric and housekeeping data, Surveyor 1 took the first picture of its landing site. The spacecraft operated for a whole lunar day, and sent back 10,732 images, including some portraits of its feet – intended to validate the Apollo LM landing-gear design – which had sunk to a depth of 2–3 cm in the very fine lunar dust. Other pictures showed the far rim of the buried crater Flamsteed P,* inside which the probe had landed, and a barely visible 'fog' on the horizon, possibly caused by lunar dust in suspension. Several hundred metres from the lander, a large crater was imaged, its rim surrounded with boulders.

The behaviour of the surficial material was observed, and its mechanical characteristics were deemed to be similar to those of damp garden soil, despite the mean dimensions of particles being smaller than 0.01 mm and the likelihood that it was anhydrous. The analysis of the texture of some nearby rocks confirmed the suspicion that the lunar maria were composed almost exclusively of basaltic rocks.

Surveyor 1 landing site details, near and far. At left is a tiny 3-metre-diameter crater not far from the probe. In the same image a field of boulders is visible near the horizon. At right is an image of the north-eastern rim of the buried crater Flamsteed P, some 25 km away; the peaks of the rim rise about 400 metres.

---

*   John Flamsteed (1646–1719), the first Astronomer Royal.

The probe was also used for some simple astronomical observations, taking images of Jupiter, Sirius and Canopus. It was not possible to take pictures of the Earth, as it was very close to the zenith and thus beyond reach of the camera.

When the Sun set on 14 June, the probe took some last pictures showing the solar corona and its own landing pads in the ghostly illumination of earthshine. Finally, the probe was put into hibernation, with the hope of reawakening it after sunrise, when the solar panels would reinitiate power production. The first attempt to rouse it on 28 June was unsuccessful, but the link was re-established on 6 July. During the week before its second sunset, Surveyor 1 took another 618 pictures. Although its condition soon deteriorated in the harsh environment, contact was sporadically maintained until 7 January 1967, when it finally expired. Surveyor 1 had been an outstanding success. It was later found on a Lunar Orbiter 3 picture, which clearly showed the 'boxy' shadow of the solar panels.

2.10 A FAILED ORBITER AND A SUCCESSFUL ORBITER

Just one month after Surveyor, the Americans were ready for another attempt to place a probe in lunar orbit, six years after the last of six failed Pioneer orbiters. However, it was not one the Lunar Orbiter probes.

Explorer 33 – also known as IMP-4 (Interplanetary Monitoring Platform) – was launched on 1 July 1966. The Delta-E launcher was a modified Thor rocket, with a more powerful engine, better upper stages, and three Thiokol solid-fuelled strap-on boosters. Since the launch of the Pioneer orbiters on Thor-Ables, McDonnell had transformed the scarcely reliable Thor into a launcher that was so successful that it still remains in service. The mission of Explorer 33 was similar to that of the previous IMP satellites (Explorers 18, 21 and 28): the study of interactions between the terrestrial magnetic field and interplanetary space, and in particular the solar wind. IMP-4 differed from its predecessors in that it was carrying one less scientific instrument (the rubidium vapour magnetometer), and that its mission was to be carried out from an unusual observation point: lunar orbit. Being there, the satellite would aid the mapping of the lunar magnetic and gravitational fields. The spacecraft had a mass of 93.4 kg, and was a 71-cm-diameter, 20-cm-tall octagon. From this central body protruded four large solar panels, a solid-fuelled orbit-insertion motor,* two 1.8-m-long magnetometer booms, and four communication aerials. The probe carried a transmitter, a small data computer, and six scientific instruments.

Paradoxically, despite fears that the probe would not enter lunar orbit because of some launcher malfunction, this manoeuvre failed because the launcher was too exuberant and gave too great a speed to the satellite. Explorer 33 was stranded in a $15,897 \times 435,330$-km Earth orbit, where it conducted a mission similar to its predecessors. It worked perfectly well for five years, during which time its apogee

---

* Identical to that used to place the first geostationary communications satellite, Syncom, into orbit in 1964.

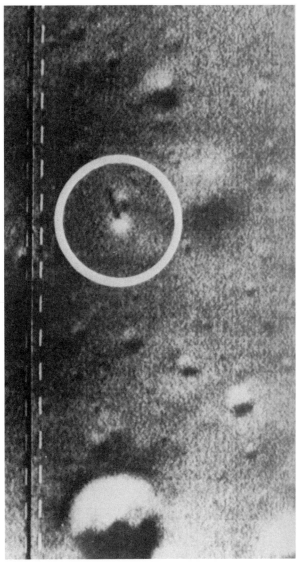

Lunar Orbiter 3 took this image of Surveyor 1 standing on the surface of the Moon. The 10-metre-long shadow of the solar panel is evident (cast in the opposite sense to those in craters).

varied between 400,000 km and 900,000 km. The last scientific data were relayed on 15 September 1971, and with a final, brief, contact six days later.

Meanwhile, the Atlas–Agena carrying the first of five Lunar Orbiters was being prepared. After the contract to build this family of probes went to Boeing, the Lunar Orbiter had materialised as a 385-kg probe, 68 kg of which was allocated to the imaging system.

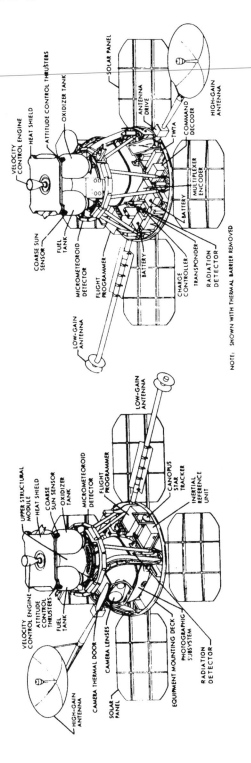

Two drawings of the Lunar Orbiter spacecraft. The small spacecraft was entirely designed around a complex imaging system.

The probe had a truncated cone shape with a base diameter of 1.4 m and a height of 2 m, around the base of which were mounted – like petals – four squarish solar panels, which took the total span to 5.2 m. At the base of this structure was the Eastman–Kodak imaging system, derived from a similar system developed for military purposes. This was an ellipsoidal pressurised shell, 90 cm long, 60 cm wide and 45 cm deep, with two quartz ports for the cameras fitted with an 80-mm focal length wide-angle lens and a 610-mm focal length lens. The images were recorded on a 70-mm Kodak film with a very 'slow', very fine grain to resist the radiation in space. Each probe carried some 79 metres of film, on which 212 pairs of images (simultaneously exposed, one by each camera) could be stored. After the film was developed by pressing it against a chemically impregnated ribbon for 3½ minutes (a procedure similar to that used in the Polaroid camera), the images were individually scanned and relayed to Earth. The photographic film was preferred over the simpler vidicon cameras because of the extreme resolution expected: $8,360 \times 9,880$ pixels for the wide-angle lens, and a stunning $8,360 \times 33,288$ pixels for the narrow-angle lens.

Alongside the imaging system were mounted the computer and solar and stellar sensors, using Canopus as a reference for attitude determination. On the same base were mounted the antennae: a low-gain antenna and a 92-cm-diameter parabolic high-gain antenna. On top of the structure were mounted four hypergolic fuel tanks and the engine system. The probe also had a micrometeoroid detector and a radiation sensor to provide warning of solar flares which could damage the photographic film.

The Lunar Orbiter 1 imaging system during ground tests.

Lunar Orbiter 1 left Earth on 10 August 1966, targeted to enter a lunar equatorial orbit from which it would take pictures of some of the candidate sites for the first Apollo landing. The flight from the Earth to the Moon, comprising the usual course correction manoeuvres, was faultless – with the exception that the stellar sensors failed to find Canopus, possibly because the structure was reflecting sunlight into the sensor. While a solution was being devised, the Moon itself was used as a reference. On 14 August, Lunar Orbiter 1's engine fired for 578.7 seconds to establish a lunar orbit with an inclination of $12°.2$, perigee at 191 km, and apogee at 1,854 km. The revolution period was $3\frac{1}{2}$ hours. After some engineering tests, the probe took twenty sample pictures covering both hemispheres, and relayed them to Earth on 18 August. These revealed that there was a fault: the system that was to move the film during an exposure to avoid blurring of a high-resolution image due to the probe's orbital motion was not working. Most of the scientists were in favour of operating the probe from a higher than planned orbit in order to conduct general mapping, but NASA ordered that the plan be followed, and so on 21 August the perigee was lowered to 58 km. The probe took a total of 413 pictures covering 262,000 square km of the near-side, and 3 million square km of the far-side. Although most of the high-resolution frames were useless, the 'wider' frames gave preliminary maps of the target areas, so the mission, which also secured the first American pictures of the lunar far-side and, as a bonus, the first two images of the Earth as seen from the Moon, was judged to have been a success.

The most interesting scientific discovery of the mission was the mascons (mass concentrations) – large areas of the surface with a gravitational field stronger than that of neighbouring areas. These were discovered by accurately tracking the probe in its orbit around the Moon. The Soviet Luna 10 probe had already hinted at their existence. Lunar Orbiter 1 indicated that the mascons were usually found inside the maria on the near-side. The micrometeoroid and radiation sensors indicated little danger for human lunar flights.

To avoid interfering with future missions, on 29 October Lunar Orbiter 1's engine was started one last time, to crash the probe on the far-side at $6°.7$ N, $162°$ E.

2.11 THE INVASION BEGINS

On 24 August 1966, the Soviets launched their second lunar orbiter. The probe was the first E-6LF. The cruise stage was identical to that of the other members of the E-6 family, but was mated to a conical mission module designed to remain attached. In the position previously occupied by the radar altimeter, there was a photographic system similar to that carried by Zond 3. Due to the sensitivity of the film to solar radiation, all of the images had to be taken within 24 hours of orbital insertion. The film was developed on board* and was then scanned. A low-resolution version of each image, yielding only 67 scan lines, was first relayed to Earth in 135 seconds. The

---

* Using a system not unlike that of the Lunar Orbiters.

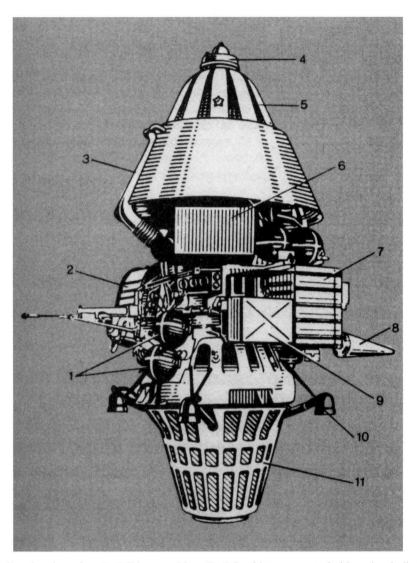

A line drawing of an E-6LF lunar orbiter (E-6LS orbiters were probably quite similar):
1, pressurisation spheres; 2, camera; 3, thermal control radiator; 4, infrared radiometer;
5, instrument compartment; 6, batteries; 7, attitude control system; 8, antenna; 9,
attitude control system electronics; 10, outrigger engines; 11, main engine.

high-resolution images, with 1,100 scan lines, were relayed later, taking up to 34
minutes each. The spacecraft could return pictures in either a three-axis stabilised
mode, or with the probe spinning on its longitudinal axis.

On 27 August, Luna 11 entered a $160 \times 1,193$-km lunar orbit with an inclination
of $27°$; but it began to malfunction immediately thereafter, when unidentified debris
became lodged inside one of the attitude control engines. Having lost its orientation,

it pointed its camera toward deep space, and so only black images were returned. By 1 October its batteries were exhausted. During its five-week mission, the probe collected data on the lunar magnetic field, the surface composition – which it analysed using gamma-ray and X-ray spectrometers – mascons, micrometeoroids, and the radiation environment of the deep space. It also investigated the radio-wave reflectivity characteristics of the lunar surface. On board was an experimental mechanical reduction gear system, called R-1, which was to be used by the Lunokhod lunar rovers then being designed. Tests showed that despite the high contact load between the transmission's components, it would work sufficiently well in a vacuum.

Just like a chess game, the next move in the exploration of the Moon went to the Americans. On 20 September, the seventh Atlas–Centaur launched Surveyor 2. The mission of the 995-kg probe, however, was complicated by an approach angle of 23° with respect to the local vertical. For both Luna 9 and Surveyor 1 this angle was close to zero, and for Luna 9 this had been a design constraint. Sixteen hours after launch, the probe corrected its trajectory – and the mission failed at that precise time, when one of the three vernier engines refused to fire and the asymmetric thrust set the spacecraft spinning once every second. This rotation was too fast for the attitude control system to cancel, and after several attempts the probe was declared lost. To collect engineering data, it was commanded to deploy its solar panels and to fire its braking rocket. It crashed on the Moon at a speed of 2.7 km/s at 5°.5 N, 12° W, with its retro-rocket still firing.

One month later, on 22 October, the Soviets launched Luna 12. Three days later the 1,620-kg E-6LF probe entered a $133 \times 1,200$-km orbit, inclined at 10°. Fearing that Jodrell Bank might once again attempt to 'steal' the images relayed by the probe, the Soviets set up a clever trick. As the Moon rose above the horizon of the Russian antenna at Yevpatoria in the Crimea, the probe began to relay its data using one of two possible telemetry frequencies. Three hours later, as the Moon was rising above Jodrell Bank's horizon, the probe began jumping from one frequency to the other, following a sequence known only to the Soviet mission controllers. As the Jodrell Bank antenna required a full day to be reconfigured to a new frequency, not a single frame was intercepted. For the first time, the Soviet Union admitted to mounting a photographic mission in support of its human lunar landing programme. Only two pictures were ever released (and were shown by Soviet television), covering an area near the crater Aristarchus* and part of Mare Imbrium (Sea of Rains). Each image encompassed an area of 52 square km with 15-metre ground resolution, but their quality (as far as it could be judged from television images) seemed far poorer than the American Lunar Orbiter pictures.

The first days of the mission were dedicated to mapping the lunar surface, after which the probe was spun on its longitudinal axis to collect data with the other instruments, which included a gamma-ray spectrometer, a magnetometer, an infrared radiometer, and a micrometeoroid detector, plus a second set of R-1

---

* Aristarchus of Samos (310–230 BC) was the first to formulate the heliocentric theory.

mechanical gear transmissions. Like Luna 11, Luna 12 measured the reflectivity of the surface to 1.7-metre-wavelength radio waves. This experiment indicated that the mean density of the surficial material – the regolith – was around 1,400 kg/m^3. The probe also detected the solar-wind-induced X-ray fluorescence of the regolith. Early in the mission, the perigee had been lowered to only 20 km, and as some fuel was left in the tanks it was decided to raise this again, but as the stabilisation system was no longer operative to keep the spacecraft properly oriented, the main engine was fired for a few seconds at a time during seven consecutive apogees, and this procedure succeeded in raising the perigee to 70 km. The mission was officially ended on 19 January 1967, after 602 orbits.

On 6 November the United States launched Lunar Orbiter 2 to reconnoitre several candidate landing sites in the northern equatorial region of the near-side. In contrast to Lunar Orbiter 1, it had a better opacised structure to avoid fouling the

The 'Image of the Century'. Lunar Orbiter 2 took this oblique view of crater Copernicus on 28 November 1966, when the spacecraft was 45 km over the lunar surface and 240 km due south of the crater. The complex of peaks in the centre of the 100-km-diameter crater rise some 300 metres above the floor. On the horizon, beyond the terraced rim of the crater, are the Carpathian Mountains.

star sensor with reflected sunlight. It reached the Moon on 10 November. The initial orbit was $196 \times 1,871$ km high, with an inclination of $12°$, but after five days of checks the perigee was lowered to 50.5 km, and on 18 November the one-week mapping mission finally began. On 23 November it took the most famous picture obtained by this family of spacecraft. On a part of the film that would otherwise have been lost during a periodic tensioning operation, it took an oblique view of the crater Copernicus, and for the first time provided people with an impression of a flight over the Moon. A radio system malfunction on 6 December prevented transmission of the last five pictures – three wide-angle and two narrow-angle. On 8 December the orbital inclination was raised to investigate the global gravitational field and mascons. After more manoeuvres – including one that minimised the time spent in the Earth's shadow during the lunar eclipse of 24 April 1967 – the probe was crashed on 11 October 1967, at $4°$ S, $98°$ E.

The year 1966 ended just as it had started: with a Soviet lander – the first lunar lander with a serious scientific payload. Luna 13 left the Tyuratam cosmodrome on 21 December. On 24 December its 'egg' – similar to that on Luna 9 – rolled on the surface of Oceanus Procellarum at $18°.9$ N, $62°.05$ W. It came to rest inclined at an angle of $16°$, inside a small but well-defined crater on a smooth plain on which the Sun was about to rise.

The Gruntomer penetrometer carried by Luna 13 was used to measure the characteristics of the lunar regolith in order to help in the design the Soviet lunar rovers of the 1970s. (Courtesy VNII TransMash.)

Luna 13 carried two 1.5-m-long booms, each with six joints actuated by torsion springs, one carrying the Gruntomer penetrometer to measure some of the physical characteristics of the regolith and the other the Plotnomer densitometer which was to use a small quantity of radioactive caesium-137 to measure its density. The penetrometer was a 5-cm-long titanium wedge that was pushed with a known force and acceleration onto the surface by a small explosive charge. The applied force was 70 N, and the cone penetrated 4.5 cm, indicating a regolith density of 800 kg/m^3. The upper structure of the regolith had previously been studied by a three-axis accelerometer mounted on the lander, which had measured the landing deceleration. Four infrared radiometers measured the regolith's temperature (which rose to 117° C at local noon), a radiation densitometer collected data on cosmic rays, and three radiation sensors collected the alpha-rays produced by the densitometer head and scattered by the terrain. Because of calibration problems, however, the densitometer's measurements were deemed ambiguous. All these experiments were used to characterise the regolith in order to prepare for the rovers that were to be sent in the 1970s. In addition, the probe had two Volga cameras to record stereoscopic images. Unfortunately, one failed to return any pictures. Nevertheless, the other – although limited in azimuth to 220° instead of 360°, and lacking the dihedral mirrors – returned five panoramas. The probe's batteries were exhausted on 28 December. The landing area had very few rocks, and was the smoothest site yet visited. As the design of the second-generation Soviet lander did not allow for a landing outside a small patch of Oceanus Procellarum, two landings were considered to be sufficient.

A portion of a Luna 13 panorama, showing the smoothest terrain yet visited on the Moon. A few boulders are visible on the horizon, and a change of texture just below the boulders indicates that the probe landed inside a crater. The object at lower left is one of the spacecraft's booms.

2.12 THE AMERICAN YEAR

The new year, 1967, did not see any Soviet missions – a clear sign that the United States had finally taken the lead. Lunar Orbiter 3, launched on 5 February, was the first lunar probe of the year. Two days later, its main engine was fired to enter a lunar orbit of 200 × 1.850 km, with the steeper inclination of 21°. The imaging mission began on 15 February, after the perigee was lowered to 48 km. Eight days later the probe began developing and relaying its pictures. Unfortunately, the last twenty-nine pairs of a total of 211 were lost due to a rewinding system fault.

The scope of the mission was no longer the search for suitable Apollo landing sites, but rather, higher-resolution photography of some of *the most likely candidates* in order to validate their suitability. The main outcome was close-up imagery of the smooth and lightly cratered area that would become the landing site of Apollo 11. Lunar Orbiter 3 also obtained a very high-resolution image of a portion of Oceanus Procellarum that was inspected in an effort to identify the shadow of the Luna 9 lander. When none was found, it was concluded that this was beyond the picture's resolution.

The orbit was altered to 143 × 315 km because of the lunar eclipse on 24 April, and then to 160 × 160 km in order to simulate the Apollo mission profile and assess

Astronaut Pete Conrad standing beside Surveyor 3 on the Moon during the Apollo 12 mission. The image clearly shows the general architecture of the lander and the deployed pantograph sampler arm. The Apollo 12 LM is in the background, just over the rim of the 200-m crater in which the probe landed.

the effects of the mascons. On 9 October the probe was crashed at 14°.6 N, 91°.7 W.

Surveyor 3 was launched on 17 April. The mass of the probe had reached 998 kg, and for the first time the spacecraft carried a second instrument, besides the camera, which itself had been improved with the addition of two mirrors in order to examine the underside of the vehicle. Also, following the cancellation of the Surveyor Block II probes on 13 December 1966, it was decided to fly some of the already developed scientific instruments on the remaining Block Is. There was a large choice: a seismometer, a micrometeoroid detector, an X-ray spectrometer to measure the composition of the regolith, and a mechanical arm to collect samples. In this case, the robotic arm, better known as the Soil Mechanics Surface Sampler (SMSS), was installed. It was mounted on the slot that had previously housed the camera that was to have taken pictures of the lunar surface during the approach. The electrical system of the arm was modified to use the camera power and command line. SMSS featured a 5-cm-wide, 13-cm-deep scoop mounted at the tip of an aluminum pantograph using four electric motors – the first to extend the arm from a minimum of 58 cm to a maximum of 150 cm, the second for the azimuth-joint actuation (112° total possible movement), the third for the elevation-joint actuation (42° total possible movement), and the fourth to open and close a moveable lip on the scoop. The mechanical characteristics of the regolith were estimated from motor current and temperature telemetric data. Better measures could have been made by a strain gauge system, but there was no time to develop it.

En route to the Moon, a trajectory correction manoeuvre was made, and finally, on 20 April, everything was ready for the fourth lunar soft landing. The retro-rocket was fired to reduce the approach speed from 2,600 m/s to 140 m/s, and was then jettisoned, and the three small vernier engines took over. A few seconds before contact, the radar altimeter lost track of the surface, and the probe landed very slowly (2.1 m/s) with the engines still burning. Before the engines could be shut off by a command from Earth, the probe had bounced several times, the first time by 20

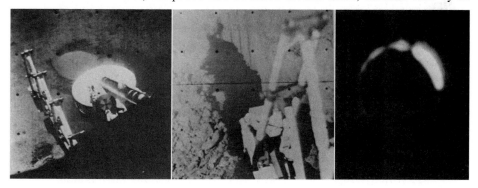

Three images from the Surveyor 3 mission. (Left) the SMSS robotic arm in operation near a footpad of the spacecraft. The circular mark left during one of the landing rebounds is evident; (centre) the SMSS digging a trench; (right) an image of Earth taken during the lunar eclipse of 24 April 1966.

metres, then by 11 metres, and since it had built up considerable horizontal velocity the landing pads excavated trenches of 30 cm length as it slid to a halt. Finally, some 40 seconds after first contact, Surveyor 3 was sitting firmly on the lunar surface inside a 200-m crater at 2°.97 S, 23°.34 W. It was sitting on a slope. (This may have been what confused the radar.) The incident was not without consequences, as the camera optics and mechanics were contaminated by lunar dust. When the robotic arm was deployed the following day, it was discovered that it too had suffered severe contamination to the actuation systems, and could thus provide only one-third of the expected maximum force during the regolith mechanics experiments. The SMSS arm operated in total for more than 18 hours, and performed some 6,000 commands, taking seven mechanics and thirteen penetration measurements, digging four trenches, and collecting some dust on a landing pad for the camera to image in order to estimate its granularity. During one experiment a bright object was carefully exhumed and revealed to be a small rock. Several commands were subsequently performed in order to collect and crush it.

Situated in a crater, Surveyor 3's view was severely limited. Nevertheless, up to sunset it took 6,326 images, including some pictures of Venus. On 24 April, from 9.48 to 14.06 GMT, the probe was the first to experience a lunar eclipse (or rather, an eclipse of the Sun by the Earth). During the totality phase – which lasted 2 hrs 26 min at the Surveyor 3 landing site – the internal temperature of the electronics boxes fell from 27° C to –123° C immediately before the end of totality. The probe took some images of the Earth's disk, framed by the halo of the solar rays refracted by the atmosphere. Unlike its predecessor, it did not survive the harsh experience of the lunar night. As will be seen, in 1969 it was visited by the Apollo 12 astronauts.

On 4 May – the same day that Surveyor 3 fell silent – the fourth Lunar Orbiter was launched. As the Apollo landing site reconnaissance was now complete, the final two probes of the family were to undertake a global photoreconnaissance of the Moon from polar orbit, covering remote areas such as the poles and the far-side in detail. To this end, an additional 18 metres of film was carried. Consideration was given, too, to taking other instruments into lunar orbit, the most important of which was a gamma-ray spectrometer to build a global compositional map. Unfortunately, this instrument was discarded at the end of 1966, and more than thirty years were to pass before a gamma-ray spectrometer was put into a lunar polar orbit – in 1998!

To accommodate the thermal requirements of the polar orbit, which could lead to several hours of continuous exposure to the Sun's heat (the dawn–sunset orbit), the bottom of Lunar Orbiter 4 was covered with five hundred quartz mirrors. The orbit insertion manoeuvre was performed on 8 May (2,705 × 6,034 km, inclined at 85°.5). In less than a month the probe took 193 pairs of images covering 99% of the near-side and 75% of the far-side, and also took the first images of the south polar region. After the imaging mission – which required more than two hundred attitude changes instead of some fifty for its predecessors in the equatorial zone – the orbit was lowered, with perigee reaching 77 km, to map the gravitational field near the poles. On 17 July, contact was suddenly lost, and never regained. The forward projection of the orbit shows that the inert probe should have impacted the Moon in October, brought down by the mascons that it was mapping.

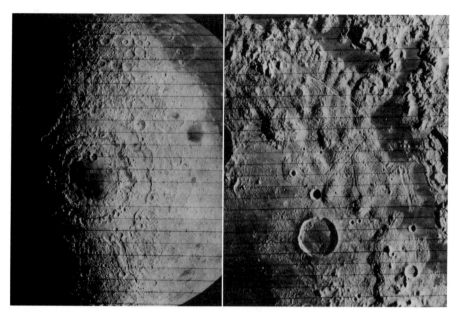

Two images of Mare Orientale taken by Lunar Orbiter 4. The image at left is a wide-angle view, showing a large part of the lunar western hemisphere. The image at right shows the complex terrain at the north-eastern quadrant of Mare Orientale.

The next lunar mission was that of Surveyor 4, which took off on 14 July, under the shroud of the eleventh Atlas–Centaur. The probe was again equipped with the SMSS, to which was added an experiment to determine the iron content of the regolith. On one of the three legs were fastened two metallic stripes, one of which was a permanent magnet. By comparing the amount of dust on each stripe, the magnetic properties of the regolith could be evaluated.

The landing site was to be at the centre of the Moon's near-side, inside Sinus Medii (Central Bay), and this required a very steep approach angle of 35°. The flight proceeded without any problem until 17 July when, 11 km above the lunar surface and 2 seconds before the braking rocket burned out, the probe's transmission ceased. No-one knows why this happened, but it may have been that the retro-rocket had some kind of structural fault which caused the propellant grain to explode. Some 150 seconds later, what was left of the probe hit the Moon at 0°.4 N, 1°.33 W. For more than twenty-five years, until Mars Observer was lost under similar circumstances in 1993, Surveyor 4 remained the last American probe to have catastrophically failed in flight.

Just two days later, on 19 July, the US launched another lunar probe: Explorer 35. Like Explorer 33, this was an IMP satellite (the sixth of the series), the mission of which was to take measurements of the deep-space environment from lunar orbit. It was quite similar to its predecessor, but with its mass increased by 104 kg in order to incorporate several more instruments: two flux-gate magnetometers with different measurement fields, three radiation sensors, a plasma sensor, a micrometeoroid detector, and an engineering experiment on the ageing of solar cells. The launcher was the same as for Explorer 33 – a Delta-E, which this time worked smoothly to dispatch the probe to its 60-hour flight to the Moon, culminating with the lunar orbit insertion manoeuvre to 800 × 7.692 km with an inclination of 147°.

Explorer 35 confirmed that the Moon does not possess a global dipole magnetic field, which means that it interacts with the solar wind by simply creating a 150,000-km long cavity in it. In contrast, Earth has a complex system of Van Allen radiation belts and a magnetospheric tail millions of kilometres in length. Furthermore, during the three days it takes for the Moon to pass through the Earth's magnetotail each month, Explorer 35 failed to detect any induced magnetic field.

Another experiment used the probe's transmitter to evaluate the electromagnetic reflectivity of the regolith. With the exception of some areas showing an anomalous signature that was then interpreted as relatively young lava flows, this showed the surface to be quite homogeneous.

As will be told later, Explorer 35 was later used in 1971 and 1972 in conjunction with the small 'particles and fields' satellites released by Apollo 15 and Apollo 16. It was finally switched off on 24 June 1973.

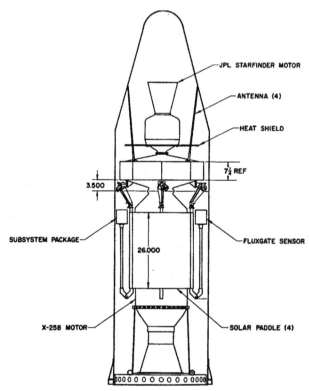

A NASA line drawing (of 1964) of an anchored IMP spacecraft like Explorer 33 and Explorer 35. The main difference between this spacecraft and the one actually flown is the Syncom motor which replaced the Starfinder motor.

On 1 August, the last Lunar Orbiter – the third American lunar probe in less than a month – was launched to complete the global photoreconnaissance begun by Lunar Orbiter 4, with a particular emphasis on the far-side. As with Lunar Orbiter 1, Lunar Orbiter 5 also had problems locating Canopus during its translunar flight, but it successfully achieved lunar orbit ($196 \times 6{,}040$ km; $85°$ inclination) on 4 August, and during the following month took 213 pairs of images – some from a 196-km perigee, and others from a perigee as low as 100 km. After the imaging part of its mission, Lunar Orbiter 5 was used to refine the map of the Moon's gravitational field and to detect micrometeoroids. In January 1968, just before the probe was de-orbited, it was used for an unusual experiment in position determination, in which pictures were taken from Earth of sunlight reflected by the quartz mirrors used for thermal control. The first attempt, on 18 January, was unsuccessful, but three days later the 150-cm Catalina telescope in California detected a moving speck 9 arcmin from the lunar limb. Despite estimates that the probe would look like a sixth-magnitude star, it actually appeared, at best, as magnitude twelve – some 150 times fainter. Ten days later, on 31 January, Lunar Orbiter 5 hit the Moon on the equator at $70°$ W.

By the end of their photoreconnaissance, the Lunar Orbiters had taken 967 wide-angle and 983 narrow-angle images, covering 99% of the lunar surface – the largest gap being a patch of several thousand square kilometres near the south pole.

Lunar Orbiter probes have again recently hit the scientific headlines in the context of the debate on scientific data preservation – a problem exacerbated by rapid technical obsolescence. NASA long ago disposed of the Ampex two-inch tape units on which JPL stored all of the original data from the lunar and planetary missions of the 1960s. Some of the last surviving Ampex units have since been located, but no governmental agency has provided the funding required to convert the Lunar Orbiter data – stored in 1,500 tapes – to CD-ROM format. Although 675 images have been scanned from the 1971 Lunar Orbiter atlas of the Moon, the scanning process has decreased the image resolution by 60%.

On 8 September 1967 Surveyor 5 was launched towards the Moon. In place of the SMSS, it carried a 13-kg alpha-ray spectrometer – a small box, 15-cm on a side, containing a few grammes of radioactive curium-242 that would emit large quantities of alpha-rays. Some of the particles would be absorbed by the regolith, and others would be backscattered. By measuring the energy characteristics of the backscattered alpha-rays, the composition of the regolith could be determined. After the course correction manoeuvre it was discovered that a fuel pressurisation valve had stuck in its open position, with the inevitable result that the helium tanks were slowly but constantly losing pressure. Following a last-minute technical meeting, it was decided to attempt a landing by firing the main braking rocket 12 seconds late, in the hope that there would be just sufficient helium for a fast landing. But to make matters worse, the landing site – which was by this time impossible to change – required an approach angle of no less than $47°$. As a final thrill, the revised flight plan was implemented in real time on 11 September. During the trip, the probe discarded all unnecessary fuel to save mass by starting the vernier engines three times. The braking rocket was started at a height of 45.7 km instead of 83.5 km and, 2 seconds before burn-out, the jettison command was sent. Due to its continuing thrust, the rocket remained in place, but the command automatically switched on the radar altimeter and the vernier engines. The retro-rocket eventually separated at a height of less than 1 km, a few seconds before Surveyor 5 landed at a speed of 4 m/s, side-slipped for a few metres, and finally stopped in a crater some 10 metres wide and 1.5 metres deep, and with walls sloping at 20°, located in Mare Tranquillitatis at 1°.42 N, 23°.19 E. At this time, the pressure in the helium tanks was just above its working minimum.

The probe took some panoramas, and then switched on the spectrometer. After taking some environmental radiation measures, the instrument was lowered onto the surface using a type of small crane. The regolith, which was rich in oxides of both silicon and aluminium, and appeared to be of basaltic composition, finally proved the magmatic origin of the lunar maria. The magnet experiment confirmed that the dust contained some iron. About 53 hours after landing, the probe tested the dust characteristics by firing its attitude control engines for 0.55 seconds to excavate a depression 20 cm wide and 3 cm deep. After sunset on 24 September, the probe was operated for a record 115 hours to monitor conditions.

During its first lunar day, Surveyor 5 relayed a total of 18,006 pictures. It was revived on 15 October, and took 1,048 more pictures on its second lunar day, and on 18 October it experienced a lunar eclipse, spending 2 hrs 28 min in the Earth's shadow. Contact was eventually lost on 17 December, during the fourth lunar day, shortly after an unsuccessful campaign of precise lunar motion determination by simultaneously tracking both Surveyor 5 and Surveyor 6, which failed because Surveyor 5's signal was by then too faint. As with Ranger 8, Surveyor 5 may be visited by a small rover some time in the future.

Two scientific instruments of Surveyor 5 in action. (Left) the alpha-ray spectrometer analysing the lunar surface; (right) the magnet experiment. The magnetised bar at left has collected much more dust than the non-magnetised bar at right.

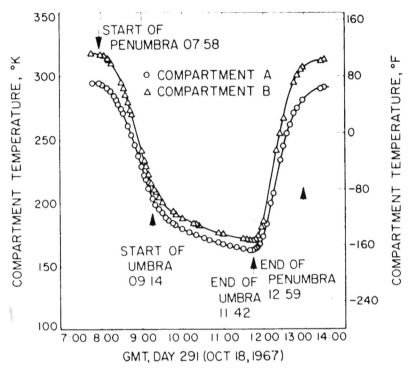

A plot of the temperatures on the outboard faces of avionics compartments A and B of Surveyor 5, recorded during the lunar eclipse of 18 October 1967. (From Vitkus, G., Lucas, J. W., Saari, J. M., 'Lunar Surface Thermal Characteristics during Eclipse from Surveyors III, V and after Sunset from Surveyor V', AIAA Paper 68-747. Copyright AIAA.)

The last of no less than eight probes launched in 1967 was Surveyor 6, which set off on 7 November. The cruise was faultless, and it landed on 10 November at a speed of 3.3 m/s at 0°.51 N, 1°.39 W, several kilometres from the wreck of Surveyor 4, and 10.5 km from the target point. The landing site was several hundred metres from a small 'wrinkle' ridge, on whose crests were several boulders – the only boulders visible in an otherwise almost featureless landscape.

Like its predecessor, Surveyor 6 carried the alpha-ray spectrometer, and had some modifications to the camera architecture. The first part of the mission was dedicated to the chemical analysis of the regolith, which seemed to be quite similar to that on the floor of Mare Tranquillitatis. On 18 November the engines were turned on again for 2.5 seconds, and the probe landed 6.1 seconds later, 2.3 m away, after a 'hop' to a height of some 4 m. This experience proved that the astronauts would be able to see through the dust during landing and take-off, and provided some stereoscopic pictures of the landing site. Unfortunately, the spectrometer came to rest on its side, and could be used only to collect environmental radiation data. On 24 November the Sun set, and the probe was turned off 41 hours later. Contact was re-established on

The effects of the Surveyor 6 lift-off. (Left) dust sprayed on a calibration target mounted on the inside of a low-gain antenna; (right) dust moved by the engine firing. The horseshoe-shaped mark left by the alpha-ray spectrometer can also be seen.

14 December, but the signal appeared erratic, and tracking had to be discontinued. Surveyor 6 took a total of 29,952 pictures (although according to another source it took 30,065 pictures).

2.13 THE LAST MISSIONS

By the time of Surveyor 6, all of the objectives of the US unmanned lunar landing programme had been accomplished, and so it was decided to land the last probe of the series on a scientifically interesting site, and to equip it with every instrument already flown: the SMSS, the alpha-ray spectrometer, the magnet bars – a second set of which was fastened on the SMSS scoop – and the camera and mirrors to image the underside. The landing site was just north of the young crater Tycho,* at the origin of the bright ray system easily seen by the naked eye in the Southern Highlands. Another scientifically interesting landing site that was considered was the cluster of what appeared to be volcanic hills near the crater Marius, in north-western Oceanus Procellarum.†

Surveyor 7 was launched on 7 January 1968 by the fifteenth Atlas–Centaur. As the landing site had to be within a circle only 20 km in diameter, two course corrections were scheduled, but the first was deemed sufficient and the second was cancelled. On 10 January, the probe landed at 40°.86 S, 11°.47 W, less than 30 km from the northern rim of Tycho. Although the site was a region of rolling hills, and was littered with rocks, it would not have posed any problem for a human landing. When dust contamination impeded the deployment of the spectrometer, the SMSS arm was used to free it. The robotic arm was used to estimate the density of the regolith, to determine the mass of a pebble by measuring the power required to lift it,

---

*   Tycho Brahe (1546–1601), the last of the great astronomers of the pre-telescopic era.
†   Simon Marius (1570–1624), a German astronomer, and co-discoverer of the satellites of Jupiter.

Four images taken by Surveyor 7. (Clockwise from upper left) hills in the Tycho area (the apparent severe slope of the hills is caused by the tilt of camera); laser beams coming from Table Mountain (left) and Kitt Peak (right); the SMSS robot arm nudging the alpha-ray spectrometer; and SMSS operations seen on a 9 x 24-cm mirror fastened on the antenna mast.

and to dig seven holes and trenches. It was also used to move the spectrometer so that it could analyse a nearby 7-cm rock, then to analyse the floor of a 1-cm deep trench, and finally to shade it from the Sun when it became too hot.

On 20 January, Surveyor 7 was used for an experiment to prepare for human landings. It had been decided to take laser retroreflectors to the Moon in order to precisely measure the distance from the Earth to the Moon by timing the round-trip time of a pulse of light. To prove the feasibility of the plan, American observatories used lasers to shoot at the probe, which took a few clear images of the night hemisphere of the Earth, showing two magnitude –3 dots, brighter than any star, corresponding to the 61-cm telescope at Table Mountain, California, using a 2-W laser, and to the 152-cm telescope at Kitt Peak, Arizona, using a 4-W laser. Curiously enough, the beam fired by the most powerful laser was dimmer than the other beam. The lasers fired by US east coast observatories were not recorded, as these were probably lost in the glow of the encroaching terrestrial dawn. At lunar sunset on 23 January, pictures were taken of the solar corona and of Mercury. Although Surveyor 7 survived the long lunar night, its systems suffered, and it finally failed on 21 February after taking 21,038 pictures. From a scientific point of view, it confirmed the compositional difference between the maria and the highlands. In particular, it found that the rocks in the maria contain three times as much titanium and iron, but much less aluminium and calcium. It also measured a mean highland rock density of 2,950 kg/m^3 – lower than the maria rock density of 3,150 kg/m^3, and much lower than the mean lunar density of 3,360 kg/m^3. Upon seeing the elemental abundances measured by the spectrometer, some scientists concluded that the

A model of a self-landing Surveyor-based rover studied by JPL in 1970. Due to the external position of the solid-fuel retro-rocket, this spacecraft would have experienced large centre-of-mass shifts during the landing. (Courtesy JPL/NASA/Caltech.)

highlands were 'alumina-rich basalt', but Eugene Shoemaker disagreed. To his surprise, he found that the data indicated that the dominant rock in the Tycho ejecta, which had been excavated from deep within the crust, was anorthositic gabbro. This feldspathic rock was clear proof that the Moon was thermally differentiated, it was not a 'cold' primordial body, it was chemically 'evolved'.

With Surveyor 7, the US unmanned lunar exploration programme came to a complete stop. The previous year, the Surveyor Block II probes had again been proposed, but to no avail. As late as 1970, JPL studied a mission reusing Surveyor technology to land a large long-range rover on the Moon. This would be braked by the same retro-rocket as that used with the Surveyors, and would land on four wheels instead of three foot pads, but it was not funded.

But what was happening in the Soviet Union? Sixteen months had passed since its previous lunar mission. It had conducted some tests of the new human-rated lunar Zond probes (of which more later), and an E-6LS probe had been launched as Cosmos 159 on a very eccentric orbit with an apogee at 60,000 km well clear of the Moon in order to calibrate the deep-space tracking antennae at Yevpatoria in preparation for piloted lunar missions. On 7 February 1968, a follow-up probe was lost when the Molniya I stage ran out of fuel before reaching Earth orbit. Luna 14 –

a twin of Cosmos 159, and the last second-generation Soviet probe – was launched on 7 April 1968. Three days later it entered lunar orbit (160 × 870 km at 42°), and its mission was announced to be to study radio transmission stability and the lunar gravitational field, and to measure the solar wind and cosmic rays. There were also new experiments on radio-wave reflectivity at two wavelengths (1.7 metre and 32 cm) and a new occultation experiment. Contrary to Western speculation, it did not have a camera. The real mission of the probe was exactly as stated by the Soviets: to improve knowledge of the lunar gravitational field, of lunar transfer orbits, and of telecommunication techniques in preparation for human missions. No image of the probe's configuration has ever been published, but it was probably similar to Luna 12.

By early 1968, both the United States and, to a lesser degree, the Soviet Union considered themselves ready to initiate the manned assault to the Moon.

3

America wins the race

3.1 THE KENNEDY MEMORANDUM

This story, too, begins on the arid steppes of Kazakhstan, at 07.06 GMT on 12 April 1961, when a Semiorka rocket lifted off from Pad 1 at Tyuratam. The upper stage carried an odd-looking spherical capsule weighing a few tonnes. Inside this capsule was a 27-year-old military pilot who was about to enter the history books: Yuri Gagarin. Fifty-two minutes later, Radio Moscow announced the successful launch of the Vostok (East) spacecraft with its human load, and shortly afterwards, at 8.55, after orbiting the Earth once, Gagarin returned to Earth, provoking the enthusiasm of the 'Communist masses'. A Soviet newspaper emphatically declared: 'It is thus proved that socialism is a better launch pad than capitalism'. In the United States, where space activities were slowly gathering momentum, the news came as a shock – not unlike a second Sputnik.

In contrast with 1957, however, the United States had a new young President, John F. Kennedy, and he assigned much more importance to propaganda and was more willing to associate the nation's image with a high-risk venture. After his inauguration in January, Kennedy had first to face the prospect of a war in South-East Asia. Gagarin's flight on 12 April, of which he had been forewarned by the CIA, presented a second major international issue. The *coup de grace* against the public image of 'the world's greatest democracy' came just five days later, when a group of Cuban counter-revolutionaries, sponsored by the US, unsuccessfully tried to overthrow Fidel Castro in the infamous Bay of Pigs operation. On 19 April – the same day on which the last invaders were being pushed back by Castro – Kennedy and the Chairman of the National Space Council, Vice President Lyndon B. Johnson, had a 45-minute meeting during which Johnson pleaded the cause of the space programme as the only venture able to restore America's stained national prestige in the eyes of the world. Johnson, moreover, was not new to such advocacy, having been the major critic of political inertia after the shock of Sputnik.

The following day, Kennedy sent a memo to the Vice President in order to focus the problem: 'In accordance to our conversation I would like for you as Chairman of the Space Council to be in charge of making an overall survey of where we stand in space. 1. Do we have the chance of beating the Soviets by putting a laboratory in space, or by a trip around the Moon, or by a rocket to land on the Moon, or by a rocket to go to the Moon and back with a man? Is there any other space programme which promises dramatic results in which we could win? 2. How much additional would it cost? 3. Are we working 24 hours a day on existing programmes? If not, why not? If not, will you make recommendations to me as to how work can be speeded up. 4. In building large boosters should we put our emphasis on nuclear, chemical or liquid fuel, or a combination of these three? 5. Are we making maximum effort? Are we achieving necessary results? I have asked Jim Webb [NASA Administrator], Dr [Jerome] Wiesner [Kennedy's science advisor], Secretary [Robert] McNamara [of the Department of Defense] and other responsible officials to cooperate with you fully. I would appreciate a report on this at the earliest possible moment.'

Johnson began to explore the feasibility of a human flight to the Moon – the endeavour which promised the most dramatic results – and he first contacted NASA, which was not caught unprepared. Since being established in October 1958, the agency had decided that its major long-term task would be to put humans in orbit around the Moon in a programme named after the Greek Sun god: Apollo. The first such mission had been proposed to the Eisenhower administration, but the ageing President, an opponent of the so-called 'military–industrial complex', and sceptical of the case for associating the nation's prestige with space, had rejected it. Kennedy himself had refused the same proposal just one month earlier, in March.

NASA informed the Vice President that it was unlikely that America could beat the Soviets with the launch of a space station, but it should be possible to send humans to the Moon before 1967 (the fiftieth anniversary of the Soviet revolution, which it was thought *they* were sure to try to celebrate with a mission around the Moon). The cost of the project was estimated as $34 billion.*

Johnson next turned to the military. Incredibly, the Army, Air Force and Navy were independently studying ambitious lunar projects. The Army expressed its wish to establish, by 1966, a large lunar base with a crew of twelve. The project would use no less than 150 heavy launchers, of two different types (Saturn I and Saturn II, of which more will be told later) at a launch rate higher than five a month! The reasons for such a base were rather vague, if not ridiculous. What was the use of a reconnaissance centre 380,000 km from Earth, at a time when spy satellites were overflying the Soviet Union every day at a height of 200 km? And how could the use of Moon-based weapons against Earth targets, three days' flight away, be 'feasible and desirable', as stated in the secret document which initiated Project Horizon (as it was called) in 1958? Briefly put, the military strategic relevance of a lunar outpost – very close to zero – was evident only to the military. A similar project, called LunEx (Lunar Exploration) was being studied by the Air Force. This included the use of

---

* The eventual cost was 'only' $24–27 billion at 1970 levels.

winged spacecraft not unlike the Space Shuttle, and the development of throttleable cryogenic engines – the RL-10 engines later to be used on the Centaur. The first expedition of the LunEx project would be carried out in 1967, to be followed by the establishment of a lunar base the following year. In contrast, the Navy was interested in developing a small lunar landing module based on the SLV (Soft Landing Vehicle) prototype built and tested by the China Lake weapon-testing centre. According to Navy plans, the SLV could carry a dog to the Moon in 1963, could return a soil sample as early as 1964, and could land a Navy pilot in 1967. The military assured Johnson of their support in a unified national venture, and halted their own projects.

On 5 May Alan Shepard took his place inside the tiny Mercury capsule on top of a Redstone rocket – the same type of rocket that had carried Explorer 1 into space – and at 14.34 GMT the engine fired and sent the capsule to a peak altitude of 187 km, after which it re-entered the atmosphere and splashed into the Atlantic 478 km downrange of Cape Canaveral, the entire flight having lasted only 15 min 22 sec.* A few days later, Johnson presented the results of his investigation to Kennedy, McNamara, and James Webb.

Then on 25 May 1961 Kennedy addressed Congress on urgent national needs: 'I believe that this nation should commit itself to achieving the goal, before this decade is out, of landing a man on the Moon and returning him safely to the Earth.' It took less than an hour for Congress to fund the project, overruling the opposition of a small minority. In contrast, polls showed that the majority of the general public – whom no-one had cared to consult – considered the project to be a waste of money. The scientific community reacted coldly to a project clearly dictated by national prestige considerations and by the logic of the Cold War as opposed to a sincere scientific curiosity. Wiesner, agreeing, suggested to Kennedy that he should not refer to the Apollo project as a voyage of scientific exploration – a suggestion which Kennedy always followed.

During the ensuing years, opposition to the project was heard again – chiefly in the scientific area – as it became clear that the same knowledge could be gained by a good unmanned-probe programme, at far lower cost and much less risk. In time, however, opposition dissipated, and was almost non-existent at the time of the first lunar missions, although it arose again later. Despite the many dissenting parties, the project obtained, over the years, almost all of the required funds, from a minimum allocation of $160 million in 1962 to a maximum of $2,967 million in 1966, to end at $601 million and $77 million in 1972 and 1973 respectively.

On a few occasions, Kennedy also cautiously attempted to initiate international cooperation. During his first meeting with Premier Nikita Khrushchev, in Vienna on 3 June 1961 (less than ten days after addressing Congress), he proposed a joint lunar programme with the Soviets, but this overture came to nothing, in part due to the eventual failure of the summit meeting. The second, and better known, occasion was

---

* This type of trajectory was followed by the second piloted Mercury, flown by Virgil (Gus) Grissom, on 21 July 1961. For their first orbital flight, the United States had to wait until 20 February 1962, when John Glenn – onboard another Mercury spacecraft, launched by the more powerful Atlas rocket – orbited Earth three times.

when Kennedy presented a speech before the United Nations on 20 September 1963, by which he time he had begun to doubt whether a race to the Moon with the Soviets really existed: 'Surely we should explore whether the scientists and astronauts of our two countries, indeed of all the world, cannot work together in the conquest of space, sending some day in this decade to the Moon not the representatives of a single nation, but the representatives of all of our countries.' These words were enough to cause a horrified reaction within the entire military–industrial complex. The most important trade magazine, *Aviation Week and Space Technology*, published a very harsh editorial affirming that 'President Kennedy has dealt his own national space programme its hardest blow', and that the speech represented 'a good example of the brittle brilliance of the White House staff members who spawn scintillating ideas without much thought on their ramification.' Moreover, the speech might 'change the motivation of several million Americans who are involved in all phases of the manned space flight effort from a patriotic sense of extreme urgency.'

Undaunted, Kennedy repeated his offer a couple of weeks later when he met the Soviet Foreign Minister Andrei Gromyko. Khrushchev made his answer known shortly afterwards: 'At this moment we have no plans for Moon flights' (which, as we shall see, was almost true in 1963), and then disdainfully concluded: 'We are quite well on Earth, too'. Over the following days, Khrushchev familiarised himself with the funding required by Soviet scientists to stage such a mission before the US, and on 1 November declared: 'We are taking the US proposal in due consideration. Wouldn't it be beautiful if a Soviet man and a US woman flew together to the Moon?' A few weeks later, Kennedy was assassinated in Dallas, shortly after visiting the main NASA facilities in Florida and Texas, and the concept of a common US–USSR lunar manned mission was quietly forgotten by both sides.

After Kennedy's death, and his ensuing 'canonisation', the Apollo programme remained substantially untouched by budget cuts and political moves that might otherwise have forced its cancellation.

3.2   WORK BEGINS ON THE LAUNCHER

Responsibility for the rocket that was to take humans to the Moon was assigned to the Marshall Space Flight Center in Huntsville, Alabama, which had shortly earlier passed to NASA's control after having been for fifteen years the Army's Redstone Arsenal (named, rather mundanely, after the characteristic colour of the local soil). Huntsville was the residence of the German engineering team led by charismatic Wernher von Braun, who were 'imported' to the United States after the Second World War.

The career of von Braun, as already mentioned, began in Berlin during the 1930s, with a group of rocket propulsion enthusiasts sponsored by the Wehrmacht, which from the beginning believed in the military potential of long-range rockets. The nature of the research soon led to the team being moved to an almost uninhabited island on the coast of the Baltic Sea, near the small town of Peenemünde, where, from 1940, engineers perfected an extremely powerful ethyl alcohol and liquid oxygen engine. The team also solved the problem of the cooling of the combustion chamber, by developing a clever system (still used to this day on every high-thrust liquid-fuel rocket engine) in which fuel – in this case, alcohol – flowed inside a tight coil sculpted on the inside of the nozzle, thus limiting its warming and increasing its operational life. This engine was later integrated on a rocket called A 4, which was first launched (unsuccessfully) in March 1942; but success came on 4 October, when an A 4 reached a height of 80 km. By this time, however, allied intelligence knew of the existence of the rocket, and a bomber raid obliterated the launching base before it could become operational. Research and development remained in Peenemünde, but the production of the rocket was moved to Nordhausen in the Harz Mountains of Thuringia. The first operational A 4 – by then renamed V2 – fell on Paris on 5 September 1944, just a few days after the city's liberation.

The real connection between the German team (and von Braun in particular) and the Nazi regime is still hotly debated to this day. Von Braun always justified himself by stating that his only interest was spaceflight, and attributed his participation in the production of such a weapon to the carelessness of youth. In fact, some say, von Braun is the only person to appear on every occasion in civilian clothing, instead of an SS uniform. But others point out that von Braun had often visited the V2 production line in Nordhausen, which was built on a concentration camp and used captives as a work force, and he could not have been unaware of the dire straits of the thousands of forced workers, an enormous number of whom were worked to death. (Some 20,000 prisoners are estimated to have died in the Nordhausen

The gargantuan F-1 engine that powered the first stage of the Saturn V.

concentration camp, in addition to 3,000 during the construction of the nearby Mittelwerk underground factory. The V2s killed a total of around 7,000 people in the United Kingdom, the Netherlands, France and Belgium.) With the war about to end, allied armies rushed to collect the secrets of the V2, and to capture its engineering team. The Americans benefited the most, as all of the main Peenemünde technicians surrendered to them, with the exception of Helmut Gröttrup, who turned to the Soviets. After moving to Texas, the team began work on several captured V2s, and later on new longer-range missiles such as Redstone, which was used to launch Explorer 1 and the first sub-orbital Mercury spacecraft, and Jupiter, used under the name Juno II, to launch Pioneer 3 and Pioneer 4.

In 1958, before coming under NASA control, the Huntsville team began to study a space launcher designed to surpass, for the first time, the capabilities of the big Soviet launcher. As the new missile project followed Jupiter, it was named Saturn. Its first version, Saturn I, could place seven tonnes in low Earth orbit.

The first stage was powered by a cluster of eight Rocketdyne H-1 engines, each one having its own fuel tank for kerosene and liquid oxygen, in addition to a small nitric acid tank used for hypergolic fuel ignition only. The second stage had six Pratt & Whitney RL-10-A3 cryogenic engines. It was designated S-IV. The original plan was to mount a Centaur, with two such engines, as a third stage, but in 1961 the Centaur was deleted from the Saturn programme (although it was adapted for the Atlas, and later the Titan III). On 27 October 1961 the first Saturn I lifted off for a test of the first stage only, as the upper stages were inert mock-ups. Two more test launches were carried out under Project High Water, which carried 95 tonnes of water on a sub-orbital flight to simulate the behaviour of liquid fuel inside the inert upper stages. The fourth launch, on 28 March 1963, successfully concluded first-stage testing. On 29 January 1964 the fifth Saturn I carried the first 'live' S-IV second stage into orbit. Preparations for this launch were witnessed by President Kennedy during his visit to Cape Canaveral on 18 November 1963, four days before his assassination. The next five Saturn I launches tested Apollo project hardware (and will be described later).

Once the Apollo project was underway, it was decided to build a derivative of the Saturn I, called Saturn IB, with a low Earth orbit payload capacity of 20 tonnes. The source of this considerable increase in performance was the new S-IVB second stage, which had a single 891-kN-thrust cryogenic Rocketdyne J-2 engine with a zero-gravity restart capability. In parallel with the Saturn IB, the development of the Saturn V – an extremely powerful rocket, able to place 140 tonnes in low Earth orbit, or to push up to 45 tonnes to escape speed – was approved.

The S-IC first stage of the Saturn V, built by Boeing, had a mean diameter of 10 metres, was 46 metres high, and carried 2,140 tonnes of kerosene and liquid oxygen – enough to keep the five Rocketdyne F-1 engines running for 158 seconds. Each of these gargantuan engines had a thrust of 6.7 MN, was 6 metres tall, and had a nozzle with a maximum diameter of 3.66 metres. Particular problems were encountered during the development of these engines, which proved to be prone to combustion instabilities – so much so that their injectors had to be redesigned several times. The S-II second stage, built by Rockwell (North American Aviation), was of the same

The launch of a Saturn V carrying the Apollo 10 spacecraft – the only lunar mission to lift off from the southernmost Pad 39B. Overall, the vehicle is 111 metres tall.

diameter as the first stage, 24.84 metres tall, and was equipped with five J-2 cryogenic engines. The third stage, built by McDonnell–Douglas, was the same space-restartable S-IVB as used on the Saturn IB.

In addition to the Saturn V, an even more powerful rocket – Nova – was studied to support the ambitious lunar missions expected to follow Apollo. This envisaged a first stage powered by eight F-1 engines with a total thrust of 53 MN, a second-stage powered by four new cryogenic M-1 engines, and an S-IVB as the third stage. The preliminary development of Nova was stopped in the mid-1960s.

Less powerful versions of Saturn were studied for simpler lunar missions: Saturn II would mate the Saturn I first stage to two or three upper stages in order to launch a basic lunar circumnavigation mission; Saturn III would consist of a first stage powered by two F-1 engines, a second stage powered by four J-2 engines, and an S-IVB; and Saturn IV would be similar to Saturn III, but with the first stage powered by four F-1 engines. Also, an extremely powerful Saturn-derived rocket was planned utilising a nuclear-powered upper stage based on the experimental Nuclear Engine for Rocket Vehicle Application (NERVA) engine, but NERVA was cancelled in 1972 without ever having flown.

3.3　THE FLIGHT IS DEFINED

As early as May 1961, the Space Task Group at the NASA Langley Research Center in Virginia began working on the details of the mission trajectory, which would constitute a major constraint to the design of the spacecraft and determine the choice of the launcher – Saturn V, Nova, Saturn III or Saturn IV.

At first, studies concentrated on the 'direct flight' technique, which envisaged a single spacecraft both for landing on the Moon and for returning to Earth. Such an approach had clear advantages – but also some obvious disadvantages. As the flight would be directly from the Earth's surface to the Moon's surface, it would require no risky intermediate manoeuvres, such as rendezvous between piloted spacecraft, in which failure might doom the occupants of one of the spacecraft. The direct flight approach was also the most logical and traditional. On the other hand, an enormous quantity of fuel was required to launch the single spacecraft to the Moon, land it, and then lift it off the Moon. It was also realised that it was a waste to land the Earth atmosphere re-entry capsule, with its massive thermal shields, on the Moon, as its shape was dictated by re-entry aerodynamics rather than lunar mission require-ments. Such an approach required either the Nova or Saturn V, depending on the mass of the spacecraft.

Another less hazardous technique under study was the 'Earth orbit rendezvous', in which, depending upon the plan, either the 'moonship' would be designed to be launched in two parts that would be assembled in orbit before setting off for the Moon, or would be designed as a single vehicle and would be placed into low Earth orbit where it would refuel from a 'tanker' before setting off for the Moon. For this solution the less powerful Saturn IB, Saturn III and Saturn IV were deemed sufficient.

At the end of 1961, when studies were well advanced, several major problems were discovered. Firstly, on a mission that was to land directly from the translunar coast, the ship would crash if its engine failed to fire. Secondly, even if the engine fired, it would be extremely difficult for an astronaut to land a 20-m-tall spacecraft piloted from its 'nose', and it would be dangerous to climb on a ladder to that height!

The solution to these problems was a revolutionary approach originally proposed by Langley mathematician John C. Houbolt in 1959: 'lunar orbit rendezvous'. This approach involved the launch of a spacecraft train from which, once in orbit around the Moon, a mission-tailored landing module would separate. Once the surface activities were over, the landing module would lift off to dock with the 'mothership' – the only part of the train that would return to Earth. However, this approach also had disadvantages. If the orbital docking were to fail, the astronauts returning from their surface sortie would be unable to return to Earth. However, this was the most economical technique, requiring mission-optimised orbital and landing modules, and the launch of a single Saturn V per mission.

After almost three years of official indifference, in November 1961 Houbolt once again proposed lunar orbit rendezvous to the Space Task Group. The last remaining resistance – that of the German team working on the launcher – was overcome the following year when the probabilities of a failed docking were precisely computed and deemed acceptable. On 11 July 1962, NASA announced its choice of 'mission mode': lunar orbit rendezvous. In order to practice the technique of rendezvous in space, a new family of piloted spacecraft was designed to fly before Apollo: the two-seater spacecraft, Gemini.

3.4   THE APOLLO SPACECRAFT

Having defined the method of reaching the Moon, and chosen the launcher, the spacecraft had to be designed. At the time of Kennedy's announcement, only twelve companies were still seeking a contract as a result of NASA's request for proposals published in 1960, namely two companies – Martin Marietta and North American Aviation – and three consortia – General Electric, Douglas, Grumman and Space Technologies Laboratories; General Dynamics and Avco; and McDonnell, Lockheed, Hughes and Ling Temco Vought. They offered a panoply of solutions: Mercury-like ballistic capsules, semiballistic (lift-generating) capsules, and lifting-body vehicles able to carry out complex manoeuvres in the Earth's atmosphere, based on the experimental aircraft then being designed by the Air Force and from which the Space Shuttle was later derived.

The commission tasked with choosing the winning proposal awarded the highest rating to the Martin Marietta proposal, but North American was the eventual winner due to its previous experience on the experimental rocket-powered X-15 hypersonic aircraft. The initial contract called for the production of twenty capsules in two different versions, sixteen 'boilerplate' test capsules (identical to real capsules

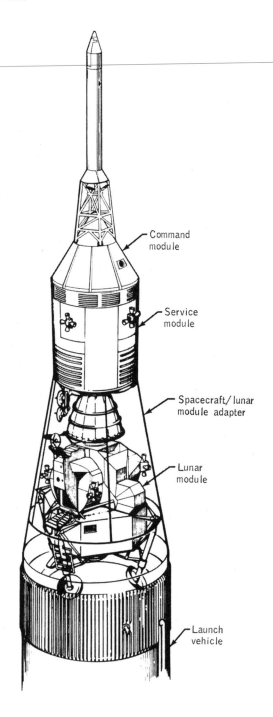

- Command
  module
- Service
  module
- Spacecraft/lunar
  module adapter
- Lunar
  module
- Launch
  vehicle

The Apollo spacecraft launch configuration.

in mass and inertia but non-habitable), ten mock-ups, five engineering prototypes, and two training simulators.*

After lunar orbit rendezvous was chosen, North American's contract was reduced to the 'mothership' only, and a competition was held for the contract for the lunar landing vehicle. Of the three-man crew, one would remain in lunar orbit while the other two went down to the surface.

As eventually built, the North American spacecraft was made of two portions: a Command Module (CM) protected by heat shields – the only part of the whole Saturn–Apollo complex designed for Earth return – and a Service Module (SM) to provide electrical power and attitude and trajectory control.

On top of the CM, during the launch and for the first minutes of the flight, was mounted a solid-fuel rocket which was to lift the CM clear of the launch vehicle in case of emergency. The 10-metre long, 1.2-metre-diameter rocket was mounted on a truss, and had two retractable lifting surfaces (canards) and an aerodynamic attitude sensor based on the X-15 'Q-ball' (Q being the mathematical symbol for dynamic pressure).

The CM was a truncated conical shape, 3.7 metres tall, with a base diameter of 2.9 metres and a mass of around 6,000 kg. On board, the astronauts had 6.2 m^3 of living and working space, and five windows of a few tens of square cm. There were two hatches: the main one being 86 cm wide and 73 cm tall, and the other a circular 76-cm-diameter hatch at the top of the cone, through which the astronauts could enter the lunar landing vehicle. The internal atmosphere (the composition of which will be discussed later) had a pressure of 355 hPa. The external surface was covered with heat shields, the three parts of which (top, sides and bottom of the capsule) were made of steel honeycomb covered with a layer of epoxy resin of a thickness varying between 3.2 cm and 6.6 cm in order to create an aerodynamic shape to provide some atmospheric trajectory control. The lift-over-drag ratio of the capsule was 0.3, compared with 4.4 for the Space Shuttle and 12 or more for modern jetliners. It had a hydrazine and nitrogen tetroxide attitude control system using twelve 411-N nozzles. Electrical power was provided primarily by the SM and (during re-entry) by three silver–zinc batteries in the CM itself.

The SM was cylindrical, some 7 metres long, and with a mass close to 25,000 kg. On one end were the straps holding the CM until re-entry, and on the opposite end was the large Service Propulsion System (SPS) rocket engine for lunar orbit insertion and trans-Earth injection. The 97.4-kN-thrust Aerojet engine used unsymmetrical dimethyl hydrazine (UDMH) and nitrogen tetroxide – hypergolic fuels in tanks that occupied most of the module. A separate system using the same propellants fed four groups of four attitude-control nozzles. Other tanks carried helium for pressurisation and the fuel cells' reactants. The fuel cells used oxygen and hydrogen and a lithium catalyser (not the best catalyser, but the lightest) to produce both power and drinking and cooling water. The SM also carried most of the antennae

---

* One of the boilerplate capsules, serial number 1227, was lost in the Bay of Biscay by the Royal Navy during an emergency landing rehearsal. The capsule was recovered by the Soviets and turned over to the US in the port of Murmansk during the first visit of US military vessels since the end of the Second World War.

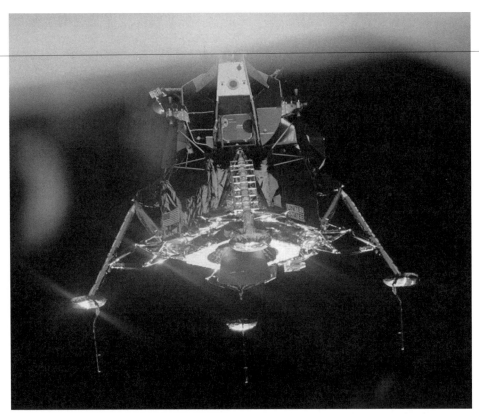

The Apollo 11 LM, undergoing visual examination in lunar orbit, shows the descent stage engine and the three contact sensors under the landing pads. These sensors will break under the weight of the module during landing.

and communication systems. For brevity, the mothership was designated CSM (Command and Service Module).

After the choice of lunar orbit rendezvous, on 7 November 1962 Grumman won the contract to build the Lunar Module (LM). This called for the construction of fifteen flight units and ten engineering prototypes. The initial concept was signed by Houbolt himself. Working on this idea, Grumman first designed a five-legged spacecraft with four large cabin windows, not unlike a helicopter. Later, however, a second version with four legs and two much smaller windows was proposed. The final version consisted of two parts: a descent stage and an ascent stage, where the crew was housed. The launch mass varied depending on the mission, but was around 15 tonnes.*

The LM comprised two parts, or 'stages'. The descent stage was a simple box structure covered by a thin mylar foil. It housed the 47-kN throttleable descent engine, four propellant tanks for UDMH and nitrogen tetroxide, helium tanks for

---

* The original name was Lunar Excursion Module – known as the LEM – but on 12 May 1966 the 'Excursion' was deleted. However, the acronym LEM is still widely used.

pressurisation, and five silver–zinc batteries. The frame was octagonal, and to four of the sides were attached the landing legs – folded during the trip to the Moon – with a 9.4-metre diagonal base when deployed. At the tip of every leg was a circular footpad, under which was mounted a 173-cm-long contact sensor covered with strain gauges designed to break under the weight of the LM at landing. To the forward leg was attached, some 75 cm over the ground, a short nine-rung ladder, and immediately above this was a platform (dubbed the 'porch') facing the exit hatch of the ascent stage. (After Apollo 9 tested LM no. 3 in Earth orbit, the contact sensor on the forward footpad was deleted in order not to hinder the astronauts' descent of the ladder.)

Beginning with Apollo 12 (LM no. 6), on one of the free sides of the descent stage was stowed a suite of experiments called the Apollo Lunar Surface Experiment Package (ALSEP), the most critical component of which was the 56-W System for Nuclear Auxiliary Power (SNAP-27) generator, producing electrical power from the heat released by the radioactive decay of a small mass of plutonium oxide. To avoid environmental contamination in case of launch failure, the radioactive kernel was transported inside an aerodynamically stable explosion-proof carbon heat shield that was capable of withstanding re-entry should it ever have to do so. Once on the Moon, the ALSEP and the generator, activated by the insertion of the kernel, would be laid on the surface about 100 metres from the LM, this distance being deemed sufficient to preclude gas contamination of the experiments by engine exhaust when the ascent stage lifted off.

The ascent stage took the total height of the LM to 7 metres. It housed two batteries – which, like those on the descent stage, provided 28 V – and the 15.6-kN-thrust UDMH and nitrogen tetroxide ascent engine. The crew's fore-aft cylindrical compartment was 234 cm in diameter, and was pressurised to 355 hPa. In order to lighten the structure, Grumman used panels so thin that they could be seen buckling under pressure; indeed, the wall thickness at some points was as little as one-tenth of a millimetre! On the rear wall of the cylindrical compartment were equipment racks. On the roof was the hatch to the CSM. On the floor, another hatch provided access to the ascent engine and its piping, and to the environmental control system which provided 48 hours (later increased to 72 hours) of autonomy. At the front of the compartment were two control panels and the place for the two astronauts, who were to remain standing during the descent and the ascent from the Moon, and were to sleep either on the floor or on hammocks. In front of the control panels were two small triangular windows that were inclined downward to provide a good view in the final phase of the landing, and there was an even smaller window above the commander's head (on the left side of the cockpit) to be used during the rendezvous. Between the two astronauts, low down, was the square cross-section hatch leading to the porch, which opened inwards and to the right (from an internal perspective).

On the outside, the ascent stage had four clusters of attitude control engines and two parabolic antennae – one for communication, and the other for the docking radar. Under the stage were the pyrotechnic devices for separation from the descent stage, which would be used as a launch pad. During launch from Earth, the LM would be housed immediately under the SM engine nozzle, in a truncated conical compartment on top of the S-IVB known as the Spacecraft/LM Adapter (SLA). Once the CSM had separated, the compartment would split open into four petals, which would detach to expose the LM ready for docking and extraction by the CSM.

From the beginning of the project, one of the biggest problems presented was learning to pilot a landing on the lunar surface at a gravity one-sixth of the Earth's. This problem was addressed three ways: an electronic simulator; a hoisting system shaving five-sixths off the weight of a simulator; and a proper flying vehicle. The hoisting system – a huge 120-m long, 75-m tall crane – was built at Langley. Bell helicopters received a $3.6 million contract to build the flying vehicle. This Lunar Landing Research Vehicle (LLRV) was publicly revealed in April 1964, and flew for the first time from Edwards Air Force Base in California on 30 October that same year. It was an ugly-looking 3-metre-tall metal truss with four legs, giving it a 4-metre-wide landing base. Five-sixths of the aircraft's weight – 1,680 kg – was lifted by a downward-pointing centrally mounted turbofan engine. The remaining sixth was countered by a pair of variable-thrust hydrogen peroxide rocket engines, and the craft's attitude was controlled by sixteen more engines. The turbofan was mounted on a universal joint to ensure that its thrust was always vertical. The pilot's ejector seat was on an open platform at the front. An entire flight would take place at a height of no more than 100 metres. Two emergency rockets were fired in case of malfunction.

Two LLRV were built and, after completing test flights and being modified with LM-like commands, were assigned to the training of the lunar landing mission commanders. In 1966 NASA ordered three more aircraft, and the whole series was renamed the Lunar Landing Training Vehicle (LLTV). These five aircraft were used extensively, and the last flight – by Eugene Cernan – took place on 13 November 1972. Three LLTVs were lost during training accidents, of which the most famous was that of 6 May 1968, when the manoeuvring engines lost pressure and the aircraft began to roll sharply; the pilot, Neil Armstrong, ejected only a second before ground impact. The second and fourth LLTV still exist. Other landings were simulated using the small vertical take-off X-14 experimental aircraft, which had an innovative digital flight control system that could imitate the behaviour of a number of other vehicles, including the LM.

3.5   LOGISTICS

In addition to the sheer technological challenge, the Apollo project required a noteworthy logistic effort, and several infrastructures had to be established for the task.

Engine test stands were built *ex-novo* or by modifying pre-existing installations at Edwards Air Force Base, for F-1 engines, and at Rocketdyne's field laboratory in the Santa Susanna hills, near its plant in Los Angeles, to simulate high-altitude working of cryogenic J-2 engines. Huntsville was equipped with stands to test the correct functioning of entire stages, and the Mississippi Test Facility (now the Stennis Space Center) was equipped with four stands to test the first and second stage at different simulated altitudes.

The grandest accomplishment was the building of what is now called the Kennedy Space Center – adjacent to the Air Force's Eastern Test Range at Cape Canaveral.* It included the Vehicle Assembly Building (VAB) and the twin launch pads 39A and 39B, and launch control facilities.

The VAB was a 160-m-tall building inside which up to four Saturn V launchers could be simultaneously assembled in a vertical position, complete with LM and CSM, in a controlled atmosphere. After assembly, huge 'crawlers' moved the launchers, along with their service towers, to the pads. The crawlers themselves were impressive machines, each weighing around 4,000 tons. They were designed to move a Saturn V from the VAB to the pad – a distance of 5 km – on eight tracks, tilting the platform as necessary to maintain the rocket in a vertical state, and travelling at a maximum speed of about 1.6 km/h. The service towers also had to withstand the hot engine exhaust gases and then be reused.

The concrete pads were built *ex-novo* several kilometres north of existing Cape Canaveral pads on a part of the coastal strip known as the Merritt Island Launch Area. Almost identical facilities provided for the refuelling and control of the Saturn V launcher and its payload during the days preceding launch. The VAB, crawlers, pads and control room facilities are still being used in modified form by the Space Shuttle.

For mission control, astronaut training and overall technical management of the programme, a new NASA centre was built in Houston, Texas – which happened (not by a chance) to be the state which the new President, Lyndon B. Johnson, had represented. The problem was also raised of how to handle the lunar rock samples that were intended to be returned to Earth. It was not possible to immediately hand over the container to scientists, as this might cause contamination of the precious material by terrestrial atmospheric agents, humidity, and so on. As it was also impossible to disprove the existence of lunar microbial life in the returned samples, it was deemed necessary to carry out some biological tests prior to their distribution. Finally, there was a documentation and identification requirement for each sample. It was thus decided to build, in Houston, a Lunar Receiving Laboratory (LRL), where the samples would be 'uncrated' in as neutral an environment as possible, photographed, documented, and finally prepared for handing over to the scientists. With time, the scope of the laboratory grew to include microbiological analysis, a preliminary sample-age analysis by measuring cosmic ray exposure, and the management of the brief precautionary quarantine imposed on the astronauts and the CM.

Another logistical problem was the transportation of the three Saturn V stages from the factories to the VAB. The first stage was built near New Orleans, and was transported along the coast. A similar method was adopted for the second stage, built at Seal Beach, California, with the barge passing through the Panama Canal.

For the third stage, built near Los Angeles, a different system was adopted. A small Californian aerospace company, Aero Spacelines, understood as early as 1961 that it could enter governmental business by designing an aircraft able to carry bulky

* From 1964 to 1973 the entire area was called Cape Kennedy.

spacecraft components. Upon securing a preliminary contract from NASA, it modified a surplus Boeing 377 four-engined propeller airliner by enlarging the fuselage to accommodate 6-m-diameter loads! This distinctive shape prompted its nickname of the 'Pregnant Guppy'. In July 1963 it was used to carry an S-IV mock-up from Los Angeles to Cape Canaveral. However, the payload volume of 826.5 m^3 was still not sufficient to contain a full S-IVB with a LM adapter, and so the 'Super Guppy' version was created. Powered by turboprops, it had a payload volume of 1,4105 m^3. It entered NASA service, and carried every S-IVB used by the Apollo missions. During the 1980s it carried its most precious cargo: the $5 billion Hubble Space Telescope.

Finally, to track Apollo capsule telemetry during launch and re-entry, to visually track it, and to maintain communications during re-entry over the ocean, eight C-135A transport planes – the military version of the Boeing 707 jetliner – were modified to the EC-135N Apollo Range Instrumentation Aircraft (ARIA) standard by adding a very large steerable antenna inside a 3-m-diameter 'bubble' in the nose. These aircraft were based at Patrick Air Force Base in Florida. On four of them was mounted an aerodynamic fairing housing the stabilised telescope of the Airborne Lightweight Optical Tracking System (ALOTS) for optical tracking of spacecraft.

3.6   FIRST TESTS

The first lunar mission hardware tests used boilerplate capsules in emergency rocket tests. These included both pad and in-flight aborts, in the latter case the boilerplate being launched by a small Little Joe II rocket. This was a solid-fuel single-stage rocket derived from the Little Joe rocket on which the Mercury spacecraft had been tested. The first Little Joe II was launched from White Sands, New Mexico, on 28 August 1963. The second rocket pushed a boilerplate to supersonic speeds, whereupon it was blown up on command in order to trigger the emergency system. On 8 December that same year, the simulated emergency was timed to coincide with the passage through the sound barrier. On the next test, on 19 May 1965, the rocket failed for real. In all cases the emergency rocket functioned smoothly. For the last test, on 20 January 1966, a real Apollo capsule (serial number 002) had to face a simulated emergency at a height of 17,000 metres, and also survived unscathed.

On 7 November 1963 and 29 June 1965, two more boilerplates were used for the launch pad abort tests, lifting off under emergency rocket thrust and then returning to Earth on parachutes after reaching a height of 1,500 metres. Two scale models of the CM were launched on sub-orbital flights by Atlas ICBMs, for heat shield tests. In the first of these Project Fire tests, the model capsule reached a speed of 40,500 km/h, simulating a return from the Moon, and survived a temperature of 11,000° C.

More boilerplates were taken into orbit by Saturn I launchers. The first was launched on 28 May 1964, and the second on 18 September. In both cases a 'live' emergency rocket was carried. The final three Saturn Is took into orbit boilerplates which deployed panels to assess the micrometeoroid threat to Apollo, in which guise they were named the Pegasus satellites. The career of the Saturn I rocket thus ended with ten successes out of ten launches. Finally, many boilerplates were used to test landing systems, water impact behaviour and buoyancy characteristics.

The Apollo programme was divided into ten mission groups: 'A' – unpiloted tests of the CM and of Saturn IB and Saturn V launchers; 'B'– unpiloted LM tests; 'C' – piloted CM tests; 'D'– full piloted tests of the CM and LM in low Earth orbit; 'E' – test of the piloted CM and LM in a highly elliptic Earth orbit (never carried out); 'F' – piloted LM tests in lunar orbit; 'G' – the first landing; 'H' – more landings of limited scientific scope; 'I' – lunar piloted orbital survey missions with a large scientific payload (never carried out); and 'J' – long-duration (at least three days) landing missions using wheeled or flying vehicles to explore scientifically interesting sites.

On 26 February 1966, in inaugurating the Saturn IB launcher, AS-201 (also known unofficially as Apollo 1) tested, for the first time, the S-IVB cryogenic stage. It boosted a fully functional spacecraft on a sub-orbital trajectory up to a height of 499 km to test the attitude control system. At apogee the SPS engine fired to accelerate the capsule to a speed close to that of re-entry from orbit. The CM was recovered 40 minutes after lift-off, 8,850 km downrange. On 5 July 1966, AS-203 (unofficially Apollo 2) put the S-IVB into orbit to test its restart capabilities. In addition, the pressure in one of the tanks was deliberately increased above its operational limits to breach the common bulkhead, in order to verify its tolerance. This caused the stage to explode, generating a cloud of fragments which re-entered the atmosphere the same day. On 25 August 1966, AS-202 (unofficially Apollo 3) took a capsule on a sub-orbital flight to a height of 1,140 km. During this flight the SPS engine was restarted three times. By then, NASA felt ready to launch its first 'C' piloted mission.

3.7   TRAGEDY

In mid-December 1966, NASA held a press conference to present the Apollo capsule, which was described as a faultless machine after more than 20,000 modifications. The plan called for a first piloted flight in February of the following year, and a landing on the Moon in 1968, as soon as the LM was ready. The crew of the first Apollo mission was also presented. The commander was Virgil (Gus) Grissom – one of original seven Mercury astronauts and a veteran of two spaceflights: the second Mercury flight – a sub-orbital 'hop' at the end of which he narrowly escaped drowning after the capsule filled with water – and Gemini 3, the first piloted test of that programme. The crew also included Edward White, the first US astronaut to accomplish an extravehicular sortie, during the Gemini 4 mission, and Roger Chaffee, a reconnaissance aircraft pilot during the Cuban missile crisis, who would be making his first flight into space.

The capsule, serial number 012, was a Block I, intended for system development; it was unable to fly a lunar mission, as it had no LM docking system. On 6 January 1967 it was mated to Saturn IB serial number SA-204 on Pad 34, and put through a lengthy series of system tests. On 27 January the astronauts boarded the capsule for a countdown simulation that was to conclude with an emergency evacuation drill. In

The cover of the AS-201 Command Module after the fire that killed Gus Grissom and his crew.

the afternoon, after some delay due to a foul smell 'like buttermilk' in the air-conditioning system, the capsule hatches were closed.

The first hatch sealed the pressurised cockpit and opened inward, in order to be held shut in space by the pressure difference between the inside of the capsule and the outside. The second hatch was part of the heat shield, and the third was part of the aerodynamic fairing mated to the abort rocket. That day, the last hatch had to be left partially open in order to provide passage for wire bundles required by the tests. The capsule was then pumped up to the launch pressure of 1,124 hPa with pure oxygen, and the simulation was started, despite some persistent communication system problems. The test was almost complete when, at 23.31 GMT (18.31 local time), Grissom was heard to shout '*Fire!*'. Chaffee confirmed this with 'We've got a fire in the cockpit!' The technicians standing near the capsule tried to open the hatches, but were driven back by flames and smoke coming from the ripped-open pressure shell. Even when the flame ceased, and they were able to return, their rescue efforts were hampered by a toxic smoke that limited visibility to a few centimetres. Four minutes after the first sign of fire, the inner hatch was opened and the rescuers found the corpses of the astronauts. Chaffee was still strapped into his couch. Grissom and

White were so tangled below the hatch sill that it was hard to tell them apart, because everything in the capsule was charred black. The bodies were removed with some difficulty, as the molten spacesuits had become glued to the couch covers.

NASA immediately named an investigating commission, which two months later explained that the fire had probably been started by faulty electric wiring close to Grissom's feet, on the left side of the capsule. From there the flames had enveloped the left side of the cockpit, burning the nylon fittings. With the rise in temperature and pressure, the fire had spread to the right. The astronauts were in the meantime desperately trying to open the hatch, which was held firmly in place by the pressure. After 20 seconds, the pressure had breached the capsule's shell, the oxygen had been depleted, and the atmosphere had become toxic due to carbon monoxide. The astronauts had rapidly lost consciousness, and had died of asphyxia. Thirty seconds later only some flammable cooling-fluid pipes continued to burn. It was a classic case of a 'flash fire' in a confined pure-oxygen environment.

The commission identified six causes contributing to the loss of the crew: the high-pressure pure-oxygen atmosphere; the vulnerable electric wires (some were in plain sight in the cockpit and could be damaged if displaced, and others ran in metallic housings with sharp edges); the presence inside the capsule of more than 30 kg of flammable materials such as nylon nets, velcro strips, polyurethane cushions on which to lay the hatch during the emergency drill, seat covers, and about 10 kg of paper; the environmental control system, using flammable fluids, which had created countless problems during development; the inward opening hatch, which had been sealed by the pressure of the fire; and inadequate fire training and equipment for the support team.

NASA had used a pure-oxygen system on Mercury and Gemini, so it was not unaware of the risks of fire. It had drawn up Apollo specifications and directives on the materials to be taken on board, on flammability tests, and on combustion characteristics. However, the testing of materials had been at the partial pressure of 354 hPa that would be used in space, rather than at the full pressure of 1,124 hPa that was used on the ground during pre-launch activities, in which combustion has completely different characteristics than in space, due to convective air motions, and – the most serious fault – some of the materials (the polyurethane cushions) rejected by NASA had been carried on board without any objection from the safety representatives of the agency. With tragic irony, the fire could have been avoided if NASA had known of a very similar accident which happened in an oxygen-filled simulator in which Soviet cosmonaut Valentin Bondarenko had died three weeks prior to the flight of Gagarin.

Unfortunately, letters to North American and reports stressing the problems of flight safety, and in particular fire prevention, had gone unheeded by North American, and this fact had not been noticed by the agency. By interviewing the astronauts overseeing the development and production of the capsules, a very tense relationship with the spacecraft builder was revealed. In particular, a degree of arrogance by the company and a remarkable inertia in accepting and implementing modifications was noted. An anonymous astronaut stated that the atmosphere was completely different with respect to that which had been established with Mercury and Gemini builder, McDonnell–Douglas. Some attributed North American's behaviour to the company's lack of previous experience in spacecraft design – this despite the fact that the contract had been won, in part, based on its presumed previous experience!

After the investigation, as a belated reformulation of that contract, key positions in CM production management were given to former employees of Martin Marietta, which had lost the contract despite its higher rating on the 'scoring' system employed.

Beginning in July 1967, the wreck of the spacecraft – which, at the behest of the astronauts, was officially given the name Apollo 1 – the heat shield, 81 boxes of documents, and other material concerning the investigation, were closed inside a pressure-tight container filled with nitrogen, and stored in a warehouse at Langley Research Center. According to some accounts, this is still at Langley, but it has also been reported that in 1990 small cracks were discovered, from which gas was leaking, and that NASA obtained the agreement of the two surviving widows to bury the wreck in the same disused Cape Canaveral missile silo as housed the debris of the Space Shuttle *Challenger*, which had exploded in the Florida sky in 1986 – 19 years and 1 day after the Apollo fire.

3.8   MORE TESTS

During the first half of 1967, while the fire investigation was underway, components of the first Saturn V began to arrive at Cape Kennedy from all across the United States. To save time, it was decided to test the complete launcher in what NASA described as 'all up' testing, and to reduce the test flights from thirteen, as originally planned, to just two. The development of the LM was not yet complete, and tests would be carried out without this component. In the first test, the third stage would boost the CSM to a high apogee, and the SPS engine would accelerate it to simulate lunar re-entry. As no astronauts were on board, the capsule was of the same standard as that lost in the fire.

After several delays – one of them just 17 seconds before lift-off – on 9 November 1967 the first Saturn V took Apollo 4 into space. The third stage restarted after two orbits to raise the apogee to 17,346 km. After S-IVB separation, the SPS engine was fired twice – the first time to raise the apogee to 18,216 km, and the second time to drive the spacecraft into the atmosphere at a speed of 11.14 km/s. Nine hours after launch the capsule was recovered intact, not far from Hawaii. Suddenly, it seemed that the tragedy of Apollo 1 had been forgotten, and that the Moon was once again within reach.

The next mission was the first flight of the LM. Although in an advanced state of development, the LM was far from being complete. The objective of the mission was to test the flight control systems, in particular the descent engine, whose trottleable injectors had been very difficult to design, and the ascent engine, which had to be so simple that it could not fail to fire.

Apollo 5 (as the first LM flight was called) was launched on 22 January 1968, on the same Saturn IB that should have taken Grissom and his crew into space. Once in a low orbit, the aerodynamic fairing substituting for the CSM and the four petals

The discarded Apollo 8 S-IVB stage. On top of the stage is a dynamical mock-up of the Lunar Module.

were deployed, and finally the LM – the only one to fly without landing legs – separated from the S-IVB. The first firing of the descent stage, simulating a lunar landing, was aborted by the main computer after four seconds, but it later proved possible to restart it twice, and also to test its throttleability. At the end of the last firing, the abort manoeuvre was tested. As the descent stage engine was shut off, the ascent stage engine was fired, and pyrotechnics separated the two stages. The ascent stage engine was fired twice – the second time for 383 seconds, to fuel depletion. In all, the tests lasted eleven hours. The two stages disintegrated on re-entering the atmosphere a few weeks later. It must be noted that the Saturn IB first stage carried a recoverable movie camera to film the S-IVB separating under the thrust of its ullage rockets and then starting its J-2 engine; this footage has been shown many times on TV.

Finally, on 4 April, Apollo 6 provided a disappointing end to the Saturn V test flights when the launcher experienced a series of problems. The first stage developed longitudinal 'pogo' vibrations, stretching and shortening five times a second because of resonances in the fuel-feed piping; the second-stage number 2 engine's thrust decreased, and then developed a non-axial component before being shut off; and the third stage did not restart once in orbit. It must be mentioned, however, that Apollo 6 also carried a recoverable movie camera in the base of its second stage, and this took the even more famous movie of the separation of the S-IC and, a few seconds later, the discarding of the interstage 'skirt', which gave the appearance of flaming in the engine plume.

The 'pogo' effect was eliminated by a partial redesign of the first-stage piping, while the problem with the second and third stages was attributed to a flexible joint which, subjected to a high flux of very-low-temperature hydrogen, to vibrations, and to the hostile high-altitude environment, could easily break down and damage the injection and ignition systems. In particular, part of the second-stage engine ignition system had actually exploded, pierced the nozzle wall, and thereby created the observed non-axial thrust component. A similar failure had probably occurred on

The Mare Foecunditatis crater Goclenius (54 x 72 km), imaged from Apollo 8, shows a complex system of grabens. The four large craters in the background are: (left to right) Colombo A, Magelhaens A, Magelhaens, Gutenberg D.

the third stage which, deprived of a working ignition system, could not restart. Other anomalies of that flight were the discovery of some incorrectly connected piping, and the detachment of debris which caused the SLA to depressurise. After the necessary modifications had been made, it was confirmed that the third flight of the Saturn V would be manned.

Before this, however, the CSM had to be tested. The first piloted mission of the programme – Apollo 7 – would essentially fly the mission that had been assigned to Grissom's crew, and so would ride into orbit on a Saturn IB, the fifth of the series. The Block II capsule used for this flight had been built two years earlier, but after the fire it was completely dismantled and then reassembled with the incorporation of several modifications. Many plastic components were rejected, the accepted components were subjected to combustion tests, and the hatch was now a single-piece outward-opening one that required 3–10 seconds, depending on whether it was opened from the inside or the outside. For the launch – the most dangerous part of the mission – a mixed oxygen and nitrogen atmosphere was adopted. This would be changed to a partial-pressure pure oxygen atmosphere once in space.

Onboard Apollo 7 were Walter Schirra, a veteran of the fifth Mercury flight and Gemini 6, and the rookies Donn Eisele and Walter Cunningham. In the 10 days following launch on 11 October, the SPS engine was fired eight times in all, as rendezvous tests were carried out using the empty S-IVB stage. The mission is mostly remembered because of some tense exchanges between the astronauts and ground controllers. Schirra refused to make a television broadcast, and he and his colleagues complained endlessly that they were suffering the symptoms of head colds. After a particularly heated dispute with the ground, one flight controller jokingly suggested timing the spacecraft's re-entry so that it descended into a hurricane! Human frailties apart, the Apollo 7 mission was a complete success, and provided the much-needed confidence in the technology developed for the lunar flights. One change ordered after Apollo 7 was to make the petals of the LM adapter detach rather than simply open out at an angle, in order to reduce the risk of one fouling the LM on a later mission.

It is not surprising that none of the three astronauts ever flew again into space; Eisele died on 1 December 1987. This was also the last launch of the Saturn IB in support of the lunar programme. The Apollo 7 CM is now on show at the National Museum of Science and Technology, in Ottawa, Canada.

According to the original plan, three more missions were scheduled before the Moon landing. The first two would test the LM in Earth orbit, and the third would put the whole spacecraft train into lunar orbit for a full dress rehearsal of procedures prior to firing the LM's descent engine for a lunar landing.

3.9   THE DAWN OF EARTH

Meanwhile, more problems – of both a technical and public-image nature – were afflicting NASA. Completion of the LM was running late, and its first piloted flight, scheduled for 1969, was uncomfortably close to the deadline of the end of decade specified by Kennedy. Problems of 'image' were also shaking the whole of the United States. The Vietnam war – which up to 1967 had received widespread public support – was now bitterly opposed, both internally and abroad.

With the civil rights protests, and with the assassinations of Martin Luther King and Robert Kennedy, the desperation of many social classes was revealing itself in rebellion and hostility. The blood-bath of the Chicago democratic convention was a stain on the image of a country proposing itself as a champion of democracy. The peace movement sprang up almost overnight, after television broadcasts began to show the realities of the Vietnam war. It was considered as not just rhetoric, heroism and a sense of duty, but as atrocities, death and destruction at the expense of innocent populations, producing the (not too incorrect) perception of the Apollo project as a manifestation of the abhorrent 'military–industrial complex'. For the first time since its beginning, the lunar programme was criticised as a waste of public money. But the critics were probably off target, because the whole programme cost no more than a year's worth of the Vietnam war – which, in 1967, was costing \$2 billion per month – diluted over a whole decade, and NASA's annual budget was never more than one-tenth of the Department of Defense's budget, or even the Department of Health's budget. The majority of public opinion, however, still supported lunar flights, and this was reinforced by a film first shown in the United States on 29 April 1968 which depicted a realistic near-future in which flights to the Moon were routine. This, of course, was Stanley Kubrick's film of Arthur C. Clarke's *2001: a Space Odyssey*.

Despite its problems, NASA was very aware that it was in a race. Intelligence services noted that the Soviets were preparing to launch a piloted mission to circumnavigate the Moon, and that they might be ready to do so as soon as the end of 1968. In July, before the Apollo 7 mission, it appeared that Apollo 8 would be a useless mission. It would be a piloted test of the Saturn V launcher, but the CM had already been tested, and the LM, which was to be tested on that occasion, was unlikely to be ready.

At the beginning of August, George Low – the NASA official in charge of the development of the CM – had an idea. As NASA Administrator James Webb was abroad, Low explained his idea to Vice Administrator Thomas Paine, who timidly approved. It was much more difficult to convince Webb, but he authorised a feasibility study, which was supportive, and on 10 August the decision was taken: if Apollo 7 was successful, the CSM on the following mission, in December, would orbit the Moon for the first time. A few weeks later Webb retired, and his place was taken by Paine.

The mission was proposed to the commanders of the next two missions. The first, James McDivitt, refused, and the second, Frank Borman, accepted wholeheartedly, despite the flight being just four months away. As training got underway in September, another event emphasised just how close to a piloted flight the Soviets apparently were: the automatic probe Zond 5 flew a loop around the Moon and returned to Earth.

The Apollo 8 crew were Frank Borman, a veteran of Gemini 7, James Lovell, a veteran of Gemini 7 and Gemini 12, and rookie William Anders. After launch by Saturn V, the S-IVB/CSM would be checked out, and if everything was working as designed, the S-IVB would carry out the historical translunar injection (TLI). In the event of problems, the mission would entirely be carried out in Earth orbit and last ten days.

The translunar coast – similar to the path that Zond 5 had flown – was also called a 'free return trajectory' because it ensured, in the event of a problem, that the Moon's gravity would return the spacecraft to Earth without any need for human intervention. In the nominal mission profile, the SPS engine would be fired over the far-side to enter lunar orbit. After some lengthy discussion it was decided that ten orbits would be carried out, for a total of 20 hours, after which the engine would be fired again to return Apollo 8 to Earth.

Another little-known but essential component of lunar flight was carried into space on 8 November, when a Delta rocket took the small Pioneer 9 probe into solar orbit. This was the fourth of five probes to continuously monitor the Sun so that NASA could alert astronauts in the event of a dangerous solar flare. On lunar missions, humans would for the first time leave the terrestrial magnetosphere, and would be exposed to potentially fatal solar radiation. A last moment of anxiety arose at the beginning of December, when intelligence suggested that after the apparently successful Zond 6 flight,* the Soviets would surely launch a cosmonaut to the Moon at mid-month,† but the Tyuratam launch window closed without any piloted Zond leaving Earth.

Then came 21 December, the day of Apollo 8's departure. The astronauts were awakened at 02.30 local time, after which they donned their space suits (essential in case of capsule depressurisation). At 05.34 the three men, saluted by astronaut Fred Haise, who was conducting embarkation operations, assumed their places inside the capsule, and two hours before lift-off the hatch was closed. In the meantime, a few kilometres away, an enormous crowd was assembling to witness the launch.

At 07.51 local time (12.51 GMT) the five F-1 engines started burning 20 tonnes of propellant each second and, after a pause dedicated to performance checking, the clamps anchoring the Saturn V to the ground released and the rocket began to rise majestically through a rain of ice slabs that had formed overnight due to the condensation of humidity on the walls of the cryogenic tanks. In the capsule, vibrations were such that radio communications were difficult to hear, and a violent shaking accompanied every movement of the engines on their gimbals. Acceleration reached 4.5 g, but the rudest moment was the separation of the S-IC stage, when acceleration suddenly fell to zero and the astronauts, although retained by their straps, were pushed towards the panel in front, only to be slammed back into their couches when the S-II ignited. Eleven minutes after lift-off, the S-IVB shutdown, having put Apollo 8 into orbit. On the second orbit, the voice of astronaut and Capcom (Capsule communicator) Michael Collins announced: 'Apollo 8, you are go for TLI'.

---

*   NASA did not know it, but the capsule crashed on landing.
†   As will be seen in next chapter, this was a real possibility.

At 2 hrs 50 min 38 sec after launch, the Apollo computer issued the S-IVB engine-firing command. The small ullage rockets fired first, and their tiny thrust settled the propellants of the stage near the bottom of the tanks, where the turbo-pump intakes were located. Immediately afterwards, the J-2 engine fired. Five-and-a-half minutes later, integrating accelerometers detected a speed of 10,830 m/s – enough to reach the Moon – and sent the shut-off signal, after which the spacecraft separated and the SLA panels were jettisoned. Borman rotated the spacecraft 180° on the yaw axis to observe the empty S-IVB and to carry out docking simulations with the LM, which in this case was represented by ballast of identical mass.

Only at this time was it possible to realise that this spaceflight was unlike any other. In contrast with the panorama shown by previous American and Soviet spacecraft, the windows of which could show only a tiny portion of Earth, which was still large in relation to the flight altitude, this time the whole blue-and-white ball of our planet could be framed by even the smallest window.

After the docking simulation, the CSM briefly fired its small thrusters to move clear of the S-IVB before the spent stage dumped its unused fuel. Many hours later, as the astronauts were closing to the leading hemisphere of the Moon, the S-IVB passed by the Moon at an altitude of 1,263 km and, perturbed by its gravity, slipped into solar orbit. During the translunar coast, which lasted slightly less than three days, the spacecraft was kept in 'barbecue mode', a slow rotation on its longitudinal axis, in order to even out the thermal stress of the uninterrupted sunlight in deep space. This rotation was cancelled only for short periods, to make television broadcasts which required the use of the high-gain antenna, and to briefly fire the

Earthrise from Apollo 8.

SPS engine, not only to make a minor course correction, but also to check the engine before attempting lunar orbit insertion.

Meanwhile, both Borman and Anders were suffering from NASA's first cases of 'space sickness'. In the particularly acute case of Borman, a possible viral infection was suspected. If this was so, the mission would have to be aborted. But after some consultations between the astronaut and NASA flight surgeon Charles Berry on an encrypted communication link, the mission was cleared to proceed.

On the morning of 24 December, the Moon was extremely close – although the orientation of the ship's rotation had prevented the astronauts from seeing it – and Apollo was slowly accelerating in its gravitational field. At 68 hrs 58 min 4 sec after launch, preceded by another countdown, communication was cut off as Apollo 8 slipped behind the Moon's leading edge, and the radios both on board and on the ground received only static.

As preparations were underway to fire the engine for lunar orbit insertion, Anders noticed something that seemed to pour like dark oil on the windows: after overflying the night hemisphere, the spacecraft was now over the dimly-lit lunar terminator.

The SPS engine was fired for four minutes to slow the spacecraft by some 850 m/s and so place it in the desired orbit with perigee at 111 km and apogee at 312 km. Now, with most of the work over, and waiting for contact with Earth to be re-established, the astronauts could enjoy the sight. The ground looked like wet sand, but its colour changed dramatically depending on the Sun's height over the horizon. The first recognisable landmark was the large crater Tsiolkovsky – discovered by Luna 3 only nine years earlier – which had a distinctive mare-like floor. In contrast with previous speculations, it proved possible to identify craters on the terminator, where the Sun was at an elevation of only 2–3°.

During the second orbit, Anders drew from a rack two Hasselblad cameras, with objectives of focal length 80 mm and 250 mm, and a movie camera, and began a plan of observation previously agreed with Harrison Schmitt, an Apollo geologist and a member of a small group of scientist-astronauts. A major problem lay in locating a usable window, as the main windows were heavily contaminated with sealant, which acted strangely in vacuum, and the other windows also had defects. Only the LM rendezvous window proved usable, despite being very small.

At perigee on the fourth orbit, the SPS engine was fired for a few seconds to circularise the orbit at 111 km. Immediately after exiting from radio occultation, Anders was presented with a view which remains one of the most enduring images of the Apollo missions. Passing from the lunar far-side – from which the Earth cannot be seen – to the near-side – it was possible to see the partially illuminated terrestrial globe 'rising' from the lunar surface. Despite having so far adhered strictly to the list of features to be imaged, this view was so spectacular that many 'discretionary' frames were devoted to it.

The main task was to photograph the lunar surface, particularly so-called 'site number 2' – an area of Mare Tranquillitatis close to the equator which, on the Lunar Orbiter images, appeared to be suitable for a landing. During the ninth orbit, after both Anders and Lovell had taken a few hours of rest and sleep, the first television broadcast from lunar orbit was made. Both Borman and Lovell entertained their audience by comparing the view of the sterile world passing below to the Earth, filled with life. Anders likened the lunar surface to beach sand darkened by a bonfire. Finally, as Apollo 8 crossed the terminator, the astronauts concluded their broadcast with Christmas wishes (in Houston, it was night-time on Christmas Eve), and recited the opening verses from the Book of Genesis.

As the three astronauts were the first humans to see the far-side, and many details of the near-side too small to be seen from the ground, they assumed the right to name a few. Anders assigned his own name and the names of his fellow travellers to three small craters on the boundary between the two hemispheres. Lovell chose one of mountains in the Montes Secchi range,* near the border of Mare Foecunditatis, and named it Mons Marylin, after his wife. None of these names were accepted by the International Astronomical Union, the official authority for the naming of celestial bodies. However, during its 1970 annual congress, a large far-side crater was named Apollo, and some small craters near it were dedicated to Grissom, White, Chaffee, Borman, Lovell and Anders. Ironically, these craters were in darkness during the Apollo 8 flight, and were thus never seen by their namesakes. Other names were made official in 1977 – and one of them was Mons Marilyn.

After ten orbits, the SPS engine was fired for the last time over the far-side. At 89 hrs 28 min 39 sec after lift-off, spacecraft telemetry was acquired even before the astronauts' voices were heard – a clear sign that the engine had fired properly, as Apollo 8 would have emerged from behind the Moon eight minutes later had the engine failed. On leaving the Moon, some pictures were taken of its full disk, showing part of both the near-side and the far-side from a perspective similar to that obtained by Luna 3, but of far better quality.

Particularly interesting was a vertical perspective of Langrenus† – a 132-km diameter crater that from the ground, being close to the limb of the near-side, appears foreshortened. In this picture, a small crater, showing a large system of white rays, not unlike the larger crater Tycho, was plainly visible over on the far-side. This crater was later dedicated to Giordano Bruno and, according to a theory first proposed in 1976, could be the only large lunar crater formed in historical times. Its position correlates with eyewitness reports related in the *Chronicle* of Gervase of Canterbury, which states that on the evening of 11 June 1178, a 'flaming torch' was seen rising from the Moon. Despite its great public appeal, this theory was recently proved wrong when it was demonstrated that the formation of such a large crater would have showered our planet with a hundred billion kilograms of ejecta. As no mediaeval chronicle – neither from Europe or the Middle East, nor from China, Japan or Korea – tells of a meteor shower this significant, the event witnessed in England probably had an atmospheric origin.

The Apollo 8 astronauts took a total of 800 pictures, of which at least 600 were close-ups of the surface, and shot more than 200 metres of 16-mm movie film.

The only problem on the 2½-day return trip was when a spurious command to the computer caused a loss of attitude, and required 45 minutes to correct, during

---

* Pietro Angelo Secchi (1818–1878) Italian astronomer.
† Michel Florent von Langren (1600–1675), a Belgian astronomer.

which an emergency navigation technique based on Earth-sighting rather than star-sighting was tested – an experience which would later prove useful to Lovell.

On the morning of 27 December, the SM was jettisoned, and shortly afterwards the CM made its first contact with the Earth's atmosphere. At 11 km/s, this was the fastest that any piloted spacecraft had attempted to return to Earth. The heat shield had been shown to be up to the job during the early test flights, but it was a tense time nevertheless. For a lunar return, a two-stage re-entry was used, with the spacecraft dipping into the outer atmosphere to shed some speed, then 'skipping' like a stone on a pond back into space on an arc that would lead to a second 'dip', this time at a speed comparable to a return from orbit. Although only 6 g, the peak deceleration seemed much worse after 8 days of weightlessness.

Some 30,000 metres over the Pacific Ocean, which was still in darkness, the ARIA aircraft and the aircraft carrier *USS Yorktown* established communication with the capsule. At a height of 9 km the drogue parachutes opened, followed by the three large main parachutes. Despite Borman's attempt to activate the parachute separation command at the exact time of impact, this was so smooth that he acted too late, and the capsule was tipped over by the wind. Several minutes passed before it was uprighted by three spherical balloons designed for this task, mounted on the nose near the LM tunnel. When dawn finally came, the recovery teams could reach the spacecraft, and a few hours later the three astronauts flew home. Lovell was later to command another mission, but neither Anders nor Borman ever flew in space again. The historic Apollo 8 CM is now on show at the Chicago Museum of Science and Industry.

3.10 LAST TESTS

With the successful Apollo 8 mission, the race against the Soviet Union was all but won. A cosmonaut flight onboard a Zond spacecraft would now take second place. Implicitly recognising its defeat, on 1 January 1969 the Soviet Union announced that the launch of a space station was imminent – a clear sign that the Moon was no longer the priority. A few days later, the Soyuz 4 and Soyuz 5 spacecraft docked and exchanged crewmembers during a spacewalk.

Kennedy's goal of landing on the Moon before the end of the decade still required some final effort, and the objective of Apollo 9 and Apollo 10 was to carry out these last tests. Apollo 9 was launched into Earth orbit on the fourth Saturn V on 3 March 1969. On board were James McDivitt, a veteran of Gemini 4, Dave Scott, Gemini 8 co-pilot, and rookie Russell Schweickart. The mission was to test the LM, the lunar suit and, most importantly, orbital docking techniques. The S-IVB placed the whole complex in an elliptical orbit and, after the LM had been extracted, performed a series of engine ignition tests, the last of which put it into a heliocentric orbit with a period of 325.8 days.

Schweickart donned the lunar spacesuit, which was effectively a miniature spacecraft, concentrating, in a 40-kg back-pack, an environmental system providing several hours of oxygen and a constant internal temperature through a water cooling system. After exiting from the front LM hatch, Schweickart travelled a few metres and stood in a pair of 'golden slippers' on the porch. The plan for him to cross over to the CM, whose hatch Scott had opened, was cancelled.

The next day, McDivitt and Schweickart boarded the LM and separated from the CM, in which Scott remained. The LM, named Spider, flew autonomously for six hours, and reached as far as 178 km from the mothership, named Gumdrop. This was the riskiest manoeuvre ever attempted in Earth orbit. If the docking were to fail, the two men onboard Spider would be doomed, as the LM carried no heat shield for atmospheric re-entry.

After six hours of free flight, during which the engines of both LM stages were tested, Scott made a perfect docking, demonstrating that one of the most complex manoeuvres was perfectly feasible. The closing-in of the two spacecraft was entirely carried out by optical means, with the CM pilot taking aim at a T-shaped target on the roof of the LM. Moving at about 30 cm/s, a male probe engaged a conical female drogue, and twelve spring latches locked to ensure an air-tight connection between the two spacecraft. Later, in a prophetic experiment, the ascent stage engine was fired and the entire spacecraft train was manoeuvred to a slightly different orbit. On 13 March, after 151 orbits, Apollo 9 returned to Earth, splashing down, for the only time of the entire programme, in the Atlantic Ocean. Of the three astronauts, Scott was the only one to fly in space again.

The ascent stage of the Apollo 10 LM manoeuvres before docking with the CM.

Spider's descent stage re-entered the atmosphere nine days later, but the ascent stage was left in an orbit ranging between 227 and 6,935 km, and did not re-enter until 23 October 1981. Gumdrop is now on show at the Space and Science Center in Jackson, Michigan.

Comforted by this string of successes, on 24 March 1969 NASA announced the crew of the Apollo 11 mission who, if everything went as planned, would make the first lunar landing. The mission commander was Neil Armstrong, a former Navy pilot who flew the X-15 rocket ship as a civilian, on one occasion attaining a speed of 6,418 km/h and zooming to an altitude of 82,500 km, then became an astronaut and flew Gemini 8; the LM pilot was Edwin 'Buzz' Aldrin (who legally changed his name to Buzz in 1979), an Air Force pilot credited with shooting down two aircraft during the Korean war, an MIT graduate, and Gemini 12 veteran; and the CM pilot was Michael Collins, an Air Force pilot born in Rome, son of an embassy military employee, and veteran of Gemini 10, after which he was temporarily grounded while he received a complex surgical operation on his spine.

The only missing flight before the landing was a lunar orbit LM test, and some – in particular, manned spaceflight director George Mueller – proposed skipping this test and attempting a landing with Apollo 10. However, two problems led to the rejection of this proposal. The LM assigned to the mission had not endured the 'slimming diet' imposed by Grumman, and was so heavy that there was little chance of it carrying out a successful landing, and waiting for the next LM would not advance the schedule at all. In addition, flight controllers were asking for very-low-altitude navigation experience in order to further assess the effect of mascons on landing precision. The original Apollo 10 mission was thus carried out. It was to be flown by the most experienced American space crew to date. The commander was Tom Stafford, a veteran of Gemini 6 and Gemini 9, the LM pilot was Eugene Cernan, who had flown with Stafford on Gemini 9, and the CM pilot was John Young, a veteran of Gemini 3 and Gemini 10.

Apollo 10 lifted off at 12.49 local time on 18 May from Pad 39B – the only time that the southernmost of two pads was used by a Saturn V. A few hours later, after the usual two Earth orbits, the S-IVB was fired for almost six minutes for TLI. At about mid-firing, because the liquid hydrogen relief valves were intermittently discharging gas, the complex began shaking so violently that the astronauts had problems seeing the instrument panels. Stafford even considered aborting the mission, but luckily he did not because, although disturbing, the vibrations did not exceed design limits. The CSM, Charlie Brown, separated from the S-IVB and retrieved the LM, not surprisingly named Snoopy. The entire manoeuvre was broadcast via a new colour camera which Stafford himself had asked to carry (the camera used on Apollo 8 was black-and-white). The S-IVB flew by the Moon at a distance of 3,111 km, and on into solar orbit. A second television broadcast was held during the second day of the mission, during which the astronauts showed the blue sphere of Earth 205,000 km away – a view that astounded the terrestrial audience. When the LM was readied on the third day, it was discovered that some insulating fibreglass had detached and formed a weightless sort of snow.

The rest of the flight proceeded smoothly through lunar orbit insertion and circularisation in an equatorial orbit at 110 km. On the afternoon of 22 May, Cernan and Stafford entered the LM and undocked. Snoopy and Charlie Brown flew in formation for some time until, over the far-side, the descent-stage engine was oriented against orbital motion and fired for about 30 seconds to enter the 'descent orbit', which had a perigee of only 14,450 metres. Some twenty minutes later, now much lower, they could better appreciate the Moon's curvature and the change in the horizon. Flying so low over the Moon was something no astronaut had tried before. Both astronauts had previously flown aircraft at a height of 15 km, but neither had done so at a speed of more than 6,000 km/h! The 'mountains' rising from the horizon proved to be the rugged rims of craters. The perigee was above Mare Tranquillitatis, just east of landing site number 2, which Apollo 8 had photographed in detail and had been chosen as the prime site for the first landing mission. Like Lovell before him, Cernan named a feature of Mare Tranquillitatis, Faye Crevasse, in honour of his wife. This name too was accepted in 1977. The low pass confirmed site 2 to be smooth, but Stafford noted that a litter of large boulders towards the far end of the target area might pose a problem (and, indeed, *did* pose a problem). Seventeen minutes past perigee, the descent-stage engine was fired to change the orbit to 352 km apogee and 22.4 km perigee in order to make a low pass over landing site number 3, in Sinus Medii, where Surveyor 6 had landed and Surveyor 4 had crashed.

When the time came for descent-stage separation and an abort simulation, there occurred the only mishap of the entire mission. The switch commutating between constant attitude and CSM radar searching mode was moved by Cernan and promptly flipped back into its original position by Stafford. As soon as the two stages separated, Snoopy began to spin around all three axes, searching for Charlie Brown. At this point, Cernan, in a live radio communication with Earth, released an epithet concerning the supposed profession of the LM's mother.* Stafford took manual control of the ship's attitude and stopped all rotation in 15 seconds. The switch-throwing issue was later determined to have been due to ambiguity in the assignment of responsibilities between the two astronauts. Both knew that the switch had to be flipped, but Stafford did not know that Cernan had already done so. To preclude such an error in the future, a clearer division as to who executed which step was drawn up.

Following a series of complex orbital manoeuvres, Snoopy again approached Charlie Brown after 8 hrs 10 min of free flight. The docking was perfect and, after thirty-one lunar orbits in some 60 hours, Apollo 10 was ready to return home. The descent stage was left in a low lunar orbit, from which it has undoubtedly decayed as a result of perturbations by mascons. After a series of engine test firings, the ascent stage was abandoned in solar orbit.

After making a grand total of nineteen live television broadcasts Charlie Brown re-entered the Earth's atmosphere on 28 May, and splashed down in the Pacific in sunlight. The astronauts were the first to shave in space, and on exiting the capsule they appeared remarkably neat and tidy. In contrast with Cernan and Young, who both participated in other lunar missions, Stafford returned to space as commander

---

* From then on, NASA imposed a stricter code on the astronauts' speech – at least when communicating with ground control.

of the joint Apollo–Soyuz Test Project in 1975. Charlie Brown is now on show at the Science Museum in London, and is the only CM in Europe.

3.11 THE SMALL STEP

Training for the Apollo 11 mission began during the spring of 1969, and continued until June. The astronauts also underwent a brief course on geology, held by Harrison Schmitt. Armstrong seemed to be particularly interested in learning this new discipline.

Despite the string of successful flights, the unknowns were still so many and so great that thoughts were given to skipping the planned launch window in mid-July and slipping the launch to the next window in August in order to provide extra time to prepare, but on 12 June the lift-off date was finally fixed at 16 July. The last word was also spoken on a longstanding debate about who would be the first to step on the Moon. Naval tradition required the captain of the ship (or, in this case, the spacecraft) to be the last to exit, and his second to step out before him. This rule was also enforced during the Gemini missions, during which extravehicular activities were carried out only by the co-pilot astronaut.

On the LM the commander's position was on the left (as with aircraft cockpits), and the exit hatch opened inward to the right, thus blocking the LM pilot in his position. Exchanging places in the bulky spacesuits, and in the very limited space of the LM, proved impossible.* NASA therefore decided that the first to step out of the module would be the mission commander. It has often been said (although never substantiated) that the choice was influenced by the fact that Armstrong – unlike Aldrin, and most of the other astronauts – was no longer a military man, and was a civilian pilot. To allow a member of the military to be the first to step onto the lunar surface could conceivably have opened the programme to criticism. The decision, of course, was not well received by Aldrin, and the two astronauts effectively became only work colleagues.

The day of launch – 16 July – was approaching quickly. The astronauts went into quarantine for ten days, and the Saturn V, CSM and LM (serial numbers 506, 107 and 5) were readied on Kennedy Space Center Pad 39A. At the suggestion of Michael Collins, the CSM was named Columbia, to honour Christopher Columbus, the discoverer of the Americas, and also as a reference to visionary nineteenth-century French writer Jules Verne, whose *From the Earth to the Moon* told of a fictitious flight to the Moon beginning from Florida that was launched by the 'Columbiad' cannon. The LM was named Eagle, with the mission logo showing a bald eagle – a symbol of the United States – swooping low over the lunar surface and carrying an olive branch in its claws symbolising peace.

As candidates for the first lunar landing, five smooth areas close to the equator had been chosen based on Lunar Orbiter images. The first area – the northernmost – was close to the eastern border of Mare Tranquillitatis and near the small crater

---

* Armstrong and Aldrin had on one occasion attempted to do so, and had damaged the simulator!

This photograph is the only clear still image of Neil Armstrong on the surface of the Moon. It also shows the American flag and, to the right, the Swiss-supplied solar wind sample collector (dubbed the 'Swiss flag').

Menzel*; the second – the site eventually chosen – was an 18.5 × 4.8-km ellipse at the southern border of the same mare, close to 6.5-km-diameter crater Moltke[†], and a few kilometres from the landing sites of Ranger 8 and Surveyor 5; the third was in Sinus Medii, close to the Surveyor 6 landing site; the fourth and fifth were in Mare Insularum (Sea of Islands), one about 100 km south of the crater Copernicus and the other near the crater Encke**.

During the early hours of the morning, the astronauts suited up and headed to the launch pad. It was estimated that the occasion had drawn almost a million people to the Cape, thus turning the area into the largest-ever camp site. The launch, at 13.32 GMT (09.32 local time) was perfect and routine. After 2 min 43 sec the S-IC stage separated at a height of more than 60 km. The S-II stage fired for about six minutes, and finally a brief firing of the S-IVB stage put the complex in the usual parking orbit. After one-and-a-half orbits of the Earth, the 'go' for TLI came, and Apollo 11 was placed on a free return trajectory. Because of the problems encountered by Apollo 10, the hydrogen valves had been modified, and this time there were no alarming vibrations. Half an hour after the injection manoeuvre, the LM was extracted from the S-IVB which, like its predecessors, entered a solar orbit after

---

*   Donald H. Menzel (1901–1976), American astronomer.
[†]   Helmut Karl von Moltke (1800–1891), the Prussian astronomer who in 1874 helped Julius Schmidt, the director of the Athens Observatory, to draft his lunar map.
**   Johann Franz Encke (1791–1865), the German astronomer who computed the orbital elements of the comet carrying his name.

flying by the Moon, this time at a distance of 3,380 km. Two small course corrections were made and, on approaching the Moon, the spacecraft was oriented so that the astronauts could photograph the partially lit lunar globe.

In the early afternoon of 19 July, above the far-side, the SPS engine was fired, and the spacecraft entered an orbit of 314.3 km apogee and 111.1 km perigee. Soon after circularising the orbit (101 × 122 km), Aldrin boarded the LM to check its systems, as Armstrong and Collins, from Columbia, searched for the landing site; when they finally located it, it was illuminated by the low Sun, which cast very long shadows and made it appear too rough for a landing. After some hours of rest, during the afternoon of 20 July Columbia and Eagle undocked and flew on in formation. Once over the far-side, the complex landing manoeuvre began. The LM was oriented so that its engine thrust was opposite to the orbital motion, then it was fired in order to produce a 15-km perigee a few hundred kilometres east of landing site 2. They were following the manoeuvres rehearsed by Apollo 10. Two minutes after Columbia re-emerged from behind the Moon, Eagle, now much lower, also established communication with Houston, but its radio signals were erratic because the parabolic antenna on top of the LM was not properly oriented.

Approaching perigee, with the ground rapidly passing under the astronauts and as a movie camera was shooting the scene for posterity, Eagle received the order to proceed with the powered descent even though communications were still erratic. When the engine was fired again to begin the descent, it became clear to Armstrong that due to the mascons, the LM would overshoot the target by some 3 km, taking it to the far end of the assigned landing ellipse. At a height of 14 km, the most nerve-wracking moment came: landing radar switch-on. For the radar to provide reliable ground distance measurements, it was necessary to turn the spacecraft with its engine facing forward and windows facing straight up. This phase of the descent would be carried out without the astronauts being able to see the ground! At a height of 12 km the radar altimeter finally sensed the rapidly closing ground, and the computer began to adjust the real trajectory to follow the planned one by firing the manoeuvring engines in short bursts.

Aldrin, in the meantime, was checking the systems by manually comparing the real data with the expected data. For this he decided to ask the computer to directly show the difference between the two measures (DH); but at that moment the computer alarm flashed error '1201', signifying an overload. This had arisen because the astronauts had forgotten to switch off the radar used for formation flying with Columbia, which was now nowhere near, and this pointless work instigated the overload error. The mission controllers therefore decided to keep track of DH themselves so that the descent could proceed. Finally, at a height of about 2,500 metres, the attitude engines pitched Eagle upright to give Armstrong a clear view of the ground ahead.

The module continued descending, guided by the computer or, with brief corrections, by Armstrong: 1,500 metres, 30 m/s vertical speed; 900 metres, 21 m/s. At that point Charlie Duke – the rookie astronaut acting as Capcom – relayed the 'go' for landing. Armstrong, in the meanwhile, was scanning the panorama. Finally, at a height of 300 metres he could see where the spacecraft was heading – and he did

Buzz Aldrin stands beside the small suite of scientific instruments deployed on the lunar surface during the Apollo 11 mission. In the foreground is the solar powered seismometer, with its white low-gain antenna pointing towards the Earth. Immediately behind it is the passive laser retroreflector. Behind the reflector the television camera can be seen to the left of the American flag. The object immediately to the left of the descent stage engine bell is the ALSCC (Apollo Lunar Surface Close-up Camera) – better known as the Gold Camera – used to take three-dimensional extreme close-ups of the regolith.

not like it. There were too many boulders – some as large as a car – and a 100-metre-wide crater. To avoid landing there, he took control from the computer and decided to manually steer to a more suitable location. Moreover, the spacecraft had crossed the 'dead man's curve': it was now so low and so slow that the abort manoeuvre would result in a fatal crash. Hovering at a height of 105 metres, he began to search for a more suitable landing site. He skimmed over the large crater and headed for a more promising area just beyond it, where he reduced the horizontal speed to zero preparatory to initiating a vertical descent, but a great deal of lunar dust was raised by the LM engine plume, making visual estimation of the module's altitude difficult.

At a height of about 20 metres, a worried Duke relayed that fuel was sufficient for only another 60 seconds. Armstrong – still uncertain of his actual height – continued a very slow descent, taking care to null any residual horizontal speed which could damage the landing legs on ground contact. Only 30 seconds were left when Aldrin announced that the contact sensors projecting below the landing pads had touched the ground, but as this was barely perceptible Armstrong kept the engine running for a second or so to be sure. At last, with only 23 seconds – or 317 kg of fuel –

remaining, the engine was shut off, and when Armstrong finally announced, 'Tranquillity Base here. The Eagle has landed', a relieved Duke replied, 'Roger, Tranquillity, we copy you on the ground. You've got a bunch of guys about to turn blue. We're breathing again.' It was 20.17.39 GMT on 20 July 1969. Eagle was on the lunar surface some 6,870 metres from the planned site, at 0.674 N, 23.472 E. The site was studded with boulders, and lines of rocks radiated from the large crater over which Armstrong had flown, some of which passed close to the LM. The site was later officially named Statio Tranquillitatis (Tranquillity Base), and three craters a few kilometres away were named Armstrong (4.6 km diameter), Aldrin (3.4 km), and Collins (2.4 km).

The first task of the astronauts was to re-programme the computer for a possible emergency lift-off. Immediately after this, they were supposed to sleep for four hours but their level of excitement was such that, for once letting their human nature show, they asked permission to exit as soon as possible.

Meanwhile, orbiting every two hours many kilometres above, Collins became the first human being to experience the unique isolation of passing alone around the far-side and, during his repeated passes over the landing area, he tried unsuccessfully to locate the LM on the ground using the spacecraft's sextant by programming it with the most recent estimates of the position as computed by Houston; he never did spot it.

During the ensuing hours, the two men in Eagle's confined pressurised cockpit suited up, put on their EVA boots and gloves, and helmet cover, and donned a Portable Life Support System (PLSS) – a back-pack weighing, on the Moon, only 10 kg.

A little more than six hours after landing, the hatch, kept sealed by the internal pressure, was finally opened, and the last remaining air escaped. As Aldrin kept the hatch open, Armstrong laid face down on the floor and, guided by his colleague, exited feet first on to the porch, where, by pulling a handle, he deployed a large panel on which were their tools and a small black-and-white television camera. The transmission was received by the Parkes radio telescope in Australia, and relayed by communication satellite to mission control at Houston. 'We're getting a picture on the TV!', announced Capcom rookie astronaut Bruce McCandless – although because the LM was facing west, the LM leg on which the ladder was mounted was in shadow, and it was difficult to discern anything in the contrasty image.

Guided by Aldrin, Armstrong put his foot on the first step of the ladder, and descended. It had been mistakenly expected that the LM would sink several tens of centimetres into the lunar dust, and the last step of the ladder was therefore mounted at a certain height which now seemed almost excessive, and Armstrong had to jump down. Once on the landing pad, he first checked that he was able to climb back, and then inspected the footpad. He noted that it had sunk just a few centimetres, and described the surface as a very fine conglomerate, 'almost like a powder'. He then announced, 'I'm going to step off the LM now.'

At 02.56.15 GMT on 21 July, Armstrong lifted his left foot from the pad of the LM and laid it down on the dust, saying: 'That's one small step for a man, one giant leap for mankind'. He then moved his left foot to test the characteristics of the

This small crater, about 60 metres from the LM, is the farthest point reached by Armstrong during the single Apollo 11 lunar walk. The white object in the foreground is the Gold Camera.

regolith, brought over his right foot, and released his grip of the ladder. He took a Hasselblad lowered by Aldrin on a rope and, passing from shadow to sunlight, began to describe the status of Eagle – as people both on the ground and on the LM urged him to pick up a contingency sample in order not to leave the Moon empty-handed should an emergency occur at just that time. In a pocket strapped to his leg he had a scoop on a folded rod. Using this, he scraped up a few small pebbles and an amount of regolith, which collected in a bag attached to the scoop. The bag went into his pocket. He discarded the scoop. Observed closely, the sample appeared neither white (as it might appear from Earth), nor grey, nor green (as described by Apollo 8 and Apollo 10 astronauts), but absolutely black!

About 15 minutes later, it was time for Aldrin to step out too. While Armstrong performed chores, Aldrin inspected the landing site. He described one rock as 'some kind of biotite' – mica – and this immediately alarmed the geologists, as mica can only form in the presence of water! When the rock was revealed to be a kind of basalt, Aldrin explained that he had only meant to say that it *looked* like a sample of biolite that he had been shown in training. They then stood together at the base of the ladder while Armstrong uncovered the commemorative plaque attached on the LM's forward landing leg. It showed the two terrestrial hemispheres, and carried the words: 'Here men from the planet Earth first set foot upon the Moon, July 1969 AD. We came in peace for all mankind', and included the signatures of the three astronauts and of President Richard M. Nixon. The LM also carried the messages of most of the world's heads of state, and medals in memory of the two deceased cosmonauts, Yuri Gagarin and Vladimir Komarov.

The time finally came for surface deployment of the first of the three instruments collectively known as the Early Apollo Surface Experiment Package (EASEP). The solar wind collector designed by Berne University, and therefore dubbed the 'Swiss flag', was a weather-vane-like strip of aluminium that was to remain unfurled during the entire period of the astronauts' extravehicular activity, to be struck by and collect the ions and atoms of the solar wind. After return to Earth, it would provide a sample of the solar wind, limited to the noble gases helium, neon and argon; it was

A selection of lunar rocks returned by Apollo 11.

impossible to discriminate other atoms from aluminium impurities, however. After setting up the solar wind collector, Aldrin spent some time on experiments to assess mobility in one-sixth gravity. Armstrong took a panorama of the landing site.

There had been a lengthy debate over the opportunity of marking the site with a flag, and the United Nations' flag was considered, as well as others. In the end, the US Senate ordered that the American achievement be marked by the 'Stars and Stripes' – after all, Soviet emblems had been delivered by Luna 2 a decade earlier. The Apollo project was, after all, born for reasons of national prestige – although at the time, most people pretended otherwise. After planting the flag, with some difficulty, and after an unplanned conversation with President Nixon which forced the astronauts to stop any activity for four precious minutes, Armstrong deployed the laser retro-reflector on the surface, and Aldrin the solar-energy-powered seismometer, the two remaining scientific instruments. The retro-reflector was a suitcase-sized and completely passive instrument with an array of quartz mirrors on its upper surface, and was to be used to measure the exact distance from the Earth to the Moon by reflecting laser beams. The seismometer would provide the first measurements of possible moonquakes, thus enabling the formulation of an hypothesis on the internal structure of the Moon. The instrument, however, had no protection from the harsh lunar night and succumbed in a little more than a month, ceasing transmissions on 25 August, having recorded only a few very-low-intensity events.

An interesting experiment carried by the first four landing missions was a camera developed by Kodak for Thomas Gold of Cornell University to take very-high-resolution stereoscopic images of a small patch of regolith, which thus appeared as it would from just some 10 cm away. Despite only forty-eight pictures being taken by the three cameras eventually used, they provided some extremely interesting fine-scale data on the regolith.

The first sampling task was called a 'bulk' sample because the objective was to fill a crate with nearby rocks and regolith. The box was then hermetically sealed.

Aldrin's last task before the end of his 'walk' was to collect two deep cores, but the regolith became increasingly compacted with depth, and so no sample was retrieved from a depth greater than 18 cm. Armstrong walked to the rim of a large crater about 60 metres from the LM – the greatest distance reached during the first lunar mission – and took photographs of its walls. The plan to take 'documented' samples, in which rocks would be photographed prior to being lifted, had to be abandoned due to lack of time, in part lost in talking to Nixon, so Armstrong hastily collected a 'suite' of rocks representative of the site.

Aldrin had had the camera for just a few minutes to take detailed pictures of the LM and the famous picture of his foot, obtained in order to evaluate the mechanical characteristics of the regolith. The only good-quality picture of Armstrong on the Moon is a view of him from the back, working on the LM.* Other images, of lower quality, were taken by a 16-mm automatic movie camera mounted behind the right window of the LM.

After 1 hr 42 min, Aldrin re-entered. The rock crates were hauled up using the rope. Armstrong's stay on the Moon had lasted 2 hrs 13 min – about 15 minutes more than expected. The walk was broadcast live on television in forty-seven countries, although the major part of the Communist bloc saw only a recorded and edited version, and China, North Korea, North Vietnam and Albania did not even mention it.

After the LM was repressurised, the astronauts could smell, for the first time, the dusty regolith on their suits. Rich in iron and other metals, upon meeting oxygen for the first time in a billion years it had oxidised and infused the cockpit with a smell like gunpowder. Armstrong answered some questions posed by flight controllers and scientists, and also provided a detailed geological description of the site. In the following hours they took some final pictures through the windows and took the rest period that they should have had before exiting. Some twenty hours after landing, the hatch was opened again for a few minutes, and the back-packs were thrown out, their landfall being recorded by the seismometer. At 17.54 GMT, small pyrotechnic charges fired to sever the contacts between the descent stage and the ascent stage, and the ascent engine was then fired to take Eagle back to orbit. Fortunately, the firing switch – which Aldrin had slightly damaged while wearing his spacesuit several hours earlier – did not create a problem.

Slightly more than three hours later, Collins succeeded in locating Eagle as a tiny dot moving over the lunar back-drop, and initiated the critical orbital rendezvous manoeuvre (as conceived by John Houbolt almost ten years previously). Closing on the LM, Collins took several photographs of the ascent stage, in some of which he caught the Earth in the background, rising over the lunar horizon behind the LM, so with the exception of himself the whole of mankind was present. After docking

---

*   This photograph was discovered in 1987, after an extended search.

(which was less than satisfactory, as Collins retracted the probe of Columbia too soon), Armstrong and Aldrin used a small vacuum cleaner to suck the lunar dust off their suits. These operations over, they transferred to Columbia with two crates containing 21.4 kg of lunar samples.

Eagle was jettisoned with its systems still operating, but with its water cooling inactive, to determine how long it could function before succumbing to the intense solar radiation – some hours, which was useful to know. Meanwhile, Columbia had fired its SPS engine to head home. Finally, on 24 July, Columbia jettisoned the SM, which re-entered like a meteor and was filmed by an ARIA plane. The CM splashed down in the Pacific Ocean and (as had Apollo 8) turned upside down.

There then began the precautionary measures designed to prevent the introduction of possible lunar microorganisms. After long studies in which national health representatives had participated, the plan still had several weaknesses. A frogman first opened the capsule's hatch for a few seconds to throw in three rubber suits designed to isolate, as much as possible, the crew from the environment. After the astronauts had boarded an inflatable raft, the hatch was closed and washed with an iodine solution, while the astronauts and recovery team continuously disinfected themselves with a different solution. After boarding the aircraft carrier *Hornet* one hour after their splash-down, and barely able to withstand the summer heat inside the suits, Armstrong, Aldrin and Collins walked a few metres to enter a mobile quarantine unit – a trailer, modified to render it completely independent of the outside world for a few days. Inside, a physician, who had voluntarily quarantined himself, performed a quick medical check on the crew. Meanwhile, Columbia was hoisted on board and connected to the trailer by a rubber duct. Respecting the quarantine as much as possible, the two sample crates and the photographs were unloaded from the capsule and separately sent to Houston, where they arrived the following day. The trailer and Columbia were taken by sea to Pearl Harbour, and then by aircraft to Houston, where they were completely isolated for three weeks.

3.12 LUNAR SAMPLES

On 25 July 1969, within hours of arriving at the LRL, the first crate of lunar samples – the crate containing the 'suite' that Armstrong had collected in lieu of the documented samples – was weighed (15.14 kg), sterilised with ultraviolet rays and a powerful germicide, washed with ultra-pure water, dried with a jet of nitrogen, and then placed in a vacuum chamber to be opened, which was done the next day after a leak in a rubber manipulating glove had been repaired.

The first view was quite a disappointment, as all of the rocks were covered with fine dust to the point that their mineralogical characteristics could not immediately be determined. Also in the crate were the solar wind collector and two deep cores. Being individually sealed, these were potentially less contaminated by the terrestrial environment and thus more suited to the search for alien life. One hundred grammes of dust and sand were used for biological tests on plants, mice and other animals.

The next day, 27 July, one of the rocks was removed and 'dusted'. Its igneous origin was evident from the readily identifiable crystals of feldspar, olivine and pyroxene. The surface, exactly as Armstrong had described it, was punctured by small apertures that were not, as some thought, vesicles created by escaping volcanic gases, but tiny impact craters – later dubbed 'zap pits'. The dust, in contrast, appeared to be composed mostly of glass particles with a large variety of shapes. The rock was analysed to measure its exposure to cosmic rays and thus the time that it had been exposed on the surface, which was discovered to be well over one million years. A small fleck was analysed spectroscopically and found to be quite similar to that analysed by Surveyor 5, which had landed 25 km from the Apollo 11 site. It was very rich in titanium, poor in volatile elements, and very poor in carbon (twelve parts per million!). After similar analyses had been carried out on several other rocks, on 12 September the first samples were distributed to the scientists who had requested them, with the agreement that no scientific results would be published before the meeting of a lunar science workshop scheduled for January 1970.

The preliminary LRL reports gave basic sample descriptions. Both igneous rocks and breccias (heterogeneous agglomerates of rock fragments in a matrix of small particles) were found to have no trace of water and very little organic material – usually about 0.0001%. All of the rocks were found to have been subjected, at one time or another, to intense shock – a discovery which confirmed the meteoric origin of lunar craters – and igneous rocks were found to be 3–4 billion years old, in good agreement with the oldest dated terrestrial rocks. From the analysis of their exposure to cosmic rays, the samples appeared to have been within 1 metre of the surface for the last 20–160 million years.

Some rocks were put on show at the Smithsonian Institution in Washington, and attracted a huge crowd – although for many they were a disappointment, as they appeared to be quite ordinary rocks.

In the same building in which the preliminary analysis of the samples was being carried out, the three astronauts, some engineers, and some laboratory workers who had come into contact with the lunar material, were segregated for three weeks. The worst case of contact was that of a photographer who handled a container of Hasselblad film that had been dropped on the lunar surface and become coated with dust. During this period, none of the confined people showed any sign of illness, every 'infected' animal survived without problems, and some plants grew faster than the normal!

At last, on 10 August the doctors decided to end the quarantine. In fact, this had never been 100% effective. Air filters were used, and all waste was burned in an incinerator, but, as Aldrin recalls, ants – potential carriers of planetary infection – could sometimes be seen! Armstrong, Aldrin and Collins departed secretly a few hours before the press conference that marked the end of their incarceration, and were immediately dispatched by NASA on a goodwill world tour, taking with them several very small lunar dust samples that were to be presented to the heads of state of the visited countries.

All three Apollo 11 astronauts retired from NASA soon afterwards. Armstrong became a lecturer at Ohio University, where he worked until 1979 and then, despite becoming a member of several corporate administration boards and participating from time to time in official ceremonies, including the significant Moon-landing anniversaries, led a reclusive life. In 1986 he became part of one of the boards investigating the loss of the *Challenger* – one of his fellow board-members being Nobel laureate physicist Richard Feynman. Buzz Aldrin had the most troubled life: two divorces and several years of alcohol disintoxication. Now in his third marriage, and evidently finally content, he guest-stars in commercial advertisements, makes public appearances, and is an advocate for renewed space exploration. Michael Collins wrote several popular books on spaceflight, and was director of the Smithsonian Air and Space Museum during the 1980s. He is now a trustee of the National Geographic Society, as is Frank Borman, commander of Apollo 8.

In recognition of the historical value of Columbia, it was the only capsule whose windows were not dissected to study micrometeoroid impacts. It is now on show in the main hall of the Smithsonian Air and Space Museum, where it is likely to remain for all time.

3.13 THE MEETING WITH SURVEYOR 3

The Apollo 11 mission fulfilled President Kennedy's challenge of 1961, but NASA had built enough Saturn Vs for an extensive test programme, and so it had enough in stock to run through to Apollo 20, and funding was already in place for the next few missions.

With the end of the euphoria, and the demise of the fear of a lunar epidemic, the next mission was prepared with a completely different approach concerning public opinion – now that the 'war' was won, it was time to 'bring the boys home'. The Apollo 12 crew consisted of two veterans: the commander was Charles 'Pete' Conrad, of both Gemini 5 and Gemini 11; the CM pilot was his friend Dick Gordon, who had flown with Conrad on Gemini 11; and the LM pilot was rookie Alan Bean, the only astronaut whose hobby was painting (although Soviet cosmonaut Alexei Leonov shares the same passion). It was the closest group of friends that NASA had ever launched into space.

The initial proposal for the landing site was an equatorial section of Oceanus Procellarum, in the western hemisphere, but one of the trajectory specialists of the site selection group, arguing that mascons were now understood, suggested trying a *precision* landing in Oceanus Procellarum within walking distance of Surveyor 3, which had been inactive for 30 months. The probe was pinpointed on Lunar Orbiter images. It was inside a small group of craters unofficially called the 'Snowman', appearing like the caricature of a human figure, and the probe had landed inside the largest crater – named Surveyor crater – which looked like the man's torso. If some of the components could be returned to Earth, the ageing of materials in the harsh environment beyond the Earth's magnetosphere could be studied for the first time. The engineers were particularly eager to examine the mechanical, optical and electronic components of the camera. From a scientific point of view, the site was criticised for being another equatorial mare, but if NASA was going to mount advanced missions to specific sites, it would need to demonstrate that the mascons

A Lunar Orbiter image of the Snowman, the Apollo 12 landing site (arrowed). Surveyor 3 had landed inside the crater that forms the man's torso (in this view, in the shadow of the 200-m crater below and to the right of the arrow).

had been tamed, and there was no better way to prove it than to land alongside an earlier craft. Conrad eagerly accepted the challenge.

To reduce the time of flight, the navigators modified the flight plan. Instead of flying on a free-return trajectory, after S-IVB separation the SPS engine would place the spacecraft into a very eccentric Earth orbit with an apogee far beyond the Moon. Near the Moon, the SPS would perform a normal lunar orbit insertion. The major weakness in this new plan was that in case of a major engine failure in the SM, the only way for the astronauts to return to Earth would be to fire the small manoeuvring thruster for a very long time. But, it was reasoned, the only way the main engine could fail was by exploding – in which case there would be no astronauts to bring back to Earth; this reasoning, as will be seen, was flawed.

To give the astronauts more time to prepare, the previous pace of a flight every two months was relaxed. As the countdown reached its final phase on 14 November 1969, a cold front accompanied by thunderstorms was passing over Florida. Serious consideration was given to postponing the mission to the following month, but meteorological aircraft finally reported cloud clearance and the absence of lightning, and the 'go' for launch was given. The lift-off proceeded perfectly for 36.5 seconds, after which the trail of hot gases acted as a conductor, and violent lightning struck between the launch tower and the rocket. The astronauts heard the master alarm sound off, and the caution and warning panel light up as the fuel cells shut down and the spacecraft's systems began to draw power from the re-entry batteries, which placed the entire mission in jeopardy. With its power supply disrupted, the equipment that processed the instrument data for transmission to the ground issued rubbish. With scrambled data on their screens, the flight controllers were helpless. But John Aaron, the controller in charge of the CSM's systems, realised what had happened because he had seen the same thing in a simulation the year before, and he promptly recommended that the signal conditioning equipment be switched to its auxiliary power supply, and when this was done the telemetry was restored.

Fortunately, the Saturn V's systems were unaffected, and it continued on course. Once the S-IVB was in orbit, the astronauts set about restarting the fuel cells. For more than two hours the spacecraft was put through a very thorough series of checks, and the 'go' for TLI was finally given. This done, the CSM Yankee Clipper (the nickname of baseball legend Joe Di Maggio) extracted the LM Intrepid from the third stage. Due to an ullage rockets failure, it proved impossible to place the S-IVB into a heliocentric orbit, and after flying by the Moon at a distance of 5,708 km it remained on a very elliptical Earth orbit with a period of 43 days.* The next day, a 9.2-second firing of the SPS engine moved the spacecraft from the free-return trajectory to the new hybrid trajectory.

On 18 November, Apollo 12 entered an 111-km-high orbit. For a few hours they snapped pictures of some interesting areas, in particular Fra Mauro, where the Apollo 13 mission was expected to land.** The following day, Conrad and Bean entered the LM, and during the thirteenth orbit headed to their rendezvous with Surveyor 3. Conrad's fear was that when the LM pitched at a height of 2,400 metres and he got his first view of where they were heading, the Snowman would not be visible, but the revised navigational techniques had worked. He was ecstatic. 'There it is, right down the middle of the road!' Seen closer, Surveyor crater appeared to be studded with smaller craters, and 'Pete's parking lot' – the expected landing point – was not that favourable, so he took control and headed for a nearby site. He allowed the LM to descend very slowly, and reduced his horizontal speed almost to zero. A few seconds later the contact sensors touched the surface, he cut the engine, and the LM fell the last few metres. Intrepid landed at 3°.01081 S, 23°.41930 W, a few metres from the rim of Surveyor crater and between it and Head crater. During the next orbit, Gordon was able to observe the LM with his sextant, to confirm that Intrepid was on target.

Five hours later Conrad descended the ladder to begin the first of the two planned walks, his image being filmed by a colour TV camera. Once at the end of the ladder, he had to comply with a $500 bet with Italian journalist Oriana Fallaci, whom he met when she was writing her book on the astronauts, *If the Sun Dies*. Fallaci had insisted that the words spoken by Armstrong on stepping onto the lunar surface were

---

* The fate of the third stage of Apollo 12 is quite interesting. In March 1971 it was perturbed on a solar orbit similar to Earth's, where it remained for 31 years. In April 2002 it cruised in the vicinity of the first Lagrangian point of the Sun–Earth system – a point about 1.5 million km from Earth along the Sun–Earth line where the gravity of the former is equal to the gravity of the latter. With a little help from the Moon, it was recaptured by the Earth, where it was discovered at the beginning of the following September, being initially classed as an asteroid. However, because its orbit was unstable, being subject to the gravitational attractions of at least three bodies (Earth, Moon and Sun), it returned to solar orbit in mid-2003.

** Fra Mauro (*d.*1459), Venetian geographer.

The Apollo 12 lunar surface three-axis magnetometer cooperated with Explorer 35 to study the lunar magnetic field. Some other experiments of the Apollo 12 ALSEP can be seen, including the seismometer (at right), can be seen at the top of the image.

not of his own devising, and that they had been imposed on him by a NASA public relations man. And so Conrad, to prove her wrong, had told her the words that he would speak after stepping onto the Moon. Jumping the 75 cm from the last step of the ladder onto the footpad, Conrad shouted: 'Whoopee! Man, that may have been a small one for Neil, but that's a long one for me.' (He was 168 cm tall, compared with Armstrong's 183 cm). Then, moving away from the LM, he took a look at Surveyor crater where, 163 metres away, was the solar panel atop the mast of the robotic spacecraft catching the Sun. The rest of the craft was in the darkness of the crater's shadowed eastern wall. Conrad then began to collect the first samples, accompanying his actions with an intermittent humming – a clear sign that his fears about the success of the mission had dissipated.

Half an hour later, Bean was also standing on the surface, his first task being to help his colleague unfurl a large umbrella-shaped parabolic antenna to be used to improve communications with Earth. (A similar antenna had been carried by Apollo 11, but had not been deployed.) But then, as Bean carried the camera away from the LM to provide a better view of operations, he inadvertently pointed it at the Sun for several seconds. After such intense illumination, the camera was useless, even though Bean tried to make it work ... by hitting it with his geological hammer! For the rest of the stay, Apollo 12 was a 'radio only' mission.

The time then came to deploy the instruments of the first Apollo Lunar Surface Experiment Package (ALSEP). Bean carried the bulky five-instrument package, as Conrad selected a place around 130 metres west of the LM that was sufficiently smooth, and far enough from the engine exhaust.

Other than the solar wind collector identical to that on Apollo 11 and already

The suprathermal ion detector (on the tripod) and the cold-cathode ion gauge (on the left, at the end of the wire) were two ALSEP instruments designed to detect traces of the exceedingly tenuous lunar atmosphere.

deployed next to Intrepid, the experiments included a seismometer incorporating four capacitor platforms, three of which could identify the direction of arrival of the quake, while the fourth could record high-frequency (up to 1 Hz) vibrations in a vertical direction. A three-axis magnetometer would characterise the magnetic field close to the surface, and study its interactions with the solar magnetic field. This last study was to be carried out in cooperation with the still-functioning Explorer 35 lunar orbiter. Two instruments – the cold-cathode ion gauge and the suprathermal ion detector – were to analyse the environment to search for a possible rarefied lunar atmosphere. The first worked for just 14 hours before breaking down, and recorded only the gases which escaped from the astronauts' suits and the outgassing of other instruments. The second instrument recorded the first trace of a lunar atmosphere. During the lunar night, the particle density of the tenuous envelope fell by three orders of magnitude, down to about 200,000 atoms per cubic cm. The instrument package was completed by a solar wind spectrometer to record the energies of the solar particles impinging on the Moon. The SNAP-27 generator – carried for the first time – caused some concern when the plutonium oxide core jammed inside the graphite shroud hinged on one side of the LM, but it was freed by a light tap with Conrad's hammer! The five experiments and the radioisotope generator were deployed at the vertices of a six-pointed star, at the centre of which was the central station, connected to the instruments with wires and carrying a small medium-gain antenna for relaying data to Earth.

After installing the ALSEP and collecting a deep sample, the first walk was over. It had lasted a total of 3 hrs 56 min.

During the first extravehicular activity, one of the most interesting and better-known samples of the entire programme was collected: sample number 12013, also known as 'rock 13' – a small, 4.45-billion-year-old breccia that was found to be unusually rich in radioactive elements. Curiously enough, rock 13 was one of the very few breccias collected by Apollo 12.

The next thirteen hours were mostly dedicated to rest, this being made slightly more comfortable by the use of two hammocks, as Armstrong and Aldrin had complained about having to sleep on the floor, although, as had their predecessors, they had to stay in their spacesuits.

Upon awakening, they set off for their second excursion. After checking the deployment of some ALSEP instruments, the first stop was at Head crater – a few tens of metres west of the LM – where Bean made the interesting observation that Conrad's footprints exposed a brighter material. This excited the geologists back in Houston, because they had hoped to find evidence of a bright ray of ejecta from the large crater Copernicus, several hundred kilometres to the north. On the rim of Head crater, Conrad rolled a small rock to test the seismometer, but no response was recorded. Conrad was also the first man to fall on the Moon, but with low gravity this was more of a nuisance than a real danger. During most of the excursion – which reached two small nearby craters at a maximum distance of 400 metres from the LM – the astronauts were almost running, so that on arriving at the rim of Surveyor crater they were exhausted.

When they had landed, the eastern side of the crater had been in shadow, and it

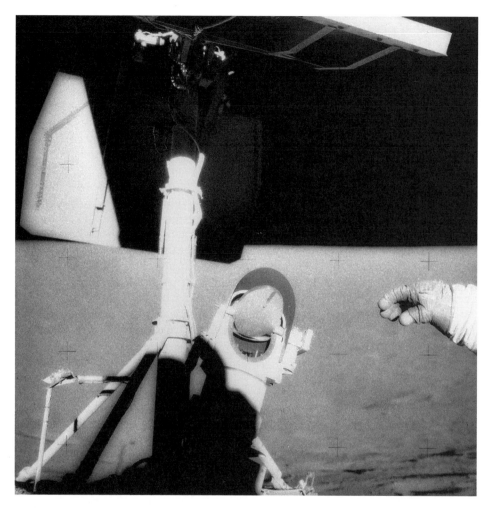

An image of the Surveyor 3 camera shows dust contamination of its mirror caused by the rather hard landing of the probe and by the nearby landing of the Apollo 12 LM. Colonies of terrestrial bacteria were found inside the camera after its return to Earth.

had looked treacherous, but by now the Sun had risen sufficiently to illuminate its floor, and venturing into it to inspect Surveyor 3 no longer seemed to be daunting. Nevertheless, NASA had insisted that they make their approach from the far rim of the crater rather than the centre, because it feared that the probe may slide and crush them. Surveyor 3 was no longer white but brown or dark orange, due possibly to its long exposure to the Sun's heat or to the dust raised by the LM. The marks left by the rebounds and the following sideward slide were still clearly visible. Moving to the side of the probe, Conrad used clippers to cut away parts of the structure, some wires, the SMSS arm and the camera and its electronics. Mischievously, Conrad had brought an automatic timer for the Hasselblad camera, so that he could take a photograph of himself and Bean beside the probe. He could imagine the reaction of those who were to see it: 'Who took this?' Unfortunately, he had put the timer in a sample bag, and did not find it until just a few minutes before re-entering the LM. With the retrieval of the solar wind collector, the excursion was over. It had lasted 3 hrs 49 min. On their roundabout route, they had collected a total of almost 35 kg of samples, including a number of cores to a depth of 70 cm. In general, the experts, listening in, had judged the astronauts' geological descriptions to be unsatisfactory. But after Aldrin's error concerning biotite, they had not wanted to risk a mistake. In fact, Conrad had resolved to describe everything simply as 'stuff'. In one of his rare specific remarks, he described olivine, a distinctly greenish rock, as 'bottle-coloured'.

Five hours later, the ascent stage left the lunar surface and headed for orbit. Over the far-side, Conrad, infringing rules, passed LM control, for several minutes, to Bean, who would otherwise never have flown it. Intrepid then docked with Yankee Clipper – and what followed was a reunion of friends. After forty-five lunar orbits, the SPS fired and the astronauts headed back to Earth. The LM was used for an artificial quake experiment by impacting it at a speed of 1.67 km/s on the Moon at 5°.5 S, 23°.4 W (or, according to a different source, 3°.94 S, 21°.21 W), 76 km east-southeast of the ALSEP. To the scientists' amazement, the shock waves reverberated for a full hour! Such a prolonged seismic signal implied that the lunar crust was brecciated to a depth of about 30 km – in essence rubble from the accretionary process. On the way home, the spacecraft flew through the shadow cone cast by the Earth, with the otherwise dark terrestrial disk being visible only by the illumination of an almost full Moon.

The splash-down on 24 November was so rough that a movie camera fell from its bracket and hit Bean on the head, giving him quite a cut, although it soon healed. A frogman supplied breathing apparatus and the new more comfortable isolation garments, rather than the unbearable rubberised suits that were worn by the Apollo 11 astronauts. Once again the crew was put in quarantine, but this time lasting only a couple of weeks, ending on 10 December. Gordon was named to command Apollo 18, but this was cancelled and he retired. Conrad and Bean each commanded a Skylab mission – the first and the second respectively. After retiring from the astronaut corp, Bean became a painter, and Conrad was the 'pilot' of the revolutionary experimental single-stage launcher DC-X. He died after a motorcycle accident on 8 July 1999. Yankee Clipper is now on show at the Langley Research Center in Hampton, Virginia. The Surveyor camera is at the Smithsonian Air and Space Museum, and the SMSS is at the Cosmosphere in Hutchinson, Kansas.

The procedure for opening the sample crates had been modified. Only one of the crates was opened in vacuum, and all the others were opened in an inert nitrogen atmosphere. The scepticism of scientists over the choice of the landing site dissipated when the samples were analysed. Other than rock 13 and the Copernicus ejecta, the lower quantity of titanium with respect to Apollo 11 samples was noted, this being probably responsible for the slightly redder colour of the western mare. Only two of the rocks were breccias, in contrast with some 75% of rock samples from Mare Tranquillitatis, although most of those had been bits of impact-fused regolith. Most of the samples were igneous, and were probably formed at some depth below the surface and excavated by the impacts that formed the Snowman craters. The analysis of the potassium and argon content revealed that their ages ranged between 2.2 and 2.6 billion years. This meant that the maria had not all formed at the same time, and that Procellarum was about a billion years younger than Tranquillitatis. This was important new information.

The most sensational discovery, however, came from the probe's camera. On several of the optical components were found colonies of live terrestrial *streptococcus mitis* bacteria! The most obvious conclusion was that contamination had occurred some time between retrieval on the Moon and arrival at the laboratory, but no trace of bacteria was found on other Surveyor hardware or geological samples. It seems, therefore, that the microorganisms survived for thirty months on the Moon, in one of the harshest environments imaginable! The camera optics also provided some interesting information on cosmic rays, due to the microscopic etching in the glass produced by the passage of the high-energy particles.

The instruments left on the surface provided very interesting data. In particular, the seismometer recorded quakes on an almost daily basis, some of which (about a quarter) were probably deep quakes, while others were produced by meteor impacts. It was also possible to divide the quakes into nine classes according to the epicentre, and note that most of the events happened in the three days nearest to perigee. The magnetometer measured a weak magnetic field quite different from the intrinsic terrestrial dipolar field. Evidently, when the igneous rocks had solidified, the Moon had possessed a field, and this had been frozen into the rock – a phenomenon called remanent magnetism. It was also possible to compare the response of the induced magnetic field with the variations of the solar magnetic field, in particular during a flare which erupted on 20 April 1970. Comparing the data from Apollo 12 with those of Explorer 35, an adaptation time of several minutes was observed, yielding a model of electrical conductivity, and thus temperature, of the regolith to a great depth. The magnetometer data also hinted that the Moon may possess a small core of 800–1,000° C, which is far too low for many characteristic phenomena of our planet, and if the Moon once had a dipolar field, this had shut down.

A few days after the splash-down of Apollo 12, the first lunar science workshop was held in Houston, during which the first results of the analysis of the Apollo 11 samples were presented. Among the discoveries announced on that occasion was the discovery of three new minerals unknown on Earth, and the identification of twenty more that were indistinguishable from their terrestrial forms – a clear sign that the Moon underwent geological processes similar to those on the Earth, and so is an 'evolved' rather than a 'pristine' body.

The search for microorganisms yielded no results, even at a fossil level, and the search for organic carbon compounds was unsuccessful, even though carbon itself was found in very low percentages. It was also discovered that the Apollo 11 rocks appeared to have an age consistently lower by as much as a billion years with respect to the regolith! How could the regolith derived from the underlying rock be older than that rock? Geochemist Paul Gast made a surprising discovery in the Apollo 12 basalts: an abundance of potassium, phosphorus and a variety of the rare earth elements. Linking their chemical symbols, he coined the label 'KREEP'. After attempting to isolate this material, he realised that it was not present as a distinct mineral. Consequently, the term did not indicate a new type of rock; it was an *adjective*, and it is more correct to say that the Apollo 12 basalts are KREEPy. By way of an 'instant science' explanation, Gast suggested to journalists that the additive might have been picked up from the ancient crust that some believed ran beneath the maria. He even predicted that it might prove to be representative of the basalt that was assumed to form the volcanic-looking parts of the highlands. However, when the KREEPy basalts were found to be rich in radioactive elements, particularly thorium and uranium, it was realised that this material could not be typical of the crust, because its radiogenic heating would have prevented the crust from cooling sufficiently to inhibit volcanism. John Wood observed that if the non-mare regolith fragments from Tranquillity Base were representative of highland rock (thrown in by impacts) then their low density meant that the highlands were a plagioclase-rich 'scum' that had floated to the surface of a 'magma ocean' made of primordial material melted by the heat liberated by impacting planetesimals. If this was so, then this explained the KREEP. As magma thermally differentiates, it spawns lava of different compositions. During crystallisation, certain elements are either accepted or rejected according to whether they fit into the regular crystalline structure; the elements that do not fit are labelled 'incompatibles'. The trace elements tend not to participate in mineralisation. These remained in the melt as the various compatible elements were extracted, and so were progressively more concentrated. The radioactives at depth helped to keep this concentrated reservoir molten, and were then locked in when it solidified. When a massive impact excavated material from great depth, sometimes it picked up some of this radioactive material. Rock 13 was such a fragment. Because such impacts are so energetic, they can melt rock and 'splash' it across the surface, where it solidifies just as if it had been erupted volcanically. Later impacts will smash it up and scatter it around, enriching the regolith with what appears to be radioactive basalt. Since the rock-dating techniques measure the relative abundances of various radioactive isotopes, regolith enriched with fragments of KREEPy rock will seem to be older than it really is. This was why the regolith at Tranquillity Base appeared to be older than the lava flow on which it resided. If NASA had called a halt after Apollo 11, having achieved Kennedy's challenge, we would probably still be puzzling over this apparent paradox.

3.14 ONE STEP AWAY FROM CATASTROPHE

With the crew of Apollo 13 began a tradition that would continue through all of the following flights: the mission commander would be a veteran, and he would fly with two rookies. The commander of Apollo 13 was James Lovell, the first man to fly twice to the Moon, having been part of the Apollo 8 crew. He should have flown with LM pilot Fred Haise and CM pilot Thomas 'Ken' Mattingly, both of whom were Air Force pilots. The launch was scheduled for 11 April 1970, and the chosen landing site was inside the highlands of the Fra Mauro region, about 100 km south of Copernicus and less than 200 km east of the Apollo 12 site. The interest in Fra Mauro was based on the belief that this region seemed to have originated in the impact which created the Imbrium basin. The LM would be named Aquarius (after the constellation), and the CSM would be named Odyssey (after Homer, and the film by Stanley Kubrick). The mission motto was *Ex Luna Scientia* – Knowledge from the Moon.

The first mishap occurred less than a week before launch. One of the children of the back-up crew LM pilot, Charlie Duke, fell ill with measles, and he exposed both of his fellow crew-members and the prime crew to the disease. Medical tests were therefore performed, and it appeared that although Lovell and Haise had antibodies, it was possible that Mattingly might develop measles during the mission. Having to choose between postponing the mission or replacing Mattingly with his back-up Jack Swigert, Lovell reluctantly dropped Mattingly. However, a new series of simulations was necessary, to harmonise the crew and to check the ability of Swigert, who had received less training. The eighth Saturn V lifted off on time, at 19.13 GMT on 11 April 1970. The second mishap of the mission occurred when the S-II's central J-2 engine shut off 132 seconds early, and it was necessary to keep the other engines running for an extra 34 seconds to compensate. After two routine parking orbits the S-IVB made the TLI burn, and shortly thereafter the SPS engine put the spacecraft on a hybrid transfer orbit without free return.

Many things had changed during the previous few months, and this new lunar mission did not attract much media attention. The war in Vietnam was entering its final and most cruel phase, and the mission of Apollo 13 was of so little interest to the general public that the 13 April television broadcast, as the astronauts showed off the spacecraft in which they were to land on the Moon, was not even aired live. A few minutes later, however, this situation was transformed.

Apollo 13 was at that moment 330,000 km from Earth, and 56 hours had passed since lift-off. It was planned that the crew, after closing the LM top hatch, should take some photographs of a very bright comet discovered four months earlier by South African amateur astronomer John Caister Bennett. As Lovell and Haise were preparing the LM for closing, mission control ordered Swigert to stir the fuel cells' liquid oxygen tanks – a manoeuvre required from time to time, to avoid stratification of the contents. At 3.08 GMT on 14 April, as Swigert selected the mixer command, there was a loud bang and the docked spacecraft were rocked by an unknown force. Swigert alerted mission control: 'Houston, we've had a problem!'. Then Lovell announced that one of the two power distribution systems, Bus B, was no longer functioning.

Immediately afterwards, Haise began a careful check of the electrical system, and discovered that fuel cell number 3 – which was supposed to provide power to Bus B – was no longer providing any power. A new alarm then flashed, reporting that the other power distribution system, Bus A, was malfunctioning. Fuel cell number 1 had failed. Only fuel cell number 2 was still generating power. Lovell, meanwhile, was checking other systems, and noticed that number 2 oxygen tank was empty and that the pressure inside number 1 tank was rapidly decreasing. But what could have caused three apparently unrelated anomalies? To further compound the situation, mission control had lost the high-gain antenna signal, as if something had caused it to point elsewhere. The connection between the anomalies was found by Lovell himself, when he looked out of the window. There was a cloud of gas emanating from a side of the SM – this could only be their oxygen. On the ground, some NASA engineers who were following the S-IVB stage with a 40-cm telescope, saw the sudden formation of a small comet where Apollo 13 was supposed to be, but which would otherwise be invisible because of its remoteness.

The chain of events appears to have been as follows. The number 2 oxygen tank had exploded, the reactant valves to two of the fuel cells had closed, the escape of gas into space had caused attitude loss so that the high-gain antenna no longer pointed at Earth, the main computer had reset after becoming overloaded, and the explosion had cracked the feedpipe from number 1 oxygen tank, which was leaking, which meant that fuel cell number 2 was doomed – and so, therefore, was the CSM.

With the CSM damaged and starved of power, there was now no prospect of going ahead with a lunar landing. The LM would now have to serve as a 'lifeboat' to keep the crew alive until they could escape in the CM and re-enter the Earth's atmosphere. While Lovell and Haise set about powering up the LM in record time, Swigert powered down the CSM's systems, using the batteries reserved for operating the CM during re-entry to shore up the flagging fuel cell while he did so. But he could not switch off the inertial platform until the 'reference' could be passed to the LM's platform, for otherwise they would be unable to navigate. The platform, to provide an orientation reference during the next days, had to be configured, and the normal procedure (using the telescope to observe bright reference stars) could not be used, as the spacecraft was enveloped in a cloud of gas and debris which prevented such observations. It was therefore necessary for Lovell to compute, by hand, the conversion between the platform orientation of Odyssey and that of Aquarius, the main axes of which, to complicate matters, were 90° from those of the mothership. Once Aquarius was ready, Odyssey was switched off, with the hope of being able to switch it back on before re-entry.

The next issue was the fact that the spacecraft was no longer travelling on a free return trajectory that would have ensured an Earth re-entry without using any fuel. The hybrid orbit in which it was travelling, perturbed by lunar gravity, would instead result in an apogee well beyond the Moon, leading to an impact with the Earth in late May, by which time the crew would be dead, and resulting in the destruction of the spacecraft.

There were two options. One was to perform a 'direct' abort by making a long engine burn in order to slow down so dramatically that the spacecraft would fall back to Earth without reaching the Moon. The other option was to return to the free-return trajectory, and loop around the Moon before heading home. The direct abort would get them back sooner, but such a major manoeuvre could only be done by the SPS engine, and would require it to fire for some five minutes. With the SM damaged, no-one wanted to take the risk that the engine might explode. The only option available, therefore, was to use the LM's throttleable descent engine to return to the free-return trajectory. Fortunately, the ability of the LM to manoeuvre the docked combination had been demonstrated by Apollo 9. The possibility of jettisoning the now useless, but massive SM to lighten the LM's task was rejected because this would expose the CM's heatshield to several days in the harsh space environment, for which it had not been designed, and because no-one knew the inertial characteristics of the LM–CM combination.

At 08.42, 5½ hours after the explosion, Lovell and Haise received the descent-stage firing command. The engine was throttled for ten seconds at minimum (10%) thrust, and for 21 seconds at 40% thrust, in order to resume a free-return trajectory. The 'barbecue' thermal control mode was restored to produce a spacecraft spin of 90° every hour or so. The crew were fortunate, at such a dangerous time, that Haise was one of the two astronauts who had specialised in the development of the LM with Grumman, and was one of the top experts on its systems. (The other astronaut was Ed Mitchell, of Apollo 14.) Relying on his knowledge, Haise computed the life support possibilities of the LM, initially designed to support two people for a maximum of 45 hours, and now required to support three people for 75–100 hours, depending on the success of a second course correction. If 100 hours, there would still be a few hours of oxygen in reserve, but after the second course correction virtually everything would have to be switched off, because the LM had only batteries, not fuel calls. The critical factor, he realised, would be coolant water, as it would probably be exhausted about five hours before re-entry. Remembering the experiment carried out by Eagle in lunar orbit a year earlier, a further few hours could be squeezed out of it, and consideration was even given to pouring urine into the cooling system. However, Haise did not identify the critical issue of carbon dioxide removal, as he had performed computations for two rather than three people. More of this later.

Sleeping inside Odyssey was very difficult, as the temperature was just above zero, and water vapour had condensed everywhere. However, after a brief rest, the time came to check the inertial platform alignment – which had been initialised 20 hours earlier – for the second engine firing to reduce the journey time by several hours. This was to be made at PeriCynthion plus Two (PC + 2) – two hours after lunar fly-by. As the spacecraft was still flying inside the cloud of gas and debris, the Sun was used as a reference. If it was within 1° of the centre of the visual field, alignment was good. As it happened, the limb fell right at the centre of the eyepiece. It was a good omen.

At 00.21 on 15 April, radio contact with Apollo 13 was lost (as expected) as the spacecraft disappeared behind the Moon. A few minutes later, as soon as the Sun rose, Haise and Swigert began to take photographs of the far-side. In contrast with previous missions, Apollo 13 flew by the Moon at a distance of 219 km, and at a higher speed, so this radio occultation lasted less than half an hour. A few seconds after the spacecraft exited from radio occultation, the Apollo 13 mission produced its only contribution to lunar science. The empty S-IVB, with a mass of 13,925 kg, impacted the surface at a speed of 2.5 km/s at 2°.4 S, 27°.9 W, on the rim of the crater Lansberg B*, 119 km north-by-northwest of Apollo 12, yielding the energy of seven tonnes of TNT. The ALSEP suprathermal ion detector was the first to detect the impact as a cloud of vaporised debris swept over the site. A few seconds later, and some 30 seconds after the impact, the seismometer detected the artificial quake

---

* Philippe van Lansberge (1561–1632), Belgian astronomer.

An impressive image of the far side of the Moon, taken from the Apollo 13 LM. CM Odyssey, with its dark windows, is seen at top.

and recorded vibrations which this time took almost four hours to dampen. (The crater produced by the impact was later identified in some Apollo 14 pictures.) As the S-IVB had been transmitting on a frequency very close to the channel used for voice communication with Aquarius, the astronauts' voices were subsequently much less noisy. (Normally, the LM would not be activated until after the S-IVB had smashed into the Moon.)

Two hours later, the time came for PC + 2 firing. This was completely successful. The mission control team now faced three different problems: to ensure that the spacecraft remained inside the narrow corridor for a safe re-entry; the removal of carbon dioxide, which would soon suffocate the astronauts; and the procedures for switching on Odyssey again before re-entry. No-one knew how to power up a CM that had been subjected to four days in the cold of space and to extreme moisture, without exhausting the batteries. Mattingly, who had not fallen ill, began to work on this problem. Concerning the trajectory, a new firing of the LM's descent engine to depletion was considered to reduce the return time by as much as 27 hours, but as this would move the re-entry area to the Atlantic Ocean or Indian Ocean, where there were no American recovery ships, it was decided to stay with the current plan.

The immediate concern, however, was the removal of the carbon dioxide, which was rising to an unacceptable level. This problem was exacerbated by the fact that the LM used cylindrical lithium hydroxide filters, while the CM – for which many spare filters were still available – used cubic ones. Fortunately, someone remembered a similar (simulated) emergency during training for a previous mission, for which a method had been devised to build an adapter. After some tests, during the early hours of 15 April the procedure was relayed to the astronauts. The cubic filter had to be placed inside a plastic bag, closed with tape, together with the fireproof cardboard cover of a flight manual, folded so as to prevent the filter from sucking the bag. Some apertures of the filter had to be closed with socks, and the main filter duct had to be connected to a lunar spacesuit umbilical that would connect the assembly to the cylindrical filter housing. In this way, two identical units were produced by the astronauts, and the carbon dioxide level – which had by that time set an alarm – began to decrease.

Some hours later – with the astronauts becoming concerned at the time it was taking to develop the procedure to switch Odyssey back on – a new course correction manoeuvre had to be carried out without using the computer, which had long since been shut off. The technique, which had been tested by Lovell during the Apollo 8 mission eighteen months previously, required only Earth-sighting as an attitude reference. To be properly carried out, the manoeuvre required that the Sun shine on the ceiling rendezvous window and the Earth remain framed by the commander's window throughout the 14-second firing. Without the intervention of the computer, the manoeuvre would require real team-work: Lovell throttled the engine and controlled the roll axis, Haise controlled the pitch and yaw axes, and Swigert used a stop-watch.

The following day was occupied with the transcription of the CM powering-up procedure. In the meantime, the internal temperature of Aquarius was falling even lower, and Haise had contracted an infection of his urinary tract, caused both by dehydration and by a misunderstood order to not use the CM hygienic waste dump, as this apparently interfered with communications. As no order to resume using it was ever issued, the crew had used all available storage, including the pockets of the spacesuits, with a resultant worsening of hygienic conditions.

On the morning of 17 April, the crew, who had taken only a few hours of rest during four days, had to carry out yet another course correction, as an unknown force was pushing the spacecraft to the limit of the re-entry corridor. What had happened was that the LM cooling system was continually dumping steam in space, and this was acting like a small rocket. The LM's descent engine could not be used because the diaphragm designed to relieve the pressure in the helium tanks once the LM had landed on the Moon had finally burst, so the small attitude-control engines were fired for 21 seconds to make the correction.

Back on course, Swigert began to switch on Odyssey's systems again, recharging its batteries using those on Aquarius. The time then came for SM separation. Swigert gave the separation command as Lovell fired the attitude engines of the LM to move away. The astronauts then rushed to the windows to see what had caused their mission to abort. First of all, the high-gain antenna pack was seen to have been bent into an unusual posture, as if it been hit by something. Then the bay housing the oxygen tanks and fuel cells came into view. An entire side-

The Apollo 13 Service Module after jettisoning, showing the effects of the oxygen tank explosion.

panel was missing, exposing a tangled mass of metal and tubing. Given the evident violence of the explosion, the fear was that it might have somehow damaged the heat shield. If so, then they were doomed.

One and a half hours later, the time came to jettison Aquarius, which re-entered over the 6,000-metre-deep Tonga trench, where the graphite shroud housing the SNAP fuel for the ALSEP should have sunk.* With three hours to CM re-entry, two causes for concern still remained: the condition of the heat shield, and the integrity of the parachutes, which had been exposed to extreme cold for several days. On the plus side, the accuracy of the trajectory was confirmed by Lovell when he watched the Moon set behind the Earth at the predicted time. As the CM entered the atmosphere, communications were cut off by the ionisation of the air, and there was considerable anxiety in Houston – and indeed, all around the globe – when this radio blackout lasted fully a minute longer than expected, during which the worst was feared, but then a signal was received and there was jubilation. At 18.07 GMT on 17 April – 142

---

*   This was the second space accident during which a radioisotope nuclear generator re-entered the atmosphere. In the first case of a Nimbus satellite which exploded during launch in May 1968, the two small generators were recovered intact, at sea. It is expected that the Apollo 13 SNAP will withstand corrosion for 870 years – ten times the half-life of plutonium-238.

hrs 54 min after launch – Odyssey made one of the smoothest splash-downs of the programme.

None of the three astronauts returned to space. Lovell had already announced his intention to retire from the astronauts corp. Haise, after recovering from his illness, was back-up commander of both Apollo 16 and Apollo 17, and went on to pilot three of five atmospheric flights of the first Space Shuttle *Enterprise* (in particular commanding the very first, on 12 August 1977) after which he worked for Grumman, the builder of the LM to which he owed his life. In 1973, after being considered for the joint Apollo–Soyuz flight, Swigert retired from NASA and pursued a career in politics. He ran for the US Senate in 1977, without success, but in 1982 he was elected. On 27 December that same year, eight days before he was to take up his post, he died as a result of a bone tumour. Odyssey was for a long time on show at the Musée de l'Air et de l'Espace, at Le Bourget, near Paris, but was returned to America in the mid-1990s and is now on show at the Cosmosphere in Hutchinson, Kansas.

After the return of the Apollo 13 astronauts, there was a hiatus in lunar flights while the cause of the accident was investigated, and precautions to preclude a recurrence implemented. As the trouble had begun in number 2 oxygen tank, the members of the commission studied all of the documentation relating to that item. It was thereby discovered that it was initially mounted on Apollo 10's Charlie Brown, before being replaced by a modified tank. During the removal, one bolt had remained stuck, and the tank was slightly damaged in the ground drain line – a component considered unessential for flight. After some modifications to bring it up to standard, the tank was mounted on Odyssey and subjected to several ground tests. During one such test, a few weeks before lift-off, the tank's cryogenic system was pressurised for the first time, and should later have been emptied of some 150 kg of mixed liquid and gaseous oxygen at a temperature of –206° C by the injection of gaseous oxygen. As the tank had suffered damage to the drain line, however, it did not empty. It was then decided to switch on the internal resistances to heat the oxygen slush, boil it, and eventually evacuate it. Unfortunately, the thermostat that would cut the resistance power supply if the tank temperature reached 26° C (due to the risk of fire) had not been dimensioned both for the 28-V internal power system and for the 65-V provided by the launch tower. As soon as the tank reached 26° C, the thermostat melted due to the unexpectedly high voltage, and remained closed. It took eight hours to evacuate all of the oxygen, but during this time the internal tank temperature rose to 600° C, and the insulation of some wires passing through the tank melted to reveal bare metal. It then required only a single spark, flying from the mixing command switch, to set the oxygen on fire and blow up the tank. An external thermometer, which could have alerted pad engineers, had a scale extending to only 28° C. The tank had been a bomb waiting to go off!

In order to limit the consequences for a mission in the event of another oxygen tank failure, the system was modified. Valves were fitted so that each tank could be isolated, and a third tank was installed, this one on the far side of the SM where it was unlikely to be damaged by any incident involving the others.

3.15 A REPEAT PERFORMANCE

During the hiatus, two lunar missions had been cancelled by Congress, and a third by NASA itself, so that lunar manned flights would now end with Apollo 17. The Soviets had taken advantage of the pause by returning their first soil samples using the automatic probe, Luna 16. Meanwhile, NASA decided that the Fra Mauro site was so important that Apollo 14 was assigned to land there. In the meantime, interest in lunar missions continued to decrease and, as mentioned, political support began to wane.

The crew of Apollo 14 were Al Shepard – the first US astronaut to fly into space, and the only Mercury astronaut to fly to the Moon – and rookies Stuart Roosa, CM pilot, and Ed Mitchell, LM pilot. For Shepard it was the return into space for which he had waited ten years. After his brief sub-orbital flight in 1961 he had hoped to fly a Mercury orbital mission, but this had been cancelled; then he had been designated commander of the first manned Gemini mission, but upon being diagnosed as suffering from Ménière's syndrome (a disease of the inner ear) he was grounded. Hoping that the disease might heal by itself, Shepard never left the Astronaut Office – even though he embarked on a business career and became the first millionaire astronaut. In 1969, after undergoing complex experimental ear surgery, Shepard recovered to normal health, and was immediately named as commander of Apollo 13. But for the first and only time, NASA top brass vetoed the crew choice and moved Shepard's crew to Apollo 14 in order to give them more time to train. Shepard's spaceflight experience was so limited (his Mercury flight had lasted 15 min 22 sec) that Conrad scornfully called the Apollo 14 crew 'the rookie-only crew'.

An Apollo 16 orbital image of the crater produced by the Apollo 14 S-IVB stage. The crater is remarkable for its system of dark rays.

Apollo 14 lifted off late in the afternoon of 31 January 1971. Conscious that their mission was to be NASA's redemption, the astronauts entered orbit resolved to carry out as perfect a mission as possible.

The problems began about two hours after translunar injection, as the CM, Kitty Hawk (named after the place of the Wright brothers' first flight), attempted to dock with the LM, Antares, on top of the S-IVB stage. Roosa eased the probe into the drogue on top of the LM, but the latches at the tip of the probe failed to engage and the two spacecraft separated again. Five more attempts were made during the next 90 minutes, but to no avail. The time when the S-IVB would dump fuel was fast approaching. Alternatives were quickly considered. One idea was to depressurise the CM and bring the two spacecraft into 'manual' contact, with an astronaut poking out of the top hatch and capturing the LM by hand! This was rightly considered to be far too risky. Finally, it was decided that, as Roosa was pushing Kitty Hawk against Antares using the attitude engines, in order to bring the twelve spring-loaded latches around the collars into contact, Shepard would command docking probe retraction in the hope that they would engage – and this solution worked. The cause of the anomaly was never identified, but it was probably due to some debris that found its way into Antares' docking system. The eventual fate of the S-IVB was once again lunar impact, this time at 8° S, 26°.6 W, 160 km south-west of the Apollo 12 seismometer. (The crater was later identified in Apollo 16 pictures, and was easily noticeable because of its unusual bright and dark rays.)

A few hours later, the first rest period began, and Mitchell – unbeknown to his fellow travellers and to NASA – tried a space experiment of his own. As he had always had a fascination with parapsychology, Mitchell had earlier contacted 'ESP' enthusiasts in Florida, who, at the beginning of the first rest period, would try to perceive a sequence of standard symbols imagined by the astronaut, tens of thousands of kilometres away. Unfortunately, the launch had been delayed by 40 minutes and so too had the rest period. The 'ground receivers' therefore attempted to catch Mitchell's thoughts 40 minutes before their actual 'transmission'; the experiment was, of course, inconclusive.

Lunar orbit insertion was perfect, and this time a slightly different descent profile was tested. Instead of separating the two spacecraft in a circular 109-km orbit and having the LM perform the burn to enter the 'descent orbit', the SPS engine was used to lower the perigee to 15 km before Antares separated. This would make more fuel available to the LM, if it had to hover for a significant time while Shepard sought a safe spot on which to land. The second mishap of the flight happened at this time. Shortly before the powered descent was to begin, ground control stations noted that the abort switch was intermittently activating, possibly because of poor soldering. If this were to happen while the engine was on, the computer would think that Shepard had aborted the descent and it would cut the engine, separate the descent stage, and initiate an emergency rendezvous. On Earth, a crisis committee prepared software fixes that were uploaded to Antares during the following orbit. No sooner had the descent begun than it was jeopardised by the approach radar. Realising that the radar had not been properly initialised following the software update, Mitchell shut it off and then turned on again, and at last, 3 km from the ground, good altitude data began to pour in. Antares landed in the Fra Mauro region, at 3.645 S, 17.471 W, about 1 km beyond the low ridge on whose summit was Cone crater, the main geological target of the mission. The landing site was much more hummocky than expected, and Antares set down with one foot inside a small crater, inclined 7° from vertical. Shepard's first words from the surface were, rather mundanely, 'We made a good landing.' Once again the landing accuracy was first confirmed by Roosa on Kitty Hawk, who by using the CM sextant could see the LM on the ground.

This time the CM pilot too had some work to do, once back in a circular orbit. Roosa had been provided with a Hycon automatic panoramic camera, yielding a maximum ground-resolution of 2 metres. Unfortunately, the camera jammed after taking just 140 pictures, and he had to resort to using a more 'modest' Hasselblad equipped with a 500-mm lens to snap a list of features to be imaged. As already mentioned, one of these photographs showed the crater created by the Apollo 13 S-IVB stage. It must also be mentioned that Roosa – who had previously served in the US Forest Service – carried, as personal objects, 500 seeds of five different trees: Loblolly Pine, Sycamore, Sweetgum, Redwood, and Douglas Fir, of which 450 subsequently germinated. Today, dozens of 'lunar trees' grow throughout the world, including a pine in the garden of the White House and another in Switzerland.

The first of the two planned excursions began six hours after landing, early on 5 February. The mood was very relaxed, and Shepard's words on setting foot on the Moon were, 'It's been a long way, but we're here.'

During this first walk, which lasted 4 hrs 47 min, the astronauts collected samples close to the LM, and deployed a laser retroreflector identical to that left by Apollo 11, a 'Swiss flag' solar wind collector, and ALSEP instruments almost the same as those on Apollo 12: a new capacitance seismometer, a solar particle detector, and the two cold and hot cathode sensors to study the thin lunar atmosphere. The three-axis magnetometer had been replaced by a small grenade launcher that was to be used after the astronauts had lifted off, to calibrate the two-station seismic network (although it was deployed too close to the ALSEP central station, and was never used for fear of damage). To probe the structure of the blanket of Imbrium ejecta that formed the hummocky plain on which they had landed, Mitchell deployed a line of seismic sensors and exploded thirteen of twenty-one small charges provided in a 'thumper'. At the end of the walk, a rest period was observed.

The second excursion began eighteen hours after the first. The objective was to climb the low ridge over which they had flown on the way in, in order to inspect the 350-metre-wide Cone crater, where the scientists hoped to find both clues of the past history of the Fra Mauro Formation preserved in the crater's wall, and samples brought to light by the impact. The 1-km walk included stops near several smaller craters. To carry samples and equipment, the astronauts were provided with a two-wheeled cart called the Modular Equipment Transporter (MET), which had a 16-mm movie camera and a magnetometer to record the magnetic field at sampling sites along the way. Shepard dragged the cart while Mitchell read the map. Unfortunately it proved difficult to find reference points and they wasted a lot of time trying to find designated sampling sites. The climb up the ridge was made more difficult by the rigid spacesuits, so that their heart rates reached 150 and they had to rest several

An image of part of the Apollo 14 ALSEP. The object to the left of the astronaut's shadow is the SNAP radioisotope generator, and the 'boxy' object is the central station with its antenna aimed towards the Earth. At the right side of the image, connected to the central station, is the grenade launcher. The white object in the background is the laser retro-reflector.

times. Slightly more than one hour after the beginning of the walk, Mitchell announced: 'Our positions are all in doubt.'

After cresting the ridge without finding Cone, they were told to halt and to sample the nearest large boulders. At the time, the astronauts could not determine exactly where they were, but later comparison between pictures taken on the surface and orbital images revealed that they had stopped within 20 metres of the crater's rim, so they had sampled their objective without realising it. The reason that they had not been able to find Cone was that, contrary to expectation, it did not have a raised rim, it was simply a hole in the ground, and worse, the fact that the near-horizon atop the ridge was the lip of the pit was not evident because it was dug into the receding slope.

Three hours after leaving the LM, the astronauts headed back, sampling some sites on the way, in one case taking a core to a depth of 80 cm. Before re-entering, Shepard had his moment of glory. He modified the contingency sample-collecting scoop by attaching a golfing six-iron to its end, and took a golf ball from his pocket. With his first attempt, using just one hand (as the spacesuit was too bulky for him to use both) he did not succeed, and with his second attempt he hit the ball badly. Using

The Modular Equipment Transporter on Cone Ridge, loaded with tools. It carried its own 16-mm movie camera. On the left of the MET is the Gold Camera, and the black dot in the background, to the right of the MET, is the LM.

a second ball, he effectively established a record when the ball travelled about 400 metres (although Shepard claimed 'miles and miles'). Following his commander's lead, Mitchell, after folding the solar wind collector, threw its pole like a javelin.

The second walk lasted 4 hrs 34 min. Although the MET mobile magnetometer took only two readings during the walk back from Cone crater, the results confirmed that the rocks have remanent fields of different strengths.

After docking with Kitty Hawk by using a new ascent trajectory requiring fewer manoeuvres than previously, Intrepid was impacted on the Moon at 3°.42° S, 19°.67 W, some 70 km west of the Apollo 14 seismometer and 100 km east of the Apollo 12 seismometer.

Finally, after thirty-four orbits, the SPS engine was fired to head home. During the trip, weightlessness experiments were performed in preparation of those designed for the Skylab space station. These included experiments in electrophoresis (mass transport in a solution in an electric field), material solidification, and liquid transfer between two containers. On 9 February, the CM splashed down only 965 metres from the expected point in the Pacific Ocean. This time the quarantine was more relaxed. The astronauts were taken by helicopter from a basic isolation facility

A close-up view of an Apollo 14 astronaut's 2-cm-deep footprint, taken by the ALSCC Gold Camera. Apollo 14 was to be the last mission to carry this instrument. (Courtesy Thomas Gold and NASA.)

onboard the aircraft carrier *New Orleans*, to a second definitive unit waiting for them at a Samoan airport. Shepard, Mitchell and Roosa arrived in Houston on 12 February, and remained in quarantine in the LRL until 27 February.

Although Apollo 14 had saved the honour of NASA by overcoming numerous problems, the geologists and other scientists were critical of the work carried out on the lunar surface. In particular, it had been expected that photographs would be taken of every sample before collection, with a gnomon (a type of sundial with a calibration strip) being placed beside it in order to be able to reconstruct the original orientation of the sample. As Shepard and Mitchell did this only a few times, the geologists had to find all the other samples in panoramic photographs, which was very laborious. In contrast to the mare-sampling missions, almost all of the 42.8 kg of rocks returned by Apollo 14 were breccias, consolidations of shattered precursors. Their analysis took longer than expected, but the results showed that the Fra Mauro Formation had been laid down as a splash of ejecta 3.85 to 3.82 billion years ago when a huge impact excavated the Imbrium basin. Mitchell's seismic tests indicated that this blanket was 75 metres thick at the landing site. The mission had achieved its primary scientific objective, but the astronauts' field technique had been poor and there had not been sufficient time on the surface to do the site justice.

The most important ALSEP observation was obtained by the suprathermal ion detector on 7 March 1971. On that day, both the Apollo 14 instrument and the identical Apollo 12 instrument recorded, after a light quake, a very rarefied cloud of gas which was probably water vapour released from the depths of the Moon. Another important observation was made by the seismometer on 13 May 1972, when it recorded the impact of a large 1,100-kg meteorite some 145 km to the north. The

Apollo 14 ALSEP was the first to fall silent. After a string of minor mishaps that partially compromised the correct working of several instruments, the central station ceased transmitting in January 1976.

None of the three Apollo 14 astronauts ever returned to space. Mitchell has long since turned to his parapsychological researches for the California Institute of Noetic Sciences; Roosa died on 12 December 1994; and Shepard died on 21 July 1998, three months before his friend and rival John Glenn returned to space on the Space Shuttle *Discovery* 36 years after becoming the first American to orbit the Earth. Kitty Hawk is now on show at the Astronauts' Hall of Fame in Titusville, Florida.

3.16 THE FIRST SCIENTIFIC MISSIONS

After the Apollo 14 mission, Eugene Shoemaker, who, in 1960, proved that Meteor Crater in Arizona marked an impact, applied stratigraphic analysis to the study of the Moon, and established the Astrogeology Branch of the US Geological Survey, declared that if America were to continue to send astronauts to the Moon without ever taking a scientist – and in particular, a geologist – it would be better to follow the methods of the Soviets who had returned their first samples using an unmanned probe. The degree to which Apollo had addressed scientific issues had increased in recent years, but the requests of the scientists often clashed with mission planning.

Even before 1967, expeditions in desert areas were organised – initially as personal ventures but later NASA-funded – to train the astronauts in the rudiments of field geology; in particular how to describe what they were seeing, and how to capture the geological diversity of a site with only a limited sample of rocks. These expeditions – organised by geologist-astronaut Harrison Schmitt and by his former teacher Leon Silver of Caltech – attracted the interest of many (but not all) astronauts. Fred Haise had been one of the best students, but Shepard and, indirectly, Mitchell, considered the courses to be a waste of time, and their eventual on-site work at Fra Mauro reflected this opinion. At the same time, Egyptian-born geologist Farouk El-Baz – editor of a huge catalogue of every single feature visible in Lunar Orbiter pictures – began training CM pilots to make geologically pertinent observations from lunar orbit. Both Mattingly and Roosa were extremely interested in this brief course. It would give them something 'meaningful' to do during their lonely vigils while their colleagues were on the surface. On each mission, the commander set the tone. Every gramme onboard the LM had to be justified, and mission planners did not find a particularly compelling need to carry a heavy geological hammer or a telephoto lens. The critical factor, therefore, was the commander's attitude to the science.

In the end, NASA – pressed by the scientific community, and faced with the number of remaining missions shrinking – decided to advance the first scientific J mission from Apollo 16 to Apollo 15. For these missions the LM would have uprated oxygen tanks and batteries to stretch the surface-stay duration to 72 hours instead of 45 hours as on H missions. Consideration was also given to equipping the LM with solar panels. The spacesuits were made more flexible, and their oxygen supply was increased to facilitate 8-hour moonwalks. The descent stage engine was upgraded to handle the increased mass of the LM. A great deal of this mass increase was due to the addition of the Lunar Rover Vehicle (LRV).

The LRV was an electric car built by Boeing in record time – less than two years from the signing of the contract in November 1969. It had a mass of 209 kg, increasing to around 700 kg with astronauts, equipment and samples. It was built on a metallic frame 310 cm long, was capable of travelling at up to 18 km/h, had a round-trip range of 65 km, and was carried to the Moon doubly folded on the left-front side of the descent stage. Each wheel was made of an elastic metallic mesh, and was connected to an electric motor which could be isolated in case of failure. The front and rear wheels could be steered independently or in unison, and a gyroscopic platform and an odometer provided the crew with information about their position. It was powered by a pair of silver–zinc 36-V batteries, had a rod-like medium-gain antenna to relay voice and an umbrella-shaped high-gain antenna for television from a camera on a mount that was controlled from Earth. Consideration was given to making it completely autonomous after the astronauts lifted off in order to further explore the neighbourhood (like the then-being-developed Soviet Lunokhod robot), but this was never implemented.

For the J missions, the SM was also modified. In particular, a large empty bay was given over to a suite of scientific instruments: photogrammetric cameras, laser altimeters, X-ray, alpha-ray and gamma-ray spectrometers, mass spectrometers, and, on Apollo 15 and Apollo 16, a small Particles and Fields (P&F) satellite to be ejected in lunar orbit. The hexagonal-cross-section P&F was 78 cm long, and had a diameter of 36 cm. On the front, the probe carried an omnidirectional antenna for data relay, and from the other end it deployed three long booms, of which two carried weights and the third a magnetometer. The only other instrument on board was a cosmic-ray telescope. The whole body of the 36-kg satellite was covered with solar panels.

To lift the heavy SM and LM, the fuel management system of the Saturn V was revised, it having been decided that, after nine successful flights, this could be less conservative.

Given this advanced capability, the issue became where to send the first mission. The primary interest was volcanism. One candidate was the complex of volcanic-looking hills near the crater Marius in northwestern Oceanus Procellarum. Another was Davy Catena – a string of kilometre-sized craters inside the larger crater Davy*. At that time, this was believed to mark a fault in the floor of the crater which had facilitated volcanism, resulting in a chain of collapse-features. The catena is now believed to have been created by the impact of a comet which had just passed very close by Earth and been ripped apart by its gravity (an event similar to the fate of comet Shoemaker-Levy 9 when it was broken up by Jupiter in 1992). However, the site selected was close to Mount Hadley** – one of the towering peaks of the Apennine mountain chain, more than 3,000 metres high, that forms the southeastern rim of the Imbrium basin – and close to Rima Hadley (Hadley crevasse), a sinuous feature only 1 km wide but 80 km long, thought by most lunar geologists to be either a channel

---

*   Humphrey Davy (1778–1829), British physicist and chemist.
** John Hadley (1682–1743), English astronomer.

etched by a sustained flow of lava or, perhaps, a 'lava tube' whose vault had collapsed. The appeal of this site over the others, was that there were multiple objectives in an area of a few square kilometres: a mare terrain (Palus Putredinis, the Marsh of Decay), a mountain (Hadley Delta) and a canyon (Rima Hadley). The LRV would enable them all to be investigated.

The crew of Apollo 15 were Dave Scott, a veteran of Gemini 8 and Apollo 9, and two rookies: Alfred Worden, CM pilot, and James Irwin, LM pilot.

A few hours after lift-off in the early morning of 26 July 1971, Apollo 15 was *en route* to the Moon. The S-IVB stage hit the surface at 0°.99 S, 11°.89 W, 180 km east of Apollo 14. Nearing the Moon, the panel protecting the Scientific Instrument Module (SIM) bay was jettisoned and went on to enter solar orbit. A few hours later, the spacecraft entered a lunar orbit with a 28° inclination – this angle being dictated by the high latitude of the northerly landing site.

The Apollo 15 lunar rover on the surface of the Moon. Visible from left: the tool storage rack, the astronaut seats (note the Hasselblad camera on Scott's chest and the combined throttle and steering handle in his right hand), the low-gain antenna, the 16-mm movie camera, the umbrella-shaped high-gain antenna, and the television camera. Note also the map of Rima Hadley on the control console.

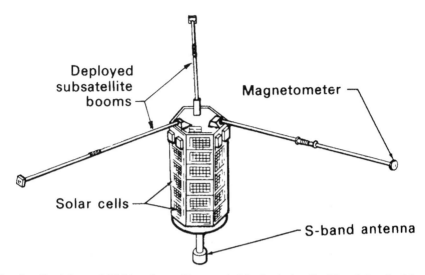

The tiny Particles and Fields sub-satellite, carried by both Apollo 15 and Apollo 16.

A hundred hours after lift-off, the LM Falcon (celebrating the mascot of this all Air Force crew) separated from the CM Endeavour (a tribute to the ship of the eighteenth-century English explorer James Cook) in a 15-km-perigee orbit. Shortly afterwards, Endeavour returned to a 101 by 120-km orbit, where Worden would carry out his SIM bay mission. Photographs were to be taken by a topographic camera using 120-mm film. It had an ingenious system (based on that developed for the top-secret KH-4B spy satellites) to reconstruct the attitude and height of the spacecraft from simultaneous photographs of the sky. It included a 76-mm objective and, on Apollo, could take up to 3,600 frames with exposure times ranging from 1/15 to 1/240 second. The SIM bay also contained a panoramic camera based on that of the KH-3 spy satellites. Whereas the mapping camera imaged an area of 175 by 175 km, the panoramic camera recorded a 28-km-wide strip that extended 160 km to either side of the ground track. The mapping camera had film for 3,600 exposures. Because the panoramic images were wider, it could take only 1,650 frames.

A laser altimeter monitored the CSM's altitude to an accuracy of about ten metres by measuring the time of flight of its 10-nanosecond pulses. The altimeter data revealed the profiles of overflown features – profiles that had always been difficult to accurately compute using only shadow data. It was thus discovered that the bottom of several near-side maria (Imbrium, Serenitatis, Crisium and Smythii) are at the same height, and that the few mare inclines are very smooth, a clear sign that they were formed by low-viscosity lava. On the far-side, the true form of the large Aitken basin was discovered.* At 1,400 km in diameter and with its floor some 6 km below the level of the surrounding terrain, this is the largest, deepest depression on the Moon. Nevertheless, for some reason it never flooded with mare lava. As already

* Robert Grant Aitken (1864–1951), American astronomer.

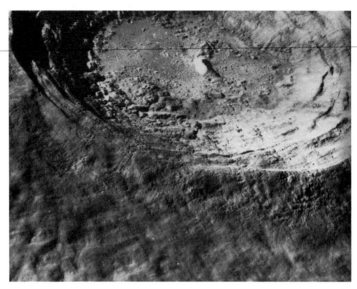

An Apollo 15 orbital image of the large crater Aristarchus.

mentioned, photographs of it were first taken by Luna 3, and it was later found in Soviet Zond 6 and Zond 8 pictures. Its rim was identified in the first maps of the lunar limb, produced by Friedrich Hayn in 1914.

The small crater Linnaeus, inside Mare Serenitatis, was imaged in detail, thus providing long-awaited definitive data on its shape and diameter, of 2,450 metres.* The size of Linnaeus had been a longstanding problem in lunar astronomy. As the crater is quite small and is surrounded by a bright halo, its actual diameter had proved very difficult to measure telescopically. In 1788 J.H. Schröter was able to see only its halo, and in 1824 Wilhelm Lohrmann estimated it as 7 km; nineteen years later, Julius Schmidt estimated it to be 11 km – before announcing that it had disappeared, possibly obliterated by a landslide or a volcanic eruption – and then as only 400 metres.

Another experiment, already carried out on previous missions, was the S-band transponder – a simple instrument which received a 2,115-MHz carrier signal transmitted from the Earth, and then retransmitted it. What was eventually received by the Earth provided data on the lunar gravitational field. Other than providing new information about the mascons, this provided the first evidence of something that had been suspected since the unmanned Ranger missions: that the Moon's centre of mass is not coincident with its centre of figure, but is displaced toward the Earth.

X-ray fluorescence was also studied, being due to the terrain re-emitting incident solar X-rays. These data provided information on the composition of the regolith, and in particular the aluminium/silicon ratio. Every maria – including the crater

---

* Carl von Linné (1707–1778), Swedish biologist.

Tsiolkovsky, on the far-side – proved to be poorer in aluminium than the highlands. Radioactive elements such as thorium, potassium and uranium were mapped by the gamma-ray spectrometer, and found to be concentrated in Oceanus Procellarum.

Another instrument providing interesting data was the alpha-ray spectrometer, measuring the flux of alpha particles – helium nuclei produced by the natural decay of radioactive elements. It proved particularly interesting to correlate spectrometric data with the presence of radon-222 – a noble gas produced by the decay of uranium and thorium. On one orbit, in fact, the spectrometer recorded an anomalous peak over Aristarchus, a bright crater in Oceanus Procellarum, in which, on 29 October 1963, James Greenacre at the Lowell Observatory in Arizona reported transient red spots. A possible link was thus discovered between the observed phenomena and the release of radioactive gases – particularly radon – following seismic events.

In addition, in the part of the orbit above the shadowed hemisphere, Apollo 15 took astronomical images of the zodiacal light using the topographic camera, and observations were also made of the solar corona.

The LM Falcon began its descent on Friday, 30 July, pursuing a carefully planned trajectory that would skim low over the mountains that form the backbone of the Apennine Range and then end with a much steeper descent than had been used for more 'open' sites. When the LM pitched and Scott got his first look at the terrain ahead, he was momentarily disorientated because the craters, which had seemed sharp and distinctive on the Lunar Orbiter pictures, were actually very subdued; but he knew that he was in the right place because snaking across the plain right where it should be was the canyon.

Scott lost sight of the ground just before touchdown because the engine lifted so much dust, so he was taken by surprise when the LM settled with two feet in a small crater and it tipped back. The Falcon had landed at 26.132 N, 3.634 E, less than 2 km from the rim of the canyon, about 6 km from the slopes of Hadley Delta, and 200 km from the hammer-and-sickle emblems carried to the Moon by Luna 2 twelve years earlier. At that very moment, the first Soviet Lunokhod was exploring the far side of Mare Imbrium. When he next passed over, Worden spotted the LM on the surface and the 150-m-diameter bright spot created by dust moved during the descent.

Only two hours after landing, the first excursion – of an unusual kind – was carried out. Scott opened the LM's top hatch (the hatch used for CM docking) and, standing on top of the ascent engine cover, poked his head and shoulders out in order to describe the landing site from this new perspective, in terms of its geological morphology. Photographs were taken of the peak named Silver Spur (after Leon Silver), which exhibited an unexpected stratification – a sign that the lunar Apennines were formed by the upraising of an already stratified terrain after a cataclysmic meteoric impact. Thirty-three minutes later the hatch was closed, and the crew took their first rest period. The modifications to the LM now provided for a more comfortable sleep, as the spacesuits, being more flexible than the previous model, could be removed and stored in lockers next to the docking tunnel. Scott and Irwin thus became the first men to sleep in shirtsleeves on the Moon – or, more precisely, in long-johns.

The first walk began fifteen hours after landing, on 31 July. The first to step out, as usual, was the commander, Scott, whose first words on the Moon were closely related to the scientific aspects of the mission: 'Man must explore, and this is exploration at its greatest!' A few minutes later Irwin joined him on the surface, and the two astronauts activated the pulley system to release and unfold the LRV from the side of the descent stage.

In less than an hour the LRV was ready, and they were heading south-west across the mare plain towards the canyon. Scott drove and Irwin navigated, but because they were not sure of precisely where they had landed, and craters made poor navigational references, they simply drove until they encountered the chasm and then followed it south towards their first objective, a small crater named Elbow on account of its location near a 100-degree bend in the canyon. It had been feared that rocks would make progress slow, but the way was clear. Nevertheless, the profusion of very small craters made for a rough ride at anything over 8 km/h.

Upon reaching Elbow, about 3 km from the LM, Scott parked and oriented the high-gain antenna towards the Earth so that the television camera could observe their activities. (The camera could not be operated while the rover was in motion.) After collecting several samples and photographing the far wall of the canyon, they drove up the flank of Hadley Delta towards Saint George, a 2-km-diameter crater embedded in the slope where they hoped to find some rocks excavated by that impact from the massif, which, it was thought, was part of the anorthositic crust. They halted short of the crater when it became apparent that its rim was not littered with large blocks. In fact, the flank of the mountain was remarkably clean. Upon spotting a rock about half a metre across, they stopped to sample it instead. After Scott had chipped a fragment off it, they rolled the rock over in order to collect some regolith from beneath it, so that the length of time that it had been sitting there could be calculated – the regolith beneath the rock had been protected from cosmic rays; it turned out that the rock had been there for 500 million years.

An operational constraint on the use of the LRV was that the astronauts could not drive farther than they would be able to walk back to the LM in the event of a breakdown. On this first test, this limit had been set at 4 km. Rather than retrace their tracks, Scott decided to head back straight across the plain, and rely on the LRV's navigational system to see them safely home. On the way they collected several more samples, including a very vesicular football-sized rock.

Although the astronauts were tired by the time that they got back to the LM, they had still to deploy the ALSEP before they could retire. The ALSEP included both old and new instruments. The third and last laser retro-reflector was deployed on the Moon, as were a new seismometer, a three-axis magnetometer, the solar wind collector, and the suprathermal and cold-cathode ion gauges. The new instruments were a dust detector, an experiment on the degradation of materials in deep space, and two thermal flux probes. These probes caused many of problems, as they had to be inserted into holes drilled to a depth of several metres. Drilling proved to be much more difficult than expected. Despite physical exertion and the fact that his tight-fitting gloves were hurting his fingers (and, indeed, bursting capillaries), Scott was able to drill to depths of 1 and 1.5 metres. The two probes were then lowered inside

In this westward-looking view of the Apollo 15 landing site on the eastern rim of the Imbrium basin, Mount Hadley Delta is to the left with Saint George on its flank, and Rima Hadley cuts across the mouth of the mare-embayed valley, with the crater Elbow at its nearby bend.

the boreholes. Each probe was equipped with four platinum sensors able to measure the temperature to an accuracy of $0°.015$. Unfortunately, the probes were not buried deep enough to be sheltered from the effects of solar irradiation, and the results were suspect. Another aspect of this experiment was the extraction of a core sample from a third drill hole, in order to provide an indication of the material that the thermal probes were embedded in. Although Scott managed to drill down 2.4 metres, it proved extraordinarily difficult to extract the tube. The core revealed that the top 45 cm of the regolith was well mixed, and that below this were as many as 58 distinct layers recording a succession of depositional events.

The astronauts' equipment now included a sort of rake used to sift the regolith to collect only rocks more than a couple of centimetres wide, and an interchangeable-point penetrometer that could be pushed into the ground by hand; the penetrometer measured the force applied by the astronaut, and the depth of penetration. Irwin excavated a short trench and, upon finding a very compacted substrate, used the penetrometer to characterise it.

The second walk began on 1 August, almost twenty-four hours after the first and after some minor problems caused by a leak of the LM's water cooling system. *En route* to Hadley Delta, they passed near the crater Dune – one of the largest of a cluster of craters with a mean diameter of a few hundred metres; they intended to sample it on their return. Beyond Dune, they began to climb the lower flank of the mountain in search of either craters which might have excavated a piece of the massif, or boulders which had rolled down from sites that were beyond the LRV's capabilities. Upon finding the flank virtually clean, they abandoned the easterly search and sampled a small crater. Upon heading back around the other way, they found a large boulder. The slope was so great that when they disembarked the LRV began to slide, so Irwin held onto it while Scott took the samples. The rock was a breccia that was encrusted with a greenish material that turned out to be magnesium-rich volcanic glass. Tests showed this glass to be the frozen spray from a 'fire fountain' which burst through the lunar crust some 3.5 billion years ago. This proved to be a very important find, because the magma that fed the eruption came from the olivine-rich mantle and, because it rose through the crust so rapidly, denying it time to interact chemically, it was virtually pure. A little farther on, they sampled a 100-metre-wide crater named Spur whose rim was littered with rocks – this was the kind of natural 'drill hole' that they had been seeking. Scott collected a small white rock which was sitting on top of a small mound as if inviting the astronaut to collect it. After removing some dust with his hand, he relayed to the ground: 'Guess what we just found!' It was anorthosite, and probably part of the original lunar crust. The 269-gramme sample was given the catalogue number 15415, but became known as Genesis Rock, and has become the most famous lunar sample. Its age was shown, by isotopical analysis, to be 4.5 billion years, so that when it formed, the Solar System was only 100 million years old! Then, some 400 million years later, it was subjected to intense shock by a massive impact.

A regolith sample collected nearby included some 'extralunar' material: a chip of a condritic meteorite. Pressed by mission control, after collecting some more samples the astronauts headed back to the LM. Making a brief stop at Dune on the way,

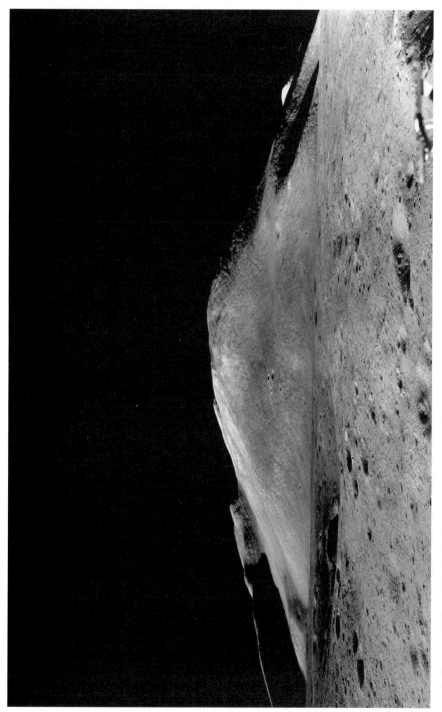

An image of Mount Hadley Delta taken during the stand-up EVA. Silver Spur (to the left) shows stratification. The crater on the right flank of Mount Hadley Delta is Saint George.

A dramatic image, taken from the rim of Saint George, of Irwin working on the LRV over the background of Rima Hadley.

they sampled a stack of boulders on its rim, one of which had vesicles up to 100-mm wide – a sign that the lava that had formed it was rich in gas.

The next day, the highlight of the third walk was a trip to Rima Hadley, 2 km west of the LM. Once there, there was just time enough to take some photographs and to collect some samples before returning to the LM. Although 350 metres deep and 1 km wide, the canyon had no real precipice, the mare just yielded to a progressively steepening slope, with large blocks of lava exposed as the regolith thinned towards the drop-off. The opposite, sunlit wall showed a stratification that implied that the mare was a succession of lava flows. There were large blocks littering the canyon floor, but not so many as to uphold the suggestion that it was a collapsed lava tube; it seemed to be an open channel along which lava flowed from a source in the mountains that form the rim of the Imbrium basin, down into the basin and onto the mare floor. The large vesicular rock that Scott collected on the first EVA was 3.28 billion years old and probably represents one of the last flows to run down the channel.

After returning to the LM, the astronauts prepared to leave. They cancelled a memorial stamp, and Scott took some time to perform his own scientific experiment. He picked a falcon's feather from his pocket, stepped in front of the LRV's camera, and spoke: 'In my left hand I have a feather, in my right hand a hammer. One of the reasons we got here today was because of a gentleman named Galileo a long time ago, who made a rather significant discovery about falling objects in gravity fields. The feather happens to be, appropriately, a falcon feather, for our Falcon, and I'll drop the two of them here and hopefully they'll hit the ground at the same time.' As soon as the two objects hit the ground in unison, Scott added: 'This proves that Mr. Galileo was correct in his findings.'

The LRV was then parked about 90 metres east of the LM, and a small statue, made by the Flemish artist Paul van Hoeydonck was laid on the ground together with a plaque commemorating the fourteen astronauts known to have died during the first ten years of the human exploration of space. Of these, six were Soviet cosmonauts: Yuri Gagarin, who had died in an air crash on 27 March 1968; Pavel Belyayev of Voskhod 2, who had died after undergoing surgery; Vladimir Komarov, Soyuz 1 pilot; and Georgy Dobrovolsky, Vladislav Volkov and Viktor Patsayev, who a month earlier had died of asphyxiation following the depressurisation of their Soyuz 11 spacecraft. The other eight were Americans: the three astronauts of the Apollo fire (Grissom, White and Chaffee), Theodore Freeman, Elliot See, Charles Bassett and Clifton Williams, all of whom had died in aircraft accidents, and Edward Givens, who had died in a car accident. The Soviets had not yet admitted to having a trainee casualty – Valentin Bondarenko – die in a fire in 1961. Another trainee, Grigori Neliubov, had died in 1966 after being expelled from the cosmonaut corp.

Three hours later the ascent stage left the lunar surface, observed by the LRV's camera, and shortly afterwards docked with Endeavour. Falcon was impacted on the surface at 26°.36 N, 0°.25 E, about 120 km west of the landing site. As this was on the other side of the canyon, it was possible to study the underlying terrain. The following two days were dedicated to orbital data collection, and finally, on 4 August, a few hours before heading home, the P&F satellite was released in a

Exposed rocky outcrops on the far side of Rima Hadley. These formations proved to be some of the most enigmatic features ever seen on the Moon, and will probably be an early objective of a robotic or human return to the lunar surface. The canyon is over a kilometre wide and 350 metres deep, but the floor is hidden by the near rim, which is marked by a line of massive boulders.

An impressive Apollo 16 orbital image of the large crater Gassendi (110 km) in Mare Humorum. The crater was a candidate for the landing site of Apollo 17.

103 × 139-km, 28°.5-inclination orbit. The small satellite returned data for six months until 3 February 1972. Other scientific data were returned sporadically until January 1973, at which time the probe was finally shut down.

During the return trip, Worden made a 39-minute spacewalk to retrieve the canisters of photographic film from the SIM bay cameras. The CM's splash down in the Pacific Ocean on 7 August prompted some concern when one of the three main parachutes failed to deploy. The shock of the water impact was rougher than usual, but not particularly dangerous. The quarantine had been abolished, as after two years no astronaut had showed any sign of illness. Apollo 15 had been a record mission: 67 hours on the Moon, seventy-four orbits, three walks lasting a total of 19 hours, 27.9 km travelled on the LRV, and 350 samples – almost 77 kg – collected, almost all of them fully documented.

Later, a small scandal surfaced when it was discovered that the astronauts had carried four hundred envelopes as part of their personal kit. These were stamped with a commemorative Apollo 11 stamp and, once on the ground, with a second stamp commemorating ten years since the first Shepard spaceflight, then cancelled at the *Okinawa* aircraft carrier's post office and sold to a philatelic collector. After discovering that these envelopes were being sold in Germany for 4,850 DM each, NASA had them confiscated, and warned Scott, Irwin and Worden that their actions would be taken into consideration when astronauts were selected for future missions. It is therefore not surprising that none of them flew in space again. Both Scott and Worden have since turned to other business, while Irwin founded a religious humanitarian association and later died of heart attack on 9 August 1991, possibly attributable to the fact that the drinking water tube in his EVA suit had not worked, and he had become severely dehydrated on the very long moonwalks. Endeavour is now on show at the Air Force's museum in Dayton, Ohio.

3.17 NEW THRILLS AND A NEW SUCCESS

The choice of landing site for Apollo 16 was again difficult. A first option was just west of the crater Marius, in a hilly terrain thought to be of volcanic origin; while a second was not far from the Ranger 9 wreck, inside the large crater Alphonsus. The most scientifically interesting possibility, however, was the young crater Tycho, possibly close to Surveyor 7. A landing in this area, however, posed several problems. The site was so far to the south that it would be necessary to use a high-inclination lunar orbit – and in this case the hybrid transfer orbit would stretch the LM's life support capabilities to the limit in case of accident. Finally, a site was chosen not far from the equator, in the Central Highlands near the large crater Descartes*. The site combined a hilly terrain with a smooth plain known as the Cayley Formation, both of which were suspected to be of very old volcanic origin, and presumably not affected by the large impact that created the Imbrium basin 3.8 billion years ago.

The veteran commander was John Young, on his fourth spaceflight, flying with LM pilot Charlie Duke, who was Capcom during some of the most tense moment of the Apollo 11 mission, and Thomas Mattingly, previously dropped from the Apollo 13 crew because it was feared that he might have contracted measles. The eleventh Saturn V, carrying the LM Orion (named after the constellation, and with reference to the astronomical instrumentation carried) and the CSM Casper (after the cartoon ghost), left Pad 39A on 16 April 1972. The flight to the Moon was routine, and the S-IVB impacted at $1°.83$ N, $23°.30$ W, some 150 km north of Apollo 12, having missed the target point by some 200 km.

On 19 April the spacecraft train entered lunar orbit, and the next day, after the SPS engine had been fired to enter the 20-km-perigee descent orbit, the CSM and the LM undocked over the far-side. Once Orion had moved clear Mattingly performed

---

*   René Descartes (or Cartesius, 1596–1650), French philosopher.

some routine checks of the SPS to prepare to return to the circular 111-km orbit in which he would perform his survey work. As the secondary gimbal-control system was activated, the spacecraft began to violently shake. After a second negative check, Mattingly called Orion to ask for advice. As none of the astronauts knew the nature of the problem, it was decided to wait a few minutes for the ground link with Houston to be re-established. The answer of the flight controllers was that the astronauts should bring the two spacecraft close together again and wait for the engineers to determine the nature of the problem, and either fix it or, if necessary, to cancel the landing and redock.

For six long hours, Casper and Orion orbited the Moon just a few metres from each other as simulations were frantically carried out on the ground to understand the anomaly. Meanwhile, lighting conditions over Descartes were slowly changing, with the rising Sun washing away the shadows that would be required to recognise landmarks and estimate the height of the LM in the final phase of the descent. But the verdict finally came. The engine could be fired with the gimbal system in that configuration.

The payload of the CSM was the same as on Apollo 15 – in particular including a second, slightly heavier (39 kg instead of 36 kg) P&F satellite carrying a particle detector and an S-band transponder to map the lunar gravitational field. High-

The Apollo 16 ultraviolet astronomy camera in the shadow of the LM.

resolution photographs were again the most interesting CM observations. Catena Davy was imaged in detail, as were several small artificial craters: those of Ranger 7 and Ranger 9, and the crater produced by the Apollo 14 S-IVB. Another target was the well-known near-side feature Rupes Recta (Straight Wall) south of Davy, which runs for 100 km roughly north–south on Mare Nubium – when viewed from Earth, it appears cliff-like, but was found to be a smooth incline of just 7°. A partial altimetric reconnaissance of the southern hemisphere – particularly the highlands near Descartes – was also carried out.

The descent of Orion toward Descartes was nominal, and the LM finally landed just a few metres from the expected point, at 8.973 S, 15.498 E. Unlike Scott on Apollo 15, Young had no plan to poke his head out of the top of the LM, but he and Duke submitted a comprehensive report on what they could see out of the windows. They had planned to go out straight after landing, but the six-hour delay ruled this out, so they retired to get some sleep.

Fourteen-and-a-half hours after landing, Young was the first to stand on the surface, and was followed by Duke. Young's first words were: 'There you are, our mysterious and unknown Descartes highland plains. Apollo 16 is gonna change your image.' After the collection of the contingency sample, the ALSEP package and the LRV had to be deployed.

The ALSEP was similar to the Apollo 15 instrument suite. The suprathermal ion gauge was replaced by a portable magnetometer, and a grenade launcher – similar to that carried by Apollo 14, but never used – was added. The solar wind collector carried a second platinum sheet. Another experiment, not concerned with lunar science, was a camera and an ultraviolet spectrometer used to observe the sky to study the distribution of hydrogen atoms in the Solar System and in deep space.*

The most important instrument was deemed to be a pair of thermal flux probes like those on Apollo 15, but using a different drill design to bore the holes into which the thermocouples were to be buried. Duke drilled to a depth of 3 metres without any problem, but then Young did not notice that the probe wires were wrapped around his feet and he pulled them out of the central station. The spacesuit did not provide enough visibility, and the wires, in the low-gravity lunar environment, retained their wound-up position and did not flatten out onto the surface. The thermal flux experiment was thus cancelled, and a procedure that might have saved it was not approved, as its implementation would require several hours during the second walk.

The disappointed astronauts then made their first trip on the LRV, a short trip of less than 2 km towards the west to sample the small craters Spook and Flag. While they had been setting up the ALSEP, Young and Duke had remarked upon the fact that the landing site was littered with breccias. This alarmed the geologists, who had predicted that they would find volcanic rocks – not the titanium-rich basalts of the maria, but a more silicic lava flow. They were going to Flag because this 300-metre-wide crater had excavated deep into the Cayley plain, and the rocks on its rim were their sampling objective. Upon finding the rim clean, they sampled

---

*   The back-up Apollo 16 camera was used on Skylab to observe comet Kohoutek around Christmas 1973.

the ejecta from the 40-metre-wide crater Plum, which was adjacent. Here again, only breccias were found. At one point, the geologists monitoring the LRV's television camera requested that the astronauts retrieve a large rock from Plum's rim, because it seemed to glint in the Sun, raising the prospect that it might be crystalline igneous material, but this 12-kg rock – the largest lunar rock ever returned to Earth, which was named Big Muley after the mission's prime geologist, William Muehlberger – turned out to be another breccia. Back at the LM, Young was filmed demonstrating the characteristics of the LRV in the so-called Grand Prix – a series of circuits travelled at maximum speed, with abrupt braking and acceleration to evaluate the grip of the wheels on the regolith.

The first excursion lasted 7 hrs 11 min, during which time the LRV travelled 4,500 metres. After a rest period, the second excursion was carried out towards the ejecta

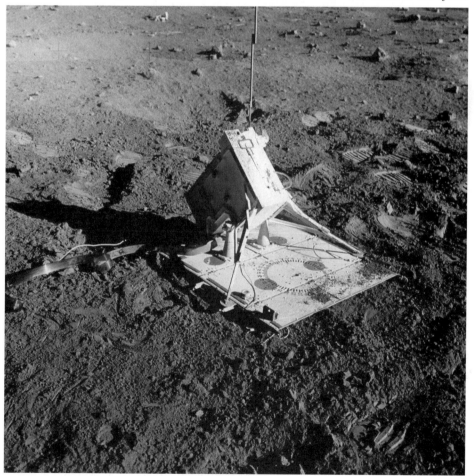

After the Apollo 16 astronauts left the lunar surface, the ALSEP grenade launcher was used to calibrate the seismometer. The launcher was operated from Earth and received commands through its own antenna.

A partial panorama of the Apollo 16 ALSEP, showing the passive seismometer. The
LM is in the background, partly obscured by the rough ground.

of a young crater possessing a system of bright rays, and aptly named South Ray. As
the crater itself could not be reached (orbital pictures and radar data showed it to be
surrounded by very large boulders), the astronauts were to sample accessible parts of
some of its rays. First, however, they drove 175 metres up the terraced north flank of
Stone Mountain, which was a lumpy mound rising 230 metres above the valley floor.
The geologist thought it might be rhyolite, a silica-rich lava. Young and Duke
collected some samples (all breccias), took photographs looking obliquely down on
South Ray less than 5 km away, and collected some rocks that had obviously landed
there after the impact that had excavated the crater two million years earlier. They
had hoped to sample a cluster of craters called Los Cincos (The Five Ones), but as
usual, orientation on the surface proved quite difficult because of an almost total lack
of landmarks. Thus, despite the astronauts approaching as close as 40 metres from
Cincos A, they did not realise that they had done so. A small 18-m-diameter crater
was instead explored. Young collected samples from the side that was potentially less
exposed to South Ray ejecta, hoping to find volcanic rocks, but these were also
breccias, and so friable that many of them crumbled as soon as they were disturbed. A
stir was created when some of the Stone Mountain rocks showed signs of hydration
(rust) when analysed on Earth, but this proved to be a case of contamination by
terrestrial humidity rather than exciting proof of the existence of water on the Moon!
  On the way back to the LM, a small crater was visited in the gap between two rays
of ejecta from South Ray, but here, too, no volcanic rocks were found. The second
excursion was over after 7 hrs 23 min, during which time the LRV travelled 11.3 km.

The third and last excursion began the next day, 23 April 1972. The objective was to drive north about 5 km to sample a 1-km-wide crater named North Ray, which sat on a slight rise just to the west of Smoky Mountain, which was thought to be a rhyolite mound similar to Stone Mountain. Despite pre-mission concern that the terrain might be too rough, they were able to make good progress, drove straight up the terraced approaches and parked right on North Ray's rim. The geologists monitoring the television noted a huge boulder which could also be seen in orbital pictures. Presuming that this had been excavated by the impact, and therefore represented the material on which the adjacent mountain rested, they requested that the astronauts sample it. The huge rock – named House Rock, due to its size – was the largest ever sampled on the Moon, and was a breccia. On their way back to the LM, they retraced their tracks, stopping only once, to explore another large boulder, named Shadow Rock because there was a hollow at its base from which they were able to sample regolith that had been shaded since the rock – a breccia from North Ray – landed some 21 million years ago. Having to compensate for the belated landing, the third excursion was the shortest, lasting just 5 hrs 40 min, during which time 11.4 km were travelled. Young and Duke spent 71 hrs 2 min on the Moon, collected 94.3 kg of samples, and travelled 26.7 km. One of the personal objects which the astronauts left on the Moon was a photograph of Duke's family.

A few hours later, Orion left the surface, and docked with Casper in orbit. More problems followed. A circuit breaker on the ascent stage was inadvertently left open, preventing it from being sent crashing into the Moon for a new active seismology experiment, and the SIM bay's mass spectrometer boom could not be retracted and had to be jettisoned. Finally, due to the SPS problem, the P&F satellite was released in a non-optimal 96×122-km, $10°$-inclination orbit, which severely limited its operational life. The probe, dragged by mascons, quickly decayed, and on 29 May it hit the surface near $10°.16$ N, $111°.9$ E, on the far-side, close to the crater Fleming.*

The two P&F satellites of Apollo 15 and Apollo 16 provided the first detailed data on the lunar magnetic field – and also many surprises. There is no dipole field of a bar magnet, but a 'leopard spotted' one having intensities ranging from ten times greater to a thousand times smaller than the terrestrial field. The study of these magnetic anomalies – called magcons, in analogy with gravitational anomalies – revealed that several of them appear to be linked with strange swirls of unusually bright regolith. Such features were found in Mare Marginis, in Mare Ingenii on the far-side, and close to the far-side craters Fleming and Gerasimovich.† One of the largest magcons was found to coincide with a near-side feature called Reiner-γ** – a strange bright swirl with a distinctive '8'-shape, which on Lunar Orbiter images was found to totally lack surface relief. The most astonishing discovery was that several of these bright swirls seem to be antipodal to large impact basins: Mare Marginis with the antipode of Mare Orientale, Mare Ingenii with Mare Imbrium, and Gerasimovich with Mare Crisium. The theory best suited to explaining the origin of

---

*   Alexander Fleming (1881–1955), British biologist.
†   Boris P. Gerasimovich. (1889–1937), Russian astronomer.
** Vincenzo Reinieri (*d.*1648), Italian astronomer.

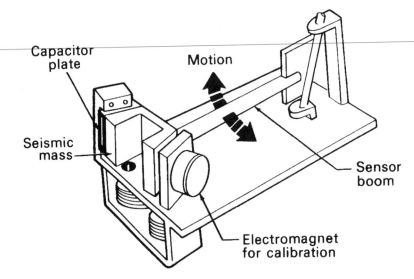

The internal workings of the ALSEP passive seismometer. It measured the displacement
of the seismic mass by the capacitance of a variable capacitor of which the seismic mass
was one of the plates.

these magnetic anomalies suggests that the Moon once possessed a real dipole
magnetic field, and that in the largest impacts, clouds of ionised gas would rise to
push the field back to the impact-site antipode, where the rocks, heated by seismic
waves concentrating there, would become magnetised. An explanation for the bright
swirls is that in these areas the local magnetic fields succeed in deflecting the solar
wind, rich in the hydrogen atoms which are responsible for the chemical reduction of
iron oxides. A more controversial theory notes that if the anomalies are taken to be
the remnants of the ancient dipole field, they point at three different positions for the
magnetic poles – a sign that the magnetic axis suffered periodical migrations billions
of years ago.

As for the geologists' predictions that the Cayley plain and the nearby mountains
were 'highland volcanism', the absence of volcanic rocks and the preponderance of
breccias demanded a rethink, and it was eventually decided that far from it being free
of the influence of the basin-forming impacts, the site was a side effect of such an
event. A comparison with terrain surrounding the Orientale basin indicated that a
splash of 'fluidised ejecta' had formed the smooth plain in the pre-existing hilly
terrain.

After the jettisoning of the satellite, the capricious SPS engine was fired, and the
lunar part of mission – cut short by one day as a safety precaution – was over. At
274,000 km from Earth, Mattingly carried out an 83-minute spacewalk to recover
the SIM camera films. During the spacewalk, a bacterial colony was exposed to the
harsh conditions of deep space as an experiment.

Apollo 16 splashed down in the Pacific Ocean on 27 April 1972. To avoid
repeating the potentially serious problem of Apollo 15, when one of the parachutes

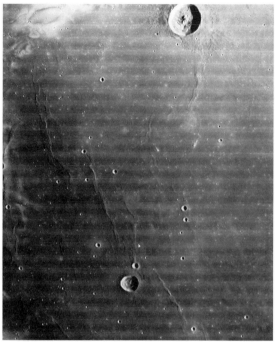

A Lunar Orbiter 4 image of Reiner-γ (the bright swirl at top left). The sub-satellites of Apollo 15 and Apollo 16 revealed that this feature matches one of the largest lunar magnetic anomalies.

did not open, the remaining fuel was not dumped overboard during the atmospheric descent. But the cure turned out to be worse than the problem. When the CM was returned to North American Aviation for post-flight inspection, the pressurised tanks exploded and injured several engineers. Casper is now on show at the Space and Rocket Center in Huntsville, Alabama.

The mission of Apollo 16 was important from the medical point of view. To avoid heart arrythmias recorded in all of the previous flights, a diet featuring large quantities of orange juice was prescribed for the astronauts. This crew was thus the first not to suffer from health problems during the flight – but the price was severe flatulence!

Duke was the only one not to fly again, and he began a new life as a Christian minister. Young became the first man to fly six times in space, on two Gemini missions (Gemini 3 and Gemini 10), two Apollo missions (Apollo 10 and Apollo 16), and two Space Shuttle mission (the first orbital flight in 1981 and the first Spacelab flight in 1983). He was to have commanded the Shuttle mission to carry the Hubble Space Telescope into orbit, but circumstances changed due to the *Challenger* accident. He is still a NASA employee. Mattingly piloted two Shuttle mission (one in 1982, and a secret military mission in 1985).

Forty grammes of regolith collected by the Apollo 16 astronauts was used in 1986 for the most unusual experiment ever carried out on lunar samples. T.D. Lin of Chiao Tung University in Taiwan prepared a small cube of concrete by mixing lunar sand with terrestrial water. According to Lin, regolith could be used to build bases on the Moon, as the cost for the transportation of construction material from Earth would otherwise be prohibitive.

3.18 THE LAST MEN ON THE MOON

On 13 August 1971 – a few days after the splash-down of Apollo 15 – the scientific community succeeded in obtaining what it had been seeking for more than two years. When the crew of Apollo 17 was named, it included, for the first time, a professional scientist: Harrison 'Jack' Schmitt, a geology graduate from Caltech with a PhD from Harvard. It was standard practice for the back-up crew for one mission to skip two missions and then fly the one after that. Schmitt would have 'rotated' from the Apollo 15 back-up crew to Apollo 18, but Apollo 18 had been cancelled, so, in order to fly him on the final mission, he was assigned the place that would otherwise have gone to Joe Engle (who was later to pilot the Space Shuttle on two of its five atmospheric test flights of 1977 and go on to fly two orbital missions). Schmitt became an astronaut during the first selection of scientist-astronauts in 1965. The main difference between this and the previous selections was that graduation in a scientific discipline, rather than extensive flying experience, was required. About a thousand candidates had been screened, but only six passed all of the tests, and of these, four eventually flew in space. Other than Schmitt, Owen Garriott flew both on the Skylab and on the Shuttle, while Joseph Kerwin and Ed Gibson flew on Skylab. The six scientist-astronauts included two physicians, three physicists or astronomers, and a single geologist. Their first assignment was fifty-five weeks at Williams AFB in Arizona, to learn to fly high-performance aircraft. It was expected that the entire crew of the LM would be able to fly it, if necessary. (The Space Shuttle 'Mission Specialists', who are not required to know how to fly the spacecraft – were still far in the future.) Schmitt's performance confirmed NASA's opinion that with adequate training anyone could fly the LM.

The crew of Apollo 17 consisted of commander Eugene Cernan, the Apollo 10 LM pilot, and the third and last man to fly twice to the Moon, Ronald Evans as the CM pilot, and Schmitt as the LM pilot. It was decided that it would be pointless to train astronauts for a mission which they would probably never carry out, so the back-up crew was comprised entirely of veterans of previous flights: Fred Haise was the commander, Worden was CM pilot, and Irwin was LM pilot.

The choice of landing site was relatively easy. After a preliminary selection, there remained a narrow valley in the Taurus Mountains, on the south-eastern rim of the Serenitatis basin, near the crater Littrow* where Worden observed what appeared to be volcanic cinder cones, the crater Gassendi** and, as a distant third, the crater Alphonsus. Gassendi was discarded, as it could present some landing problems. Alphonsus was rejected because it had less to offer than the Taurus–Littrow site. Although landing in a narrow valley surrounded by mountains would have been

---

*   Johann J.V. Littrow (1781–1840), Austrian astronomer.
** Pierre Gassendi (1592–1655), French philosopher and astronomer.

History's most reproduced image was taken by Harrison Schmitt after Apollo 17 translunar injection.

thought impossible early in the programme, every mission since Apollo 12 had landed with an accuracy of better than 250 metres, and the worst case – Apollo 11 – had been only a few kilometres, so for this final mission the landing ellipse was shrunk to suit the target. Nevertheless, if Cernan found himself coming in seriously off-track, he would have to abort.

The TLI procedure had to be modified. A winter launch to Taurus–Littrow would require S-IVB ignition over the Atlantic instead of the Pacific Ocean, and this in its turn would necessitate a night-time launch, which would create serious problems for the recovery teams in the event of an abort. Based on the experience of eight perfect lunar launches, however, the night launch was accepted.

On 6 December 1972, everything was ready for lift-off in front of a crowd of VIPs, including top politicians and old and new project engineers. That evening, the countdown had reached T–30 seconds when the event sequencer discovered that the third-stage tanks had not been pressurised. This inconvenience solved, Apollo 17 left Pad 39A just after local midnight, some 2 hrs 40 min late, at 05.33 GMT on 7 December. The launch illuminated the Florida sky, the trail of fire was visible 1,000 km away in North Carolina, and the water around the Cape became alive with hundreds of fish that were drawn to the most intense fishing lamp ever made. The launch was perfect, and a few hours later the LM was extracted from the S-IVB, which then headed for the Moon, where its impact at 4°.33 S, 12°.37 W, about 150 km east of Apollo 14, had the energy of 11 tonnes of TNT, and was recorded by all four ALSEPs.

Shortly after the extraction of the LM, Schmitt took the most famous photograph of the entire Apollo programme, and surely one of the most reproduced in the history of photography. It shows the 'full' terrestrial disk. At the centre, Africa is plainly visible, as is Madagascar and the Arabian peninsula. At the bottom, Antarctica, snow white, shows between the white clouds, and at the top the Mediterranean Sea is visible at the limb. All around the Earth there is only the darkness of space. This picture (which many have seen so many times that it no longer seems significant) remains a fitting symbol of the Apollo programme, for although it does not provide us with new scientific certainties or a new frontier to colonise, it stirs an awareness of our planet by reminding us that all of our history, our lives, our passions, and the people we love, are confined to that blue-white ball in the void. It has also been argued (probably correctly) that the environmental conscience born in those years was due to photographs such as this.

After a single course correction, the spacecraft entered lunar orbit on 10 December. The next day, the SPS engine was fired to adopt the descent orbit with a 27-km perigee, and the LM Challenger separated from the CSM (on this mission named America) to head towards the valley in the Taurus–Littrow region. Half an hour later, Cernan and Schmitt cleared the peaks of some 2,200-m-high mountains and, after a perfect descent, landed at 20.18809 N, 30.77475 E, less than 200 metres from the target point. This was the farthest point from the geometric centre of the near-side ever reached by an Apollo landing.

Evans, in the meantime, had taken the CSM back to an orbit between 100 km and 129 km, and had begun his own programme of observations. The scientific payload was quite different from that of the two previous missions. The P&F satellite, as well as the X-ray, alpha-ray and gamma-ray spectrometers had been left off, and were replaced by a far-ultraviolet spectrometer to investigate the presence of hydrogen atoms near the Moon, an infrared radiometer to measure the surface temperature and, indirectly, the depth of the regolith, and an antenna for a bistatic radar (that is, with the transmitter in lunar orbit and the receiver on Earth) to directly measure the depth of the regolith, and for lunar altimetry. One of the instruments that was not replaced was the topographic camera. This time, to choose the features to be imaged, pictures taken by the Soviet Zond 8 probe were used.* Onboard America were also five mice (*perognathus longimembris*), to be used for an experiment on the effects of radiation on the nervous system.

Four hours after the landing, the first walk began. Cernan was on the surface first, followed by an excited Schmitt – the first scientist to practice his own profession on another celestial body. The significance of having a geologist on scene became clear as soon as Schmitt, even before descending the ladder, described in detail, and with an insight beyond his predecessors, both the large-scale details, the panorama, and the small-scale details. On this occasion, mission planners had become so confident that the collection of a contingency sample was deleted.

The ALSEP, deployed during the first hours of the walk, was quite different from previous ALSEPs, and neither a solar wind collector nor a seismometer were carried. The only instruments not replaced were a grenade launcher like those of Apollo 14 and Apollo 16, and the heat-flux probes, which were to be buried at a depth of 2.5 metres. Among the new instruments were two gravimeters, one on the LRV to measure the local gravity at each sampling site, to investigate the structure of the valley floor, and the other fixed for a novel experiment to try to record – for the first time ever – gravitational waves (an exotic type of wave predicted by the theory of relativity, but so far never detected), a lunar regolith electrical characteristics sensor, and a neutron probe – a 235-cm-long probe to be temporarily inserted into the hole left by core-sampling and then retrieved before lift off to study the effects of cosmic rays on the regolith.

The three most interesting instruments were the ejecta and meteoroid detector, the seismic profiling experiment, and the atmospheric composition experiment. The first detected meteorites and measured their masses, directions and speeds, to compute the contribution to the regolith by extralunar particles. The second used a set of geophones to record seismic waves propagating through the valley floor from explosive charges distributed by the astronauts and detonated following lift-off. This would provide a measure of primary and secondary wave speeds through the material, to investigate the stratigraphy to a depth of up to 2 km. The third experiment provided one of the biggest surprises of the whole Apollo programme by revealing that the Moon produces a very thin atmosphere – comprised mostly of helium and argon – over its sunlit hemisphere, with a total mass of some 30,000 kg. If this atmosphere were to have the same density as sea-level terrestrial atmosphere, it

---

*   This was the beginning of space cooperation with the Soviets, which two-and-a-half years later would culminate with the Apollo–Soyuz Test Project.

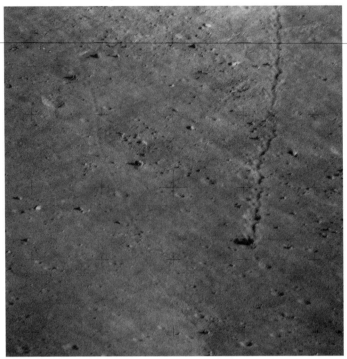

The slopes of North Massif, showing two boulders which rolled downhill.

would fit inside a cube with sides of 30 metres! The most interesting observations were those of the behaviour of argon during a whole lunar day. Concentration plots revealed two maxima at sunrise, and two minima – an initial shallow minimum, almost flat during the day, and a very deep and steep minimum during the night, shortly before dawn. It appears that the lunar terminator is swept by a thin argon breeze before the heat of the Sun initiates rock outgassing. For more than ten years the observations of the Apollo 17 instrument (which failed after just nine months) were the only detailed observations of the lunar atmosphere, although other components – including sodium and potassium – were discovered by terrestrial observers in the 1980s.

After deploying the ALSEP, Cernan and Schmitt boarded the LRV and made a short trip, stopping for half an hour near the small crater Steno, one of the Central Cluster group of craters. The only problem was that as Cernan had been preparing the LRV, the geological hammer in his shin pocket had snapped off one of the rear fenders. During the trip, the wheel threw up a 'rooster tail' of fine dust, much of which fell on the vehicle, coating it and its crew. There was some concern that the dust might damage the hermetic seals of the spacesuits. After 7 hrs 12 min, the first of three walks ended. Overnight, a makeshift fender was fashioned from a stiff map, and it was taped on first thing the following day.

The second walk was aimed at exploring the slopes of the South Massif. Orbital images showed the paths left by some boulders which had slid down. If it proved possible to sample those boulders, the stratigraphic history of the mountains could be studied. The trip on the LRV lasted a full 73 minutes, during which time 8 km was covered. Cernan and Schmitt reached the foot of the massif by travelling on a brighter terrain which was probably material brought down by a land-slide. For more than an hour, samples were collected from the base of the massif. It was discovered that the grey-blue boulders had originated higher on the mountain, and the grey-brown boulders lower down. All of the samples proved to be breccias, but many of the inclusions were crystalline. Schmitt's attention was caught by a bright clast in one boulder, which proved to be pure olivine, and one of the oldest rocks collected on the Moon, its age being only slightly less than Genesis Rock.

A brief stop followed near the small crater Lara (named after the character in Pasternak's *Doctor Zhivago*) located at the base of a 75-metre-high scarp that ran across the floor of the valley, and then an unplanned stop was made to collect some bright material noted by Schmitt. This material is believed to belong to one of the very bright rays that cross the near-side, originating from Tycho – the youngest of the large craters on the Moon. An analysis of the sample dated this at 'only' 109 millions years, when Earth was mid-way through the Cretaceous period, and when dinosaurs ruled the planet!

The next objective was to be a highlight, and so it proved to be. When Worden had inspected this site from orbit during Apollo 15, he had reported a crater on the valley floor which had a distinctive dark 'halo'. The geologists suspected it of being a volcanic cinder cone. Cernan and Schmitt were to sample it. After parking by a large boulder on the rim, Cernan oriented the antenna toward the Earth to provide a television broadcast, while Schmitt took panoramic images of the 110-m-diameter crater. Suddenly, the attention of the geologist was captured by something: 'Oh, hey, wait a minute . . . There is *orange* soil.' Cernan soon took a look for himself, and confirmed: 'He's not going out of his mind, it really is!' It was a small patch of an unmistakeable orange colour. 'How can there be orange soil on the Moon?' Cernan wondered aloud, and then provided his own answer: 'It's been oxidised!'

'It *looks* like an oxidised desert soil,' Schmitt agreed. 'If ever there was . . .' he began, but then stopped and laughed. 'I'm not going to say it.' But then he laughed again and finished his interpretive remark: 'If ever there was something that looked like fumerolic alteration, this is it.' It appeared that the photogeologists had been correct, this was a volcanic crater. If the orange hue was due to iron oxide – rust – then there must have been water vapour seeping from deep within the Moon. Just as Scott and Irwin had 'found what they were looking for' in the form of the small piece of anorthosite on Spur's rim, Cernan and Schmitt had found what they had been sent to find here on Shorty's rim. They scraped away the surficial material to trace the extent of the exposure, which proved to be a streak radiating out from the crater, took samples of both the orange material and the adjacent material, and then took a core sample whose colour changed with depth from orange through purple to 'the blackest black' Schmitt had ever seen. As they were packing up to leave, Cernan noticed more orange material exposed inside the crater. But subsequent analysis revealed that the 'orange soil' was not oxidised, it was composed of tiny glass spherules. It was volcanic glass, but Shorty was not a cinder cone, it was an impact crater. The glass was laid down by a 'fire fountain' 3.5 billion years ago and then

Harrison Schmitt unpacks the LRV on the rim of Shorty. Soon after this photograph was taken, orange soil was found a few metres in front of the boulder on the rim behind the rover.

covered by a lava flow, through which Shorty had punched a hole some 30 million years ago. The glass beads were orange because of the iron and titanium ratios (just as the glass encrusting the boulder sampled by Apollo 15 was green due to the high magnesium content). Upon emerging from the fire fountain, the ilmenite-rich magma droplets had crystallised to form the jet-black material if they had had time to cool slowly, and vitrified (to orange glass) if they had cooled rapidly.

After a brief stop to sample large boulders on the rim of 600-metre-wide Camelot, Cernan and Schmitt headed home to the LM, which was pressurised after an EVA lasting 7 hrs 37 min.

The third and last walk took place on the morning of 14 December. This time the targets were other mountains: the North Massif and the Sculptured Hills to the east. The first stop was some 4 km from the LM, where orbital pictures showed a large boulder, the tracks of which could be followed on the North Massif for hundreds of metres. The boulder was the primary target for the exploration and Schmitt was given all the time that he needed to analyse it. It was a huge rock more than 10 metres long, and had fractured into five parts as it came to rest some 800,000 years ago. It was a rock with a complex history, and Schmitt had to inspect its entire length before he could see the 'big picture'. The southern end had a texture that looked to Schmitt to be either a recrystallised breccia or an anorthositic magma that had captured a lot of inclusions which were breccias. The northern end was a blue-grey rock similar to those they had seen at the base of the South Massif. Whereas the southern end was extremely vesicular, this end was free of vesicles. At this point, he realised that the two rocks had been welded together by extreme heat, and he set off in search of the 'contact'. Upon locating it, he was amazed to find that it was 1 metre wide, and there were fragments of the blue-grey rock within the suture zone. Later analysis of the samples that they collected indicated that the blue-grey rock was a breccia, and the other end was an impact melt within which the breccia, massive though it was in its own right, was merely an inclusion. The melt, which resembled a magma except for the fact that it had so much material caught up within it, could only have been formed by a vast impact, and its location on a mountain on the rim of the Serenitatis basin suggested that it had been *that* impact. Cernan took some panoramic photographs as Schmitt collected samples, in the process taking another of the most famous Apollo images, showing the East Massif, the valley, the northern end of the boulder, and Schmitt. A recess in the rock can be seen where regolith is forming, as well as the traces left in it by the astronauts' scoops.

The next stop was again on the slopes of the massif, where a slightly smaller boulder was sampled. This, too, proved to be a large breccia. A characteristic of this boulder was a streak of dark material across a side which indicated that it, too, had once been very hot, and a vein of molten rock had been forced into a crack. There then followed an exploration of the flank of one of the Sculptured Hills. Here the astronauts relaxed, and spent a little time kicking rocks. The last stop was near a small crater called Van Serg, which the geologists had suspected of being another volcanic cinder cone, but turned out to be an impact crater, the shock of which had compressed the thicker than usual regolith to form loosely consolidated breccias – so called 'instant rock' – in abundance.

Another classic Apollo image: Schmitt and the Station 6 rock.

After the return to the LM, the time came to store the last samples, including an unplanned core collected by Schmitt near the LM after he found an unused sample tube. Finally, some ceremonies were carried out. Schmitt collected a rock – sample 70017, the 'Goodwill Rock' – which was to be cut up and given to the students of seventy countries, and Cernan read the plaque attached to Challenger's front leg showing the globe of Earth, the globe of the Moon, the sites of the six Apollo landings, and the words: 'Here man completed his first explorations of the Moon, December 1972 AD. May the spirit of peace in which we came be reflected in the lives of all mankind.' After storing the last 62 kg of samples – of a total of 110.4 kg collected during the mission, some 18 kg more than expected – Schmitt re-entered the LM and Cernan, preparing to do likewise, said: 'I believe history will record that America's challenge of today has forged man's destiny of tomorrow ... God speed the crew of Apollo 17.' A few hours later, at 05.54.37 GMT on 14 December 1972, Challenger's lift-off was recorded by the LRV's camera, which, for the first time in documenting such an event, was rotated upward to follow the ascent until the LM disappeared from sight.

The LM's ascent stage was crashed into the South Massif, a few kilometres from the descent stage, in order to calibrate the seismic profiling experiment. The fall should have been visible to the LRV's camera, but the impact, at 19°.96 N, 30°.5 E, took place 10 km from the camera, and was not recorded. The experiment provided some extremely interesting data. It showed that the rock of which the massif was an outcrop was also at the floor of the valley, 1 km below the surface. In this rock, seismic waves travelled at almost 4,000 m/s. Over this first layer, up to a depth of 150 metres, was a second layer probably made of less compact basalt, inside which waves

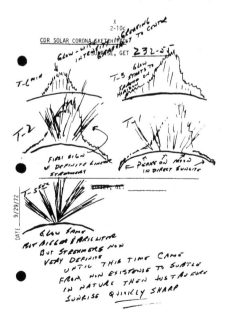

The Apollo 17 observations of streamers at sunrise, probably caused by suspended dust.

The crater Tsiolkovsky, with its mare-like floor, is one of the most intriguing geological features on the Moon, and was to be the target of an ambitious Apollo landing on the far-side. This photograph was taken during the Apollo 15 mission.

travelled at a speed of 1 km/s. This picture was completed by two layers of regolith, the deeper of which was up to 150 metres thick, and the other only a few tens of metres thick, in which the wave speed was just 270 m/s. The following day the explosive charges that had been distributed around the valley during the three walks, were detonated. This provided a new measure of regolith profiles. The explosions of the nearest charges was observed by the camera on the LRV.

During two days of orbital work, one of the most mysterious lunar phenomena was recorded. The astronauts were observing sunrise and making sketches of the appearance of the corona when, two minutes before dawn, some streamers appeared, resembling those that appear through breaks in clouds. These streamers, with time, changed in shape and intensity, and became extremely obvious a few seconds before dawn. The swiftness of the changes clearly proved that it was a local phenomenon, probably caused by forward-scattering of sunlight by very fine dust, possibly suspended by the strong electric fields associated with sunrise.

At 23.35 on 16 December, the SPS engine was fired for the last time, and Apollo 17 headed home. During the return trip, 290,000 km from Earth, Evans made the usual spacewalk to retrieve films and other experiments, including the five mice, one of which had died. In a telephone call relayed to the spacecraft, President Nixon congratulated the astronauts on their successful mission and, prophetically as it turned out, ventured that it would be many years before human beings returned to the Moon. The mood at splash-down on 19 December 1972 in the Pacific Ocean, 300 km from Pago Pago, was therefore a mixture of joy at a successful mission and sorrow that the programme was over.

None of the three astronauts returned to space. Cernan is now an active lecturer on space, Evans retired to manage a real estate business and died of a heart attack on 7 April 1990, and Schmitt was a US Senator for New Mexico between 1976 and 1982 and is now a space consultant. The CM America is on show at the Johnson Space Center in Houston.

3.19 THE CANCELLED MISSIONS

Human exploration of the Moon was doomed as soon as it began. Two months after the Apollo 11 mission, NASA made public its plan for the future. First, the Apollo Applications Program (AAP) would place a 100-tonne scientific laboratory – Skylab – in low Earth orbit. Then, from 1974, the Space Shuttle would begin flying, assuring (so it was thought) much cheaper access to space for humans and hardware. After that, in 1985, NASA envisaged a fifty-crew-member space station, a lunar base served by an atomic-powered shuttle in the style of *2001: a Space Odyssey*, and the first human outpost on Mars. However, the Nixon Administration – which had the luxury of being in office when Apollo 11 achieved Kennedy's challenge, without having taken the risks of either proposing it or assuring funding – decided not to initiate such a risky and expensive project, particularly as the Vietnam war was still absorbing a tremendous amount of the public purse and because several new highly technological military programmes were being proposed. NASA's great adventure was therefore rejected. As partial compensation, however, lunar missions up to Apollo 20 were still scheduled and funded, but NASA needed a Saturn V to launch its Skylab space station, and, having no funds to build a new rocket, had to use one that was already built and paid for, and therefore on 4 January 1970, as the scientists were gathering in Houston for their first 'rockfest' to discuss their Apollo 11 results, NASA cancelled Apollo 20.

On 2 September 1970, NASA cancelled two more missions, saving, depending upon estimates, from as little as $20 million each (which Harold Urey derided as 'chickenfeed') up to $450 million each, equivalent to the cost of a week or so of war in Vietnam. Apollo 15 was to have been the final H mission. This was cancelled, and recast as the first J mission. Having made this final cut, Thomas Paine resigned as NASA's administrator a fortnight later.

The development of a small flying platform to enable the astronauts to aerially explore the surface near the LM was interrupted, and the Mobile Laboratory (MoLab) project – a family of pressurised wheeled vehicles to be carried by an unmanned Saturn V, to explore the surface with much greater autonomy than was possible using the LRV – was cancelled. (The mission over, the MoLab was to remotely drive to the next landing site, collecting interesting samples on the way.) Even the simple and relatively cheap modification of making the LRV autonomous in order to remotely drive it after the astronauts had left, was never implemented.

Ultimately only six out of thirty-two sites selected between 1967 and 1969 were visited by humans. Other very interesting sites – including Tycho, Aristarchus, Copernicus, Alphonsus, Davy Catena, and the Marius hills – were to remain unexplored. An extremely ambitious mission, both from a scientific and technological point of view, was a landing on the far-side – preferably inside the crater Tsiolkovsky. This project's greatest advocate was Harrison Schmitt. The greatest problem was communications, as the presence of the Moon between the astronauts and the Earth would block direct transmissions. The solution, found by experts in celestial mechanics, was simple and elegant: the colinear Lagrangian point 64,500 km beyond the Moon where the gravitational attraction of the Earth and Moon are equal. A telecommunications satellite in a 'halo orbit' around this point would be visible from both the Earth and the lunar far-side, and so be able to act as a relay. However, corrective manoeuvres would be required every few months to maintain a satellite in such an instable orbit. Schmitt's enthusiasm led him to propose the modification of two unused Tiros meteorological satellites, which could have been launched together on a Titan rocket. But this time technological and operational risks were judged to be too high, and a landing on the far-side was never approved.

3.20 EPILOGUE

The Apollo programme over, hardware developed for it was used for several years more. The last Saturn V was launched on 14 May 1973 to place a modified and habitable S-IVB stage into orbit, to become America's first and only space station: Skylab. Between May 1973 and February 1974, three crews visited the station, riding in Apollo capsules launched by Saturn IBs. There were tentative plans to have it refurbished by the Space Shuttle, but it re-entered the atmosphere in 1979, before the much-delayed Shuttle was ready, and disintegrated over Australia.

The last two unused Saturn Vs are now tourist attractions at the Kennedy Space Center in Florida and at the Johnson Space Center in Houston. The last Apollo capsule – Apollo 18 – was launched on the last Saturn IB on 15 July 1975. On board were Thomas Stafford, Vance Brand and Donald K. Slayton, a veteran astronaut chosen in 1959 for the Mercury flights.* A few hours before the beginning of the Apollo 18 mission, Soyuz 19, carrying Alexei Leonov and Valeri Kubasov, had lifted

---

* Because of a slight heart defect Slayton had never flown, and was for several years responsible for nominating the Apollo crews.

off from Tyuratam. Later, using an adapter carried on the S-IVB where the LM had previously been, the two spacecraft docked for several days of joint activities. At that time, few people were aware that the Soviets had *two* crews in space, as Soyuz 18B was about to end its two-month stay on the Salyut 4 space station.

The Saturn family of launch vehicles proved to be the safest ever flown. Ten Saturn Is, nine Saturn IBs and thirteen Saturn Vs were launched without a single loss. Over the ensuing years there were several proposals to reuse some of the technology developed for the Apollo programme, such as the J-2 engines, but none were ever accepted.* One of the most recent proposals was to build new Apollo capsules be used as provisional six-place lifeboats for the International Space Station, instead of the Russian-built Soyuz capsules, as a stop-gap until a small winged craft can be introduced as a crew-transfer vehicle, but this was not pursued.

The saddest note concerning the Apollo heritage is that despite the effort involved in their installation, on 30 September 1977, upon being denied even a small amount of funding to continue receiving and processing data, NASA had to switch off the remaining ALSEPs.†

Of all of the instruments on the Moon, only the laser retro-reflectors are still being used. During the first 25 years of operation those left by Apollo 11, Apollo 14 and Apollo 15, and the Soviet counterpart on Lunokhod 2, have been used for more than 8,300 distance measurements by the McDonald Observatory in Texas, Mount Haleakala in Hawaii, Pic du Midi Observatory in France, Lick Observatory in California, the Crimean Observatory, and from sites in Australia and Germany. Measures are usually taken by mounting a laser at the telescope's focus and pointing it towards area of the Moon where the mirror is known to lie. The coherence of the beam is such that on the lunar surface, 380,000 km away, it has a diameter of 6 km, forming a cone with a vertex angle of around 3 arcsec. Complex filtering operations are then performed on the light coming from the Moon to detect a returning photon every ten seconds or so. By measuring the time difference the exact distance of the Moon can be established. The first experiments, performed in 1969, yielded an accuracy of several metres, but current measurements have an accuracy of only a few centimetres. Among several discoveries resulting from these experiments are microvariations of the length of the terrestrial day caused by atmospheric and oceanic motions, and the rate at which the Moon is receding from Earth: 3.8 cm per year. Laser measurements have enabled a better formulation of the Moon's extremely complex motion, extending the computation of ephemerides and identification of eclipses as far back as 1400 BC. It has also been determined that the whole lunar surface is vibrating at a very low frequency. This could be evidence of a recent large asteroidal impact, possibly confirming the theory according to which the crater Giordano Bruno formed in recent times, but it is also explicable in

---

* It is worth noting that all of the technical drawings of the Apollo project are preserved, although, at Nixon's insistence, the tooling was disposed of. Saturn V blueprints are stored on microfilm at the Marshall Space Flight Center, and other documents take up more than 80 m³ of space in the US federal archives.

† During the first 924 days of operation they had recorded a total of 815 meteoritic impacts – mostly during the months of November and December 1974 and June 1975.

terms of turbulent motion of magma at the boundary between the small core and the mantle.

About 74% of the 380 kg of lunar samples returned by the Apollo crews is currently stored inside the Lunar Sample Vault in Houston, and another 14% is in a special storage area in San Antonio. The remaining 12% is on show in museums all around the world, or has been destroyed in experiments or in rock cutting. Every sample, however small, has received at least a quick partial analysis, with the exception of two cores collected by Apollo 16 and Apollo 17, which remain sealed awaiting analysis by a future generation of instruments for which pristine samples would be valuable. Other samples – such as the Apollo 15 Genesis Rock – have been completely cut up, and survive only as small fragments. It must be added that analysis of lunar regolith samples has enabled the identification of a class of meteorites in terrestrial collections which have a composition that proves that these rocks were hurled from the Moon by a large impact, thus realising, by natural methods, what JPL Director Pickering once proposed to do using atom bombs! To this day (March 2003) twenty-four of these 'zero cost lunar samples' have been identified.

When Apollo 11 set off for the Moon, there were three theories for how the Moon formed: 1) that it split away from the Earth when our planet was new and still molten, 2) that it formed in the same part of the Solar System at the same time as the Earth, and was captured by it early on, and 3) that it formed elsewhere in the Solar System, and was captured much later. To general consternation, the Apollo samples ruled out *all three* theories! It was not until 1986 that a concensus was arrived at in which the Moon formed from debris left over from the collision of a Mars-sized body with the Earth very soon after the Earth formed from the solar nebula.

4

How the Soviet Union lost the race

4.1 THE FIRST LUNAR FLY-BY PROJECTS

At the end of the 1950s, with the approach of the flight tests of the Zenit spy satellites and Vostok spacecraft designed to take cosmonauts in orbit around the Earth, Sergei Korolyov began to consider the next steps in space exploration. He therefore began a study into the feasibility of a piloted lunar fly-by launched by new versions of 8K71, possibly incorporating some form of in-orbit assembly. This project eventually included the launch of at least five spacecraft of three different models. First, the 9K spacecraft – an empty rocket stage – would rendezvous with three or four 11K 'tanker' satellites (possibly guided by a cosmonaut piloting a Vostok used as a space tug), which would then empty their tanks into the 9K. The last launch would be that of the two-crew 7K spacecraft, designed to dock with 9K and, under its thrust, to circumnavigate the Moon. The project was extremely ambitious, as it required a succession of orbital encounters, a little known and potentially risky manoeuvre, and the untested transfer of liquid fuel from one craft to another in weightlessness.* The main problem with this lunar project, however, was that it did not interest the military, which had total control of the Soviet space programme. Korolyov therefore proposed two more versions of the 7K – the first being a piloted spy satellite, and the second an orbital interceptor carrying a cannon with which to eliminate US spy satellites. After studying spacecraft configurations with aerodynamic lift which could fly, Korolyov – whose team had shortly before absorbed the engineers of several recently-closed aeronautical design bureaux – opted for a ballistic capsule with an unusual 'headlight' shape, a semi-lifting solution with a lift-over-drag ratio of 0.3 (analogous to the Apollo capsule).

To test the ability of the 8K71 to carry out rendezvous in space, Korolyov ordered that Vostok 3 and Vostok 4, and later Vostok 5 and Vostok 6, be launched just a few

---

* The first fuel transfer between two spacecraft was not attempted until 1978 between the Soviet cargo vehicle Progress 1 and the Salyut 6 space station.

days apart. Vostok 4 entered orbit only 6.5 km from Vostok 3; and Vostok 6 entered orbit less than 5 km from Vostok 5. As this type of vehicle had no manoeuvering capability, a rendezvous was impracticable. Nevertheless, these missions demonstrated the feasibility of the orbital encounter between the components of the lunar train.

At the same time, however, the foundations were laid for the eventual demise of Soviet piloted lunar projects. On 13 May 1961, the rival team led by Vladimir Nikolaevich Chelomei (which included the son of Communist Party Secretary General Nikita Khrushchev) was authorised to adapt an extremely powerful ICBM into a heavy UR-500 (Universalnaya Raketa – Universal Rocket) launcher – later renamed Proton – in order to send an LK-1 (Lunnyi Karabl – Lunar Ship), carrying a single cosmonaut, to circumnavigate the Moon.* That same year, Korolyov was authorised to develop a family of rockets, simply called N (Nossitel – Launcher), more powerful than the Proton, to place payloads of 40–80 tonnes in Earth orbit. In 1962, the configuration of the new launcher was finally determined. It was to be a monoblock rocket, without boosters or auxiliary rockets, just like the Saturn family. Korolyov hoped to use the largest of these rockets to land a Soviet cosmonaut on the Moon in the second half of the decade. Once again, however, the military was not interested in such a stunt, and for a long time the N family was left without a payload.

A mock-up of the Soviet 7K Soyuz spacecraft.

---

*   The initial idea was to use the UR-500 to deliver to the heart of the United States the largest nuclear bomb ever built, weighing 25 tons and yielding 58 MT – jokingly named by Soviet scientists, 'the zarine of nuclear bombs', in keeping with the Russian tradition of gargantuan and useless weapons.

A dispute over the engines for the N family of launchers led to a bitter split between Korolyov and Valentin Glushko which once again emphasised the rivalry between the Soviet design bureaux. The latter advocated the use of corrosive and toxic combinations such as hydrazine and nitrogen tetroxide or even fluorine and hydrogen, while the former preferred the traditional combination of liquid oxygen and kerosene, although at some time in the future he hoped to use liquid oxygen and hydrogen for upper stages, as the Americans were doing. Korolyov's resentment soon surfaced, as he remained convinced that during the 1930s Glushko had plotted to have him imprisoned. In fact, he had never been satisfied with Glushko's engines, even those mounted on the 8K71. Seeking a complete change, he turned to Nikolai N. Kuznetsov, an aircraft engine designer.

In parallel, the design teams of Chelomei and of the Ukrainian Mikhail Yangel proposed their rivals to the N launchers: the UR-700 and the R-56, respectively, each of a configuration similar to the 8K71. The former was an uprated version of the UR-500 including a cryogenic upper stage, designed to place up to 130 tonnes into low Earth orbit or to send an LK-700 direct landing capsule to the Moon. The latter was the only Soviet proposal to include rocket engines as powerful as the American F-1.

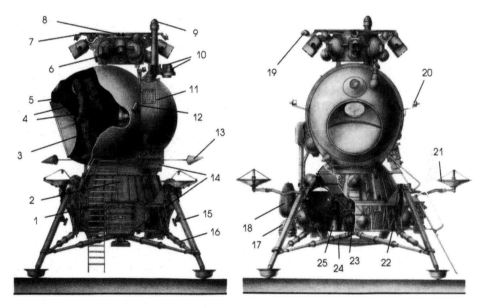

The Soviet LK lunar lander: 1, descent stage; 2, stage E; 3, cabin for one cosmonaut; 4, life support system; 5, system for landing observation; 6, attitude control engines; 7, thermal control system radiator; 8, docking system; 9, docking system targeting sensor; 10, sensors; 11, instrument module; 12, camera; 13, omnidirectional antenna; 14, batteries; 15, landing leg with shock absorber; 16, strut with shock absorber; 17, landing radar; 18, external instrument module; 19, low gain antenna; 20, docking system antenna; 21, television antenna; 22, downward-firing engine; 23, main engine; 24, deflector; 25, back-up engine. (Courtesy RSC Energiya (from the web site).)

Despite Kennedy's challenge, the Kremlin was not actively considering a lunar landing at that time, so the ambitious proposals being offered by the various design bureaux seemed destined to remain on paper.

4.2 WORK BEGINS

At the beginning of 1964, Korolyov had a meeting with Khrushchev during which he proposed a human mission to the Moon using a direct flight approach and requiring two N-1 rockets (the base model of the family) placing up to 75 tonnes in Earth orbit. Intrigued by Korolyov's presentation, and noting that the Americans – whose Apollo project had begun three years earlier – were apparently serious in their endeavour of landing a man on the Moon by the end of the decade, the Central Committee of the Communist Party issued an edict of 3 August 1964 ordering that work should begin on a project to put a Soviet citizen on the Moon before the United States. Recognising that they were starting rather late, they also approved Chelomei's simpler lunar circumnavigation project, which required the construction of twelve LK-1 spacecraft. This was a semi-lifting conical capsule, geometrically similar to Apollo but 30% smaller, and carrying a single cosmonaut. And, like Apollo, it was to be mated with a cylindrical propulsion module with an engine for trajectory control during the flight, and other auxiliary services; power was to be generated by two solar panel wings.

A first change was made in mid-October that same year, when Khrushchev was ousted from power and replaced by Leonid Brezhnev. As a result, many of the Chelomei space projects that had received Khrushchev's personal support were cancelled – but not the circumlunar mission, which was supported by a commission of the Academy of Sciences. One year later, on 25 October 1965, as the first LK-1 was nearing completion, Korolyov finally gained complete control of all lunar exploration projects, and stopped work on Chelomei's spacecraft.* He decided to launch his own 7K spacecraft on Chelomei's Proton rocket for the circumlunar mission. As the 7K, in contrast with LK-1, did not include a sufficiently powerful internal propulsion system, he proposed to install the fifth stage of the N-1 as the fourth and last stage of the Proton, to enter parking orbit and to accelerate the spacecraft to the Moon. At the same time, he decided to employ lunar orbit rendezvous instead of direct descent for his landing mission, and increased the N-1 payload to 95 tonnes.

Korolyov modified his 7K spacecraft to produce three main versions: an Earth orbital version to be launched on his Semiorka rocket to be used as a ferry to space stations; a lunar circumnavigation spacecraft to be launched by the Proton upgraded with one of his own upper stages; and a version to remain in lunar orbit

---

* Determined not to be squeezed out, Chelomei talked the military into funding the Almaz (Diamond) military space station, of which three were launched in the 1970s: Salyut 2, Salyut 3 and Salyut 5. He also developed a large pressurised module to transport cargo to these space stations: TKS, derived from the LK-1 lunar spacecraft, of which eight were launched between 1976 and 1985.

while a specialised craft landed on the Moon, with the two being launched together on an N-1.

Unfortunately, on 14 January 1966 – only three months after gaining control of the Soviet lunar programme – Korolyov died. He was succeeded by his deputy, Vasily Mishin, who, however, had far less charisma and much less political leverage.

4.3   THE BIRTH OF SOYUZ

In its Earth-orbit version, Korolyov's 7K spacecraft consisted of three in-line modules: a spheroidal 2.3-m-diameter, 3.1-m-long Orbital Module (OM), pressurised, and equipped with a hatch and a docking system; the bell-shaped re-entry capsule (SA – Spuskayemyi Apparat), 2.2 metres long and with a diameter of 2.3 metres, housing up to three cosmonauts; and the cylindrical Service Module (SM), 2.3 metres long, housing fuel tanks, engines, two solar panels with a total span of 9 metres and an area of 11 m^2, and batteries. In this version, it was called 7K-T, or, more simply, Soyuz (Union).

The lunar circumnavigation version was called 7K-L1, and its mass was reduced to some 5,500 kg by the elimination of the OM. The SM had a mass of 2,250 kg, and housed star sensors, the 4-kN-thrust KTDU-53 engine, and the attitude control system, including its thrusters. The re-entry module had a mass of 3,100 kg, and could carry up to 150 kg in a volume of 2.5 m^3, pressurised with pure oxygen at 506 hPa (0.5 atmosphere). The 7K-L1 was 4.5 metres long, and had a maximum diameter of 2.72 metres. It had the same solar panels as the 7K-T. On the front of the capsule was the landing parachute system, a conical adapter for mating with the launcher's emergency escape system, and a steerable parabolic high-gain antenna with a diameter of about 1 metre for communication in deep space. Fourteen 7K-L1s were built: a ground test prototype; two prototypes without heat shields for flight tests, similar to American boilerplates, and designated 7K-L1P; seven unpiloted test spacecraft; and four human-rated spacecraft able to carry two cosmonauts or to fly automatically. In its unpiloted version the spacecraft was called Zond – a name previously used for completely different deep space probes – and the piloted version would be called either Rodina (Home Country) or Ural (Ural Mountains) – or even Akademik Korolyov (Academician Korolyov). To save mass on these flights, the cosmonauts would not wear pressure suits and no back-up parachute was included.

In its third version – called LOK (Lunnii Orbitalnii Karabl – Lunar Orbital Ship) – the SM was replaced by stage I of the N-1 launch vehicle, which was longer, and carried more fuel. This version was powered by fuel cells, and could sustain two cosmonauts for fifteen days. In this form, the mass at launch was 8,400 kg.

For the lunar landing – project N-1/L3 – a landing module was also designed, called LK (Lunnyi Karabl – Lunar Ship). This consisted of an upper 2.3-metre-diameter spheroidal module, pressurised to carry a single cosmonaut, and a lower 'pancaked' sphere, also called stage E, which carried fuel cells, tanks, and the Isayev-designed engine to be used for both landing and lifting off from the Moon (in contrast with Apollo). The 20.5-kN-thrust engine was powered by monomethylhydrazine and nitrogen tetroxide. In case of engine failure, two smaller lateral engine nozzles – each with a thrust of 10.25 kN – would take over. To the lower sphere was attached the landing base, equipped with four legs, which, after serving as a launch platform, was to be left on the lunar surface. Around the base, housings were provided for scientific instruments to be deployed on the surface. A fully fuelled LK had a mass of about 12 tonnes.

4.4   THE FIRST TESTS

The first UR-500 – a two-stage version with no lunar launch capabilities – was launched on 16 July 1965 and placed the 12-tonne Proton 1 scientific satellite into Earth orbit.

The first Soyuz spacecraft was launched on 28 November 1966 under the name Cosmos 133. The attitude control system malfunctioned from the start, and after two days it was decided to fire the re-entry engine blindly, without any possibility of knowing whether the capsule was properly oriented. Although this returned the spacecraft to Earth, it re-entered over communist China, with which relations were worsening every day, so the self-destruct system was activated.

The second test fared even worse. During the countdown on 14 December, the emergency escape system fired, and the spacecraft's fuel tanks were ripped open, exploding like the rest of the launcher.

The third attempt was almost a success. Cosmos 140, launched on 7 February 1967, orbited the Earth for two days without any particular problems. After that, the

A Proton is carried to the launch pad with a Zond spacecraft under its shroud. Protons carrying Zonds had an emergency rocket and four stabilising surfaces for use in the event of a launch abort.

heat shield was slightly damaged, and the descent took place over the Aral Sea, which was frozen, and the capsule smashed through the ice and sank to a depth of a few dozen metres.

The launcher used for these flights was a new version of the old Semiorka with two upper stages, the first of which was the stage I of Molniya. Later, this launcher too was given the name Soyuz.

After three successful launches and one failure of the two-staged Proton, the next launch was the first of the four-staged lunar version. This 55-metre-long rocket could place 21 tonnes in low Earth orbit or send 5.7 tonnes to the Moon, and slightly less to the planets. The first stage burned nitrogen tetroxide and unsymmetrical dimethylhydrazine, with a unique architecture: to the central oxidiser tank were attached, like petals on a flower, six cylindrical UDMH tanks, each of which had an RD-253 engine. The second stage was separated from the first by a metallic truss, and carried four engines with a total thrust of 2.3 MN. The third stage had a single 583-kN engine and four smaller manoeuvring nozzles. The fourth stage – which was derived from stage D of the N-1 launcher – had a single 86-kN engine. To launch the piloted Zond spacecraft, the aerodynamic fairing was topped by a solid-fuel emergency escape system and four lattice stabilisers.*

On 10 March 1967, a Proton carried the first 7K-L1P boilerplate Zond into space, under the cover name Cosmos 146, to test the deep-space navigation system. It was not equipped with heat shields. After a whole day in low Earth orbit, stage D placed the spacecraft in a highly elliptical orbit. Its fate is still a mystery. It probably disintegrated on entering the atmosphere, without having been tracked by Western radars while in deep space. Despite some minor systems problems, the flight was successful. Unfortunately, the Soviets then incorrectly assumed that few problems were left to be solved, and that even these would be solved easily. On 8 April, the mission was repeated by Cosmos 154. This time the stage D ullage rockets were jettisoned before firing, and the spacecraft was left stranded in parking orbit.

At about this time, Brezhnev, ordered the launch of a spectacular human mission to celebrate May Day in the year of the fiftieth anniversary of the Russian Revolution. Two Soyuz were to be launched – the first carrying a single cosmonaut, and the other, three cosmonauts, and after docking two cosmonauts were to transfer from one craft to the other by spacewalking. In addition to providing a 'first', this would test the procedure by which it was intended that a cosmonaut transfer between the LOK and the LK in lunar orbit.

On 23 April 1967, Soyuz 1 was launched, carrying Vladimir Komarov, who was flying in space for the second time, and wearing only lightweight clothes. The Soviets' rush to complete the development of Soyuz is not at all surprising when considering the tremendous advances made by the American space programme with the Gemini missions of 1965 and 1966 – they were ready to attempt anything to take advantage of the pause after the Apollo fire of 27 January. But for the Soviets too, urgency and improvisation would prove fatal. Problems began as soon as Komarov

---

* Like Soyuz, Proton too is still in service in both three- and four-staged versions. By the end of 1999, the four-staged version had had 204 successful flights, eleven partial failures and twenty-one total failures.

entered orbit, when one of the solar panels failed to deploy; the optics of the main attitude sensor became fogged; the back-up attitude determination and control system – designed to sense the orientation of the spacecraft with respect to the terrestrial magnetic field – refused to yield to manual control; and only one of the two radio systems was working. The launch of the second Soyuz was cancelled, and everything possible was tried to bring the unfortunate cosmonaut home. On the seventeenth orbit, Soyuz 1 was manually oriented and its engine commanded to fire, but to no avail. On the nineteenth orbit, the engine finally fired for the necessary 146 seconds. During re-entry, a serious problem developed with the parachute system. The drogue parachute deployed properly, but the main parachute failed to unfurl from its housing. The back-up emergency system then deployed, but the lines of the two parachutes became entangled. The capsule smashed into the ground at a speed of 500 km/h, disintegrated, and burst into flames. Amidst the wreckage, recovery team found only a few remains of Komarov.

Four months later, the docking manoeuvre was carried out automatically by two Soyuz spacecraft: Cosmos 186 and Cosmos 188.

4.5   THE ZOND FLIGHTS

The off-nominal flight of Cosmos 154 and the death of Komarov pushed the first circumlunar mission by a cosmonaut from November 1967 (the fiftieth anniversary of the Russian Revolution) to 1968. Mission planners expected to succeed with four to six robotic flights before proceeding to the human flight.

The first automatic lunar Zond attempt took place on 27 September 1967. One of the engines of the Proton's first stage did not fire, because pad engineers had not removed a plastic tubing cover. One minute after lift-off the emergency escape system fired to carry the Zond to safety. The rocket impacted about 65 km from its launch pad. On the next occasion, on 22 November, the engines of the second stage did not fire. The spacecraft was found 285 km from the pad. The retro-rockets that were supposed to slow the descent had fired too soon, and the parachute had dragged the capsule for 600 metres, severely damaging it.

A new launch, on 2 March 1968, was more successful, and after placing Zond 4 in parking orbit, stage D was reignited to put the spacecraft into a highly eccentric orbit with a 200-km-perigee and an apogee at more than 350,000 km. Although the spacecraft reached the Moon's orbit, it had been decided to orient the apogee away from the Moon in order to avoid gravitational perturbations during a test of deep-space navigation, tracking, and communications. Inside the spacecraft was a proton sensor and a communication system that was used by a group of cosmonauts as an experimental relay to communicate with flight controllers a few hundred metres away.

The re-entry was to be carried out in two phases. After passing over Antarctica, the capsule would first encounter the atmosphere on a shallow trajectory over the Gulf of Guinea, where a Soviet ship was in place to track its telemetry. Generating lift, the bell-shaped spacecraft would skip off the upper atmosphere on a high arc that would cause it to re-enter somewhere over the Aral Sea. In case of failure, the capsule would perform a ballistic re-entry over the Gulf of Guinea – an eventuality for which Soviet engineers had included a self-destruct system in order to avoid the spacecraft falling into the wrong hands.

Throughout the flight, Zond 4 had problems with its attitude determination star sensor, which failed to detect Sirius. This complicated not only the course correction manoeuvres, but also the re-entry, which was a straight-in ballistic descent into the Atlantic Ocean, 200 km from the coast of Africa. Although this was within reach of the tracking ship, the self-destruct system automatically activated at a height of about 10 km. The Soviet engineers were, understandably, quite upset by this result.

For a very long time, the Soviets, initially, and then the Russians, stated that Zond 4 was aimed at the Moon, but the exact time of launch is known, and it can easily be verified that no lunar launch window was open. Another (and no less fanciful) version of the flight was provided by the US Space Command, according to which Zond 4 was lost in solar orbit.

The next attempt, on 22 April, *was* aimed at the Moon, but a failure detection system 'detected' a non-existent problem with the Proton's second stage and shut it down, the escape system was activated, and the capsule was recovered 520 km from the launch pad.

The next Zond did not even leave the pad. On 14 July, one of the stage D tanks ruptured under pressure, causing the structural failure of the whole stage. Three engineers were killed, but the number of fatalities could have been much worse had the lower stages, already loaded with toxic fuels, been damaged. According to all sources, the Zond escaped without a scratch, but was scrapped.

Two months later, the Soviets tried again, and it seems that the problems that had marred previous flights had been solved. For this mission another lifting re-entry attempt was planned, with a first atmospheric contact over the Indian Ocean and a landing in Kazakhstan. In view of the possibility of a ballistic entry, this time no self-destruct system was mounted, and eight recovery ships and two tracking ships equipped with Kamov Ka-25 helicopters were deployed to ensure coverage of a 2,500-km-wide area. Zond 5 was launched on 14 September 1968. After 67 minutes in a low parking orbit, stage D raised the apogee towards the Moon. In the re-entry module were numerous biological samples: two turtles, *Testudo horsfieldi Gray* – the first multicellular living creatures to reach the Moon – flies, plants, worms, and 237 *drosophila melanogaster* fly eggs. Other scientific instruments were dedicated to Earth photography (but, strangely, not lunar photography), the study of deep-space radiation, the solar wind and cosmic rays, and photometric measurements of several bright stars.

At about 325,000 km from Earth, on 17 September, the probe fired its engine to move the apogee beyond the Moon. As usual, the Jodrell Bank radio telescope succeeded in receiving the probe's telemetry, and the next day announced that it had just flown by the Moon at a height of 1,950 km, and was now returning to Earth! Only at this point did the Soviets admit to having launched a new mission to the Moon. Although, the British radio telescope received a human voice coming from

Zond 5 is integrated on the Proton stage D. (RSC Energiya image from Semenov, Yu. P. (*editor*), 'Rakyetno-Kosmiceskaya Korporatsiya 'Energhiya' imieni S. P. Karaliova 1946-1996'.)

the craft, this did not cause a commotion, as it was clear from the beginning that a tape recorder was on board.*

The probe's attitude control system had meanwhile suffered an incredible string of faults. The thermal protection material had sublimated under the heat of the Sun and fogged the star sensors; the Earth sensor had been mounted in an incorrect position due to a typing error in the technical documentation; and the back-up system had been mistakenly switched off by ground controllers. With only the Sun sensors for attitude determination, it took no less than 20 hours to tilt the spacecraft from one side to the other in order to align the thrust vector for a course correction about 143,000 km from Earth. Then on 21 September, it hit the atmosphere at the wrong angle, and re-entered at almost 11 km/s over the Indian Ocean. It survived the heating to 13,000° C and a 16-g deceleration and, after producing a sonic boom that was probably heard over hundreds of kilometres, at 16.08 GMT, six minutes after entering the atmosphere, it made a perfect night-time splash-down. Several Soviet ships converged on the area. The capsule – in obviously poor shape – was recovered with fishing nets by the ship *Borovichiy*, and was then transferred to the *Vassili Golovin*, which took it to the port of Mumbai, in India, where it arrived on 4

* In the case of the Vostok prototypes, the tape contained typical Russian cooking recipes in order to avoid creating more 'phantom cosmonaut' rumours.

October. An Antonov An-12 transport plane flew it back to Moscow. When the capsule was opened it was discovered that the animals had survived. The recovered films included several images of Earth taken from a distance of 85,000 km before re-entry. Of the 237 *drosophila melanogaster* eggs, only seventeen had produced adult flies, although a few larvae and pupae were still alive. The next generations of these insects showed a marked increase in spontaneous mutation – up to ten-fold. The turtles lost up to 10% of their body mass, and showed some variation in the quantity of glycogen and iron in the liver. The Zond 5 capsule is now on show at the small museum of the Energiya corporation – the former Korolyov design bureau.

The following month, both the Soviets and Americans carried out their first piloted missions since their respective accidents of 1967. The Americans flew Apollo 7 in Earth orbit from 11 to 22 October. The Soviets sent up Soyuz 2 unpiloted on 25 October, and Soyuz 3, piloted by Georgy Beregovoi, the next day. Beregovoi closed twice, but he inadvertently had his craft inverted and so no docking was carried out.

Zond 6, carrying animals and seeds, left Earth on 10 November. On board was also a camera with a 400-mm lens, a cosmic ray sensor, and a micrometeoroid detector. In particular, it was planned to study the Leonid meteor stream, which reaches maximum activity around mid-November, and which, just two years earlier, had produced the most impressive meteor storm of the twentieth century. During the first days of the flight, the only problem was that the high-gain antenna failed to deploy, which severely limited the amount of data that could be returned. On 12 November (the same day on which NASA announced that Apollo 8 would be sent to the Moon) the probe corrected its trajectory and on 14 November it passed 2,420 km over the lunar far-side. The camera imaged the surface in two photographic sessions – the first at a distance of 11,000 km, and the second at 3,300 km. At this point, telemetry showed a marked decline in the temperature of the hydrogen peroxide used in the attitude control engines. To prevent this from freezing, the orientation of the spacecraft was modified to expose the tanks to the heat of the Sun. Unfortunately, the seal of the re-entry capsule hatch, which was also exposed to the Sun, deformed, and began to leak air. Over the following days the cabin pressure declined from 506 hPa to 33 hPa, and the biological samples died. Furthermore, the cloud of gas that enveloped the spacecraft was ionised by solar ultraviolet, acted as a conductor, and caused a short circuit which damaged the altimeter!

Two course correction manoeuvres were carried out to place the spacecraft exactly central in the 100-km-wide entry corridor in the Earth's atmosphere. On 17 November the SM separated, but the high-gain antenna on the capsule's nose remained firmly attached. Some time later, after flying over Antarctica, Zond 6 hit the atmosphere over Madagascar at 11 km/s. The design of the re-entry module generated some lift, and the capsule skipped off the atmosphere into space at the much slower speed of 7.6 km/s. At this moment, the high-gain antenna finally separated. The capsule then dipped back into the atmosphere. After a deceleration of 7 g, the parachute opened, and everything appeared ready for landing. At a height of 5,300 metres, however, the damaged altimeter sent the parachute separation command, and the probe was destroyed in the impact with the ground. The landing site was inside the perimeter of the Tyuratam cosmodrome, only 16 km from the pad from which Zond 6 had been launched a week earlier. Onboard the wreck remained the data recorder and the photographic films – but also about 10 kg of TNT for the self-destruct system. Two men were put to work to find the explosive charge amid the charred remains of the spacecraft. A few days later it was found, and was detonated by Soviet army explosive experts in the steppes of Kazakhstan. Luckily, the photographic films – including the first true colour images of the Moon taken from a spacecraft – were found in good condition, and the Soviets were able to depict the mission as being completely successful.

As Apollo 8 was being prepared for lift-off on 21 December, Soviet cosmonauts and space engineers asked the Communist Party to approve the launch of a piloted Zond as soon as possible. There was no time to lose, as the Tyuratam launch window was open for only three days, from 15 December to 17 December. Two crews were training for the mission. The first consisted of Valeri Bykovsky, a veteran of Vostok 5, and rookie Nikolai Rukavishnikov, while the second (back-up) crew consisted of Pavel Popovich of Vostok 4 and Vitali Sevastyanov. On 1 December a Proton with a Zond capable of carrying cosmonauts was mounted on the launch pad, but over the following days a long string of problems developed with the capsule. The Kremlin refused to approve the launch, the window closed, and Soviet hopes of being the first to fly to the Moon dissolved with the perfect launch of Apollo 8.

On 1 January 1969, the Soviet Union announced to a world still astonished by the Apollo 8 images of the Moon, and even more so, of the Earth from deep space, that the objective of its space programme had always been the construction of large space stations. Nevertheless, nineteen days later, a new Zond, reusing the descent capsule which had survived the failure on 22 April 1968, was launched toward the Moon carrying the usual biological samples. Five minutes after launch, one of the second-stage engines failed, followed three minutes later by the third-stage engine. The escape system saved the unlucky capsule, which came down in Mongolia, a Soviet ally; even though it landed in a narrow valley in 3,000-metre-high mountains, it was recovered within three hours.

The next Zond – carrying four turtles – was launched on 8 August 1969, a few days after the return of Apollo 11. The flight proceeded without any problems, despite the failed deployment of the high-gain antenna, which became entangled with a wire. The Moon was passed at a distance of 1,985 km (or 1,200 km, according to another source), the far-side, which was almost completely lit, was imaged in colour during two photographic sessions, and the Earth was also photographed. This time the re-entry was perfect, and Zond 7 landed about 50 km from the planned site, near Kustanai, in Kazakhstan, on 14 August.

Three months later, on 28 November, a Zond in a configuration designated 7K-L1E was launched. It had been stripped of its life-support systems in order to accommodate a group of cameras to observe the behaviour of the propellants in the tanks of stage D in weightlessness, preparatory to a simulated lunar orbit insertion burn, but the first stage of the Proton exploded.

Consideration was given to the launch of a Zond carrying a cosmonaut in April

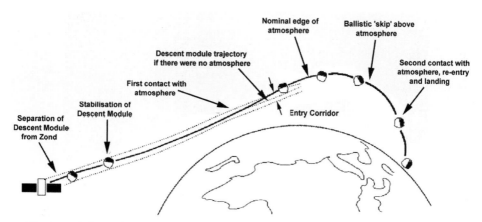

The intended Zond re-entry profile. Zond 7 was the only spacecraft of the series to perform it completely successfully.

1970 to celebrate Lenin's hundredth birthday, but nothing came of this proposal. Curiously, if this mission had been carried out, the spacecraft would have travelled along a trajectory similar to the one involuntarily followed by Apollo 13 around the same time.

The last lunar Zond was launched on 20 October 1970, by which time the public was no longer interested. On board Zond 8 were turtles, flies, onions, microbes, and various types of cereal, including wheat. It was optically tracked by Soviet telescopes until it was 277,000 km from Earth, and an onboard laser was used for telescope pointing. The Moon was flown by at a distance of 1,110 km, and dramatic pictures were taken of the leading hemisphere, featuring Oceanus Procellarum and Mare Orientale. The probe also tested a different landing technique. It flew over the Arctic rather than Antarctica, and made a straight-in ballistic re-entry over the Indian Ocean, and was recovered by the oceanographic ship *Taman* less than fifteen minutes after splash-down on 27 October. The flight of Zond 8 was the last of the Soviet circumlunar missions.

The scientific results of the series of probes were quite controversial. Despite all four spacecraft receiving a similar dose of radiation, which inflicted chromosomal damage on all of the seed samples, only the turtles onboard Zond 5 were affected.

The final Zond was launched on 2 December 1970 in the guise of Cosmos 382. It was a re-run of the 7K-L1E mission to observe the behaviour of propellants in the stage D tanks in weightlessness. This time the Proton functioned perfectly. Cosmos 382 is still in orbit.

4.6 THE N-1 LAUNCHER

In parallel with the circumlunar Zond flights, the decree of 3 August 1964 also authorised the second lunar programme, with the objective of landing a cosmonaut on the surface of the Moon before America could accomplish the task.

A view of Mare Orientale taken by Zond 8. Oceanus Procellarum can be seen at right. (Courtesy Vernadsky Institute, Moscow.)

The most important part of the project was the N-1 launcher – a huge five-stage rocket (compared with the American four stages), slightly smaller than the Saturn V. It was 105 metres tall, had a launch mass of 3,025 tons, and was designed to place up to 95 tonnes into low Earth orbit.*

The first stage (stage A) was a truncated cone 30.1 metres in length, with a base diameter of 16.8 m. It utilised a spherical 10-metre-diameter kerosene tank and a 12.5-metre-diameter oxygen tank. On the base were thirty NK-33 engines, each developing 1.51 MN of thrust, mounted in two concentric rings of which the outer ring had twenty-four nozzles and the inner ring had six nozzles. Four of the outer ring engines were equipped with large graphite paddles for thrust vectoring (the first stage of the Saturn V had only five F-1 engines, and they were gimballed for thrust vectoring). Finally, the first stage carried four lattice aerodynamic stabilisers. The problem of controlling thirty engines was solved using a complex control system called KORD (KOntrol Roboti Dvigatyelyei – Control System for the Working of Engines). In the event of a problem with one engine, the KORD would shut down

---

*   The Apollo SM was also counted as a stage, as it had a task similar to that of the Soviet stage D.

both the failing engine and that diametrically opposite it, in order to maintain symmetric thrust. The N-1 was intended to be able to carry out its mission with a maximum of four engine failures. Unfortunately, the KORD system proved unable to react fast enough, and it contributed to the eventual failure of the entire programme.

The second stage (stage B) was connected to the first stage with a metallic truss, was 20.5 metres long, and was equipped with eight NK-43 engines, each generating 1.76-MN-thrust. The third stage (stage V) was 11.5 metres long, and was equipped with four NK-39 engines, each generating 0.4 MN of thrust. The first three stages carried the lunar spacecraft into Earth orbit, where the fourth and fifth stages would take over. The fourth stage (stage G) was equipped with four NK-31 engines, each developing 0.4 MN of thrust, and would be used to head for the Moon.

The N-1's stage D was derived from the Molniya stage L and from the GR-1 (Globalnaya Rakieta – Global Rocket) third stage, and was equipped with a single Melnikov RD-58 liquid oxygen/kerosene engine.* Korolyov adapted this stage for use as the Proton's fourth stage, a role in which it still serves.†

The lunar mission plan called for the L3 lunar complex – consisting of the LOK, a small nitrogen tetroxide and UDMH attitude control module, the LK, and stages D and G, with a total of 95 tonnes – to be placed into a 200-km parking orbit with an inclination of 51°.6, where it would remain for up to a day. Stage G would carry out translunar injection before separating to leave the course correction and lunar orbit insertion tasks to stage D. Once in a lunar orbit with perigee at 14 km and apogee at 110 km (quite similar to the last Apollo missions), one of the two cosmonauts would exit from the LOK and make a short space walk to reach the LK. The LOK would then separate, and stage D would initiate the powered descent. At a height of 4 km, stage D would brake the descent and be jettisoned, to fall to the surface. At this time the cosmonaut would have had just a few seconds to decide whether to carry on or to abort the mission and head for LOK docking. A minute later, the LK would touch down on the surface. As it did so, four small downward-firing rockets would ignite to ensure that it settled firmly on even the most rugged terrain. The LK was capable of 48 hours of autonomous operations, but the first mission would last no more than 4–6 hours. The landing site would probably have been in the equatorial zone.

---

* The GR-1 was Korolyov's answer to a military request for a space-based nuclear bombing system – a criminal system because of its destabilising potential – but the missile was never built, as the military preferred Yangel's proposal. In one of the best deceptions during the Cold War, two full-scale mock-up GR-1s were displayed on Red Square on 9 May 1965. NATO assigned the non-existent rocket the name SS-10 (Scrag). The orbital bombing project – known in the West as FOBS (Fractional-Orbit Bombardment System) – was tested using inert warheads during the 1960s and 1970s, but was finally abandoned in the 1980s. As the name suggests, instead of sending a warhead on a high ballistic arc over the north pole, FOBS warheads could fly either 'depressed trajectories' over the pole or fly around the southern hemisphere to circumvent US radar defences by approaching from the south.

† Many commentators in the West have speculated on why the Proton fourth stage should be called stage D, as this is the fifth letter of the Russian alphabet.

The tail of the huge N-1 first stage undergoing static tests. The N-1 first stage was the only Soviet lunar launcher component tested in flight, its engines failing four out of four times. (Courtesy TsNIIMash.)

Once on the Moon, the cosmonaut would depressurise the cockpit and exit to the ground where, after planting the Soviet flag, he would collect soil samples and deploy scientific instruments, including a small automatic rover and a coring drill. The semi-rigid Krechet suit – which would have permitted a surface excursion of about 90 minutes – included a device which the cosmonaut would have worn as a sort of metallic hoop on his hip, to prevent him from falling on his back and becoming stranded if he were not able to get back on his feet. Once the historic mission was over, the LK would re-ignite its main engine and leave its landing frame on the Moon. After rendezvousing with the LOK, it would dock using a system called Kontakt, which acted in the same manner as velcro. The LK had a disk-shaped honeycomb structure on its roof. This would be pierced by a wedge-shaped probe on the nose of the LOK. Once the two craft were docked, the lunar cosmonaut would carry out another space walk to return to the LOK with his samples of moonrock. After abandoning what was left of the LK, the stage I (LOK) engine would fire to return the spacecraft to Earth, where it would carry out a lifting re-entry (as used for the Zond probes), and land in the Soviet Union. In fact, Mishin preferred a splash-down in the Indian Ocean, and he had Zond 8 demonstrate this mission profile, but the cost of establishing a recovery fleet was deemed too expensive – and in any case, there was concern that the re-entry capsule, which had been designed to float only in emergencies, was insufficiently buoyant to survive several hours at sea in the event of an off-course descent.

A relatively familiar site at Tyuratam: two N-1 lunar launchers on the pad.

4.7 FOUR FAILURES

Construction work on the two N-1 pads, separated by just 500 metres, began at Tyuratam just one month after the beginning of the programme, in September 1964. From the start, it was clear that it would be impossible to land on the Moon in 1967, and the date slipped first to 1968 and then to 1969. Although it was never publicly acknowledged, the Soviet Union *was* in a dramatic race against the United States.

The many problems of the project, both managerial and technical, then began to emerge – the managerial problems most of all, because not only had the Soviets to make up time for their late start, they had set in motion two competing programmes (or three including the robotic probes programme, of which more will be told in next chapter). This was an enormous waste of resources, and in what could only have been a bureaucratic nightmare, twenty-six ministries and in excess of 500 state industries participated in the programme! Funds were never at the same level as for Apollo, although it is difficult to compare them. Allocated Soviet funds amounted to 2.9 billion roubles – about $4.5 billion in 1970. The equivalence between research roubles and research dollars produces a result of around $10.1 billion. It would be very interesting to know the percentage of the Soviet Union's Gross National Product spent on the project.

Technical problems were also noteworthy. Apart from the disadvantage of the latitude of Tyuratam (46° N) compared with the latitude of Cape Kennedy (28° N), and the need to launch at 51° in order to preclude an abort into China, with the higher inclination reducing effective payload, the Soviets adopted a deceptively simple approach that eventually led to failure. As a Saturn V required the correct firing of just eleven engines of very high technology to take astronauts into lunar orbit, the N-1 required the firing of forty-six engines! Furthermore, in an effort to shorten development time, Korolyov chose not to test the thirty-engined first stage as a whole. Although NASA opted for an 'all up' test of its Saturn V, it had at least subjected the S-IC to static firing. In addition, the Soviets did not build anything like the Super Guppy aircraft, and the components of every launcher required no less than 165 special trains to travel from Samara, where production was carried out, to Tyuratam. Finally, because the public was kept uninformed of the space programme, Soviet engineers lacked the strong backing, sense of pride and occasional criticism on which Western engineers could rely.

During the second half of 1966, a group of eighteen cosmonauts began using a modified Mil Mi-4 helicopter to begin training for lunar flight. The first crew was also named: Alexei Leonov – the first man to carry out a spacewalk (which he did in 1965, during the Voskhod 2 mission) – would land on the Moon, and rookie Oleg Makarov would pilot the LOK in lunar orbit.

The first N-1 – an engineering mock-up – was mounted on Pad 1 on 25 November 1967, and went through a series of tests up to 12 December, when it was rolled back to the assembly building. The first real N-1 was mounted on the same pad on 7 May 1968. Inside its aerodynamic shroud was an L1S (a Zond spacecraft with a modified propulsive module). This was to enter lunar orbit, carry out a photoreconnaissance mission, and return to Earth. Under the Zond (as on the first Apollo missions) was a dynamic mock-up of the LK. The launch was scheduled for September, but after a month of tests it was discovered that the first-stage tanks had a manufacturing defect, and were cracking. The N-1 was therefore removed from the pad, and the first stage was disassembled.

During the following months the mock-up was used in another round of tests and launch simulations – because the launch team included conscripted soldiers, a continuous training process was necessary. A new N-1 first stage was mounted on the pad in November, to test the engine control system. It must have been an amazing sight, as the second pad was then occupied by the engineering mock-up. In January, the first stage was returned to the assembly building, from which it re-emerged a few days later, complete with upper stages and a new Zond under its

An American KH-4 CORONA spy satellite image of the N-1 launch facilities, taken shortly after the catastrophic failure in July 1969. The explosion has devastated the pad at right, where one of the two lightning towers has collapsed. The best CORONA pictures show details as small as the N-1 interstages. Better images were probably taken by KH-8 Gambit satellites, but it is unlikely that these will be declassified in the near future. (Courtesy Dwayne A. Day.)

payload fairing. On the revised plan, four N-1s were to be launched in 1969: in February, April, June and November.

Finally, after a month of tests, the engineers felt confident enough to attempt launch. At 09.18.07 GMT on 21 February 1969, the huge launcher left Soviet soil.

Less than ten seconds later, engines number 12 and 24 were erroneously shut down by the KORD system. Twenty-five seconds after lift-off, the engine thrust was deliberately reduced as the launch vehicle went supersonic in order to minimise aerodynamic stress, and 40 seconds later it was brought back to maximum rather faster than expected. The first stage started to vibrate, and the oxygen pipes of one of the engines ruptured, causing a fire near the base of the rocket before the KORD could close the fuel valves. Several nearby engines exploded, and the surviving engines were eventually shut down 70 seconds after lift off. The Zond was saved by the emergency escape system as the uncontrollable N-1 impacted 48 km from the launch pad. (One of the very few false rumours that have emerged since the fall of the Soviet Union maintains that the N-1 impact caused ninety-one fatalities.)

A few months earlier, the Soviets had lost their chance to beat America to fly the first circumlunar mission. With this N-1 failure, they lost the race to land a man on the Moon. But the programme had an inertia, and so it kept going.

A second N-1 was rolled out to Pad 1 (the mock-up was being used to bring Pad 2 into operation). On 3 July – less than two weeks before the lift-off of Apollo 11 – this was launched in darkness. This time, at a height of less than 200 metres, one of the engines ingested a foreign object that had inadvertently been left in the tanks or engine piping, and exploded, damaging nearby engines. The Zond was lifted clear and landed about 2 km away as the N-1 fell back to the pad just 18 seconds after lift-off. The explosion – comparable to a small atom bomb – was so immense that the

The re-entry capsule – the only component common to all of the versions of the Soviet 7K spacecraft. (Courtesy ESA.)

145-metre-tall service tower was lifted from its semicircular track and pushed back, and one of the 180-metre-tall lightning towers collapsed. Pad 2 and the engineering mock-up were also badly damaged, and the shock wave was so violent that cars as far away as 2 km were turned upside down! US intelligence learned of the accident at the beginning of August, when they received images taken by a KH-4 CORONA spy satellite that had overflown Tyuratam soon after the devastation.

The reconstruction of the launch complex took two years, during which time three American crews walked on the Moon. Mishin, meanwhile, had been authorised by the Kremlin to initiate a new project which, although coming second to Apollo, promised to eclipse it. The L3M project would require the launch of two N-1F – a new modified version of the N-1 using a cryogenic upper stage Sr instead of stages G and D. The first launch would place one of these stages in lunar orbit, and the second would send a 23-tonne LM with a three-men crew. The landing module and the Sr stage would dock in lunar orbit before heading to the surface. Early missions would remain on the surface until lunar sunset, but as facilities were expanded later crews would remain for up to three months! At the end of the mission, the LM would eject a Soyuz-like capsule for re-entry. Provided political and economical support were in place, the first landing was expected before the end of the 1970s. In contrast with N-1/L3, this mission had the blessing of the Soviet Academy of Sciences. The N-1 programme was thus modified to act as a pathfinder for N-1F. The third and fourth N-1s were completed in a configuration similar to the new rocket.

On 26 June 1971 (when the author of this book was almost ten days old), the third N-1 was ready for launch on Pad 2. Instead of a Zond, it carried a mock-up of the L3 complex, so there was no possibility of carrying out a meaningful lunar mission. It had been hoped to launch it at night on 20 June, so that its progress could be observed by the Soyuz 11 crew, who were at that time onboard Salyut 1, but when it was finally launched the space station was over the Pacific Ocean. At a height of 250 metres the launcher began an unplanned roll manoeuvre – probably because of aerodynamic interference between the rocket and the engine exhaust. The stress collapsed the interstage truss and the upper stages separated from the first and second stages and fell back and exploded. The first stage continued firing, and impacted 20 km from the pad 55 seconds after lift-off, producing a crater 30 metres wide and 15 metres deep. But a far more tragic accident was to follow: on 30 June, a pressure equalisation valve on the Soyuz 11 descent module inadvertently opened shortly before re-entry, and the three cosmonauts – Vladislav Volkov, Georgy Dobrovolsky and Viktor Patsayev – died due to the depressurisation of their capsule.

The fourth launch of the N-1 – with a revised first stage with a cylindrical rather than a conical tail with a diameter reduced by 1 metre, and fitted with small vernier rockets for better roll-axis control – was attempted from Pad 2 on the morning of 23 November 1972. It carried a real LOK and an LK mock-up. It was planned to put the LOK in lunar orbit for 3.7 days, where it would take pictures of future landing sites. The flight went well for 107 seconds, at which time the inner ring of six engines of the first stage were deliberately shutdown in preparation for staging, but the vibrations resulting from this sudden change ruptured pumps and pipes and the first stage exploded. Although disappointing, the first stage had worked properly to within 40 seconds of handing over to the second stage. From an engineering flight-test point of view, progress was being made.

4.8　THE END OF THE LUNAR PROGRAMME

After these four failures, the engineers working on the N-1 prepared two new N-1F launchers for 1974. If both of these succeeded, the N-1F could have entered service as soon as 1976, the first human flight could have been the seventh N-1 flight, and the first landing could have been attempted on the next flight.

However, on 18 May 1974, a few months before the first test was due, Defence Minister Sergei Afanasyev briefly visited Mishin's team to announce that Mishin was to be dismissed, and that his place was to be taken by Glushko – Korolyov's arch-rival. Afanasyev then wished the team good luck, and left. Glushko promptly halted the development of the N-1, and the already-built engines were abandoned. Kuznetsov, however, continued development, and accumulated more than 100,000 seconds firing time on a single engine.

After the cancellation of the N-1, the two existing launchers were destroyed. Parts of the spherical tanks were used to roof Tyuratam city gazebos, and parts of the structure ended their life as car shelters or hen houses. But a better fate awaited the NK-33 first-stage engines – 150 of which secretly escaped this kind of *damnatio memoriae*. For twenty years these engines were left, covered with plastic, in a fuel plant in Samara, and after the fall of the Soviet Union many of them were sold to America. The KVD engines (Kislorodno–Vodorodnyi Dvigatel – Oxygen–Hydrogen Engine) of stage Sr were mounted on the third stage of the Indian GSLV rocket, and therefore flew in space for the first time in April 2001 – more than thirty-five years after being designed.

What followed was a prime example of how resources were wasted during the Soviet era. Glushko's team began work on a new lunar project and a new heavy launcher, Vulkan, to use its Vesuvius cryogenic engines to place up to 43 tonnes into lunar orbit. The project envisaged the construction of an autonomous lunar base called Zvezda (Star), to support up to six cosmonauts for as long as one year. It would be built up incrementally.

Before the construction of the base, some automatic probes, built by Lavochkin, would be launched to map the chosen site in detail and to characterise the regolith to a depth of 10 metres. After this, a 31-tonne piloted lander, a habitation module, a large pressurised rover and a nuclear power plant would be launched in quick succession; six months later a second lander and another habitation module would be added; and finally, three months later, an industrial module with facilities for the production of oxygen from lunar regolith, would complete the base, which would have a habitable volume of 520 m^3.

When the Supreme Soviet rejected this ambitious project, Glushko began work on the development of yet another heavy launcher called Energiya (Energy), with a payload capacity similar to that of the N-1F. Work on the new rocket lasted more than ten years and it then flew only twice – in 1987 and 1988 – the last time carrying into space the Buran (Snow Storm) space shuttle. Lunar missions were also studied for this new launcher – which for the first time surpassed the Saturn V as the most powerful rocket ever built. The project envisaged the simultaneous launch of two Energiyas – the first carrying a three-man LM capable of twelve days of independent operation, and the other carrying a five-man orbital module. The whole lunar mission, including the surface stay, would have lasted up to a month.

In the 1970s and 1980s the Soviet space programme came to maturity, surpassing in some fields (mainly human spaceflight) the American programme. However, until the collapse of the Soviet Union it remained subject to the contradictions which had doomed the lunar project. In an economy that was otherwise supposed to be planned from the top, the absence of a central national space agency left the field open to the bitter competition between rival design bureaux. In this way, two competing space station programmes were carried out: the first – Almaz (Diamond) – by the Chelomei design bureau, and the other – DOS (Dolgovremennaya Orbitalnaya Stanziya – Long-Duration Orbital Station) – by Mishin at the Korolyov bureau. Nevertheless, these technologies were later integrated to form the Mir (Peace) space station, the assembly of which began in 1986.

This was similar to the course followed by the Buran project, which was started without the real need for a space shuttle! While Glushko, backed by the Politburo, of which he had become a member, built the Buran, which was almost a copy of its American precursor, Chelomei, without authorisation, and in an almost clandestine manner, studied a smaller shuttle tailored to imagined military requirements. Chelomei built a full-scale model of his own shuttle, but was severely punished when this became known. Buran flew only once in space as if to prove a point, and was then abandoned.

Hopes of resurrecting the Energiya launch vehicle were dealt a hard and possibly fatal blow in May 2002, with the collapse of the roof of the building that had been constructed for N-1 assembly, in which the remaining hardware was being stored; eight people died in the accident.

4.9   LUNAR MODULE TESTS

The story of the Soviet lunar programme would not be complete without mention of the LK tests as the T2K flights. For these tests, which were the counterparts of NASA's Apollo 5 mission, Soyuz launch vehicles were used to place into orbit lightened, legless LKs.

The first LK, under the cover name Cosmos 379, entered Earth orbit on 24 November 1970. Three days later, stage E (the combined descent and ascent stage) fired, varying its throttle to simulate a lunar landing, while in reality raising the apogee from 232 km to 1,210 km. Meanwhile, a recorded tape was played for a communication test. After a 24-hour 'stay', the landing platform was jettisoned and the engine re-ignited in a simulated lift-off, and the apogee reached 14,035 km. Soon thereafter, the batteries were exhausted. The spacecraft remained in orbit for another 13 years before re-entering on 21 September 1983.

Cosmos 398 was launched on 26 February 1971, performed a similar mission, and re-entered on 10 December 1995, being rediscovered by the world media in the days before re-entry.* Cosmos 434 was launched on 12 August 1971, made the longest engine firing of the series, and re-entered on 22 August 1981, at which time the nature of this family of satellites – which were previously suspected of being failed lunar probes – was finally revealed. In 1978, after the spy satellite Cosmos 954 fell over Canada and contaminated a large area of forest with its nuclear reactor, it was discovered that the days of Cosmos 434 were also numbered. It was then asked of a Soviet space official whether this satellite also carried nuclear fuel. He replied that nothing was to be feared from Cosmos 434, as it was 'simply' a lunar module prototype!

A series of T1K flights were also planned to test the LOK in Earth orbit, but these were never carried out. Five LK and a single LOK survive today. The 7K-T Soyuz spacecraft is still in service in its Americanised TMA version serving as the 'lifeboat' for the International Space Station.

Up to May 2003 almost 100 piloted and automatic Soyuz have been launched. Piloted flights suffered four accidents: Soyuz 1 and Soyuz 11 (both fatal), Soyuz 18A (failed launcher separation and emergency landing), and Soyuz T-10A (launcher explosion on the pad seconds before lift-off, and manual firing of the emergency escape system).

---

*   The author remembers having seen a pass of Cosmos 398 on 25 June 1995, when the satellite appeared like a very fast-moving magnitude 1 star.

5

The end of the race

5.1 THE SOVIET THIRD GENERATION

With the launch of Luna 14 in 1968, the second generation of Soviet lunar probes came to an end. These probes used the Molniya launcher, and shared a common propulsion module on which an orbital module or a small lander was mounted. During the early 1960s work had begun on the third generation of Soviet lunar probes, designed to support human lunar landings, but finally destined to substitute for them.

These new probes had a common spacecraft bus, known as the stage KT. With a dry mass of 1,100 kg, it consisted of four 88-cm-diameter tanks for hypergolic hydrazine and nitric acid propellants, and the main engine for course corrections during the flight from the Earth to the Moon, for orbit insertion and subsequent manoeuvres, and for landing if required. To this module were attached four large cylindrical tanks (increasing the total mass of propellant to 3,900 kg) that were jettisoned in lunar orbit if the probe was to land on the Moon; and, in this case, four short legs with a 4-metre-wide separation. On top of this bus was mounted the mission-specific hardware, to create three different probes:

- *E-8 rover*
  Four folding ramps were attached to the landing base, and almost all of the available room was taken up by the eight-wheeled 8EL vehicle called Lunokhod (Lunar Walker), designed to autonomously explore the surface for up to three months.

- *E-8LS orbiter*
  The common bus was not equipped with landing legs, and carried a Lunokhod instrument module to study the lunar environment from orbit. For this kind of mission much less fuel was required, and the mass thereby saved was used to accommodate an increased supply of nitrogen gas for attitude control.

- *E-8-5 automatic sample return mission*
  On the common bus, which had legs, was mounted a robotic manipulator with a small drilling unit to retrieve a sample of the lunar surface, a hermetic capsule in which to return the sample to Earth, and a liquid-fuel rocket to lift the capsule from the lunar surface. To simplify the design of the ascent stage, this could give a velocity increment to the capsule in a vertical direction only, and this constrained sample-collection missions to the eastern hemisphere of the Moon, near 60° E, and near the equator from which the Earth could be reached using a vertical ascent. (A similar constraint on the descent rather than the ascent phase had dictated the landing sites of the E-6 lander missions.) The probe could return only about 100 grammes of regolith to Earth.

The orbiter and the rover had a solar panel to sustain operations, but the sample-return variant ran off batteries and so had a limited life. This version, first proposed in early 1967, was the only one entirely designed by the Babakin team. Studies of the other versions were begun by Korolyov during the early 1960s. A variety of other possible missions for the E-8 were proposed, including collecting samples from a depth of 10 metres, prior to the establishment of the Zvezda lunar base.* Because of the mass of the probes (more than 5,700 kg at launch), they could be carried only on launchers of the Proton family.

The initial plan was to use an orbiter to carry out a detailed photoreconnaissance of possible landing sites for a human mission, to be followed soon afterwards by two rovers. These would land near the chosen site and explore the neighbourhood for at least a month. After validating the site, the rovers were to act as beacons for an unmanned LK. After this landed, the rovers were to examine it to check for possible damage. The unmanned LOK that had accompanied the LK was to carry out orbital photography and then return to Earth. Finally, one month later, the piloted mission would land, and the cosmonaut could choose which of the LKs he would use to return to orbit. It was also planned to use the Lunokhods as manned rovers, and cosmonaut Valeri Bykovsky had trained on one of them.

Although this extremely complex plan was never implemented, the automatic mappers and rovers were built and were later supplemented by the sample-return version.

5.2   A FALSE START

The final approval for the new series of lunar missions was given by the Soviet Council of Ministers on 8 January 1969. The plan outlined an extremely ambitious timescale for 1969: four E-8-5s, to be launched between June and September, and

---

*   There was even a proposal to use the E-8 bus to transport a small capsule to Mars, to be released to make an atmospheric entry shortly before the main spacecraft entered orbit. Of course, the motivation was to beat the Americans to such an achievement.

An E-8-5 sample-return probe at the Paris Air Show in 1971. The horseshoe-shaped object above the robot arm joint is the rarely-used camera. Note also the outrigger arm carrying attitude control engines. (Courtesy Patrick Roger.)

three Lunokhods in February, October and November, in addition to two probes to Venus and two probes to Mars.

The first probe, a Lunokhod, was launched on 19 February, but the Proton shroud collapsed due to launch vibrations just a few seconds after lift-off, and the payload was destroyed. Forty seconds later, the launcher self-destruct system was activated. The probe's remains were carefully sifted to locate the radioactive polonium isotope intended to keep the rover's electronics warm, but it was never found.* The recovery team found many mechanical components to be still in good condition, and these were later reused in several mock-ups.

---

* According to a rumour, conscript soldiers arrived on the impact site before the recovery team and took the polonium to their huts to provide warmth during a particularly cold winter!

On 13 June it was the time for the launch of the first sample-return probe. If it succeeded, the Soviets would be able to claim a major propaganda triumph by securing a lunar sample a month before NASA launched Apollo 11, at a fraction of the cost and without risking any cosmonaut's life; but it was not to be, because stage D never fired, and the probe burned up in the atmosphere.

The next attempt was more successful ... but only just. On 13 July 1969, only ten days after the second N-1 had blown up on the launch pad, and as thousands of campers, pressmen and VIPs were descending on Cape Kennedy for the launch of Apollo 11, a Proton rocket lifted off and TASS announced the launch of Luna 15. The first telemetry intercepts – carried out, as usual, by Jodrell Bank – showed that something unusual was happening. The probe was on a very slow translunar trajectory, probably designed to maximise the payload – a clear sign that a new type of heavy probe had been introduced.

Soviet space experts were interviewed by Western journalists, and this time some details were revealed – in order, no doubt, to steal Apollo's thunder. It was thus discovered that the probe would land on 19 July, one day ahead of Armstrong and Aldrin's landing, and that it was then to return to Earth carrying something: a lunar sample, of course. On 17 July, as Apollo was less than half-way between the Earth and the Moon, Luna 15 entered orbit (133×286 km, $45°$ inclination). Consultations between NASA and the Soviets followed, as it was feared that the Luna probe's transmissions might somehow interfere with the piloted mission. The Soviets replied that because the two spacecraft used different frequencies, no such risk existed. Meanwhile, the probe was progressively lowering its orbit (its perigee was 96 km on 18 July, and 85 km on 19 July). However, something was clearly wrong, and the landing had to be delayed by one day. In the evening of 20 July, as Apollo 11's Eagle landed, the Soviet spacecraft was in a breathtaking orbit carrying it from an apogee of 109 km down to a 16-km perigee. Finally, as Armstrong and Aldrin were about to end their stay, Luna 15 prepared to land. The main engine was started during the fifty-second orbit at 15.47 GMT on 21 July, and four minutes later hit the Moon in Mare Crisium (Sea of Crises) at $17°$ N, $60°$ E (or, according to other sources, $12°$ N, $59°$ E, or $17°$ N, $49°$ E), while still travelling at about 480 km/h. Unfortunately, the impact was insufficiently energetic to be detected by the EASEP seismometer at Tranquillity Base. The Soviets were forced to admit that they had experienced some problems in the attitude control system, and that they had underestimated the orbital perturbations due to mascons. The debut of the third-generation series could not have been more embarrassing.

Two more E-8-5s followed, but neither left Earth orbit. Cosmos 300 was launched on 23 September 1969, and re-entered four days later. Cosmos 305 was launched on 22 October, and re-entered after only two days. On the first occasion a faulty valve on stage D caused a slow depletion of the liquid oxygen tank, and on the second the attitude control system of stage D malfunctioned. Such failures in a 'mature' stage were frustrating.

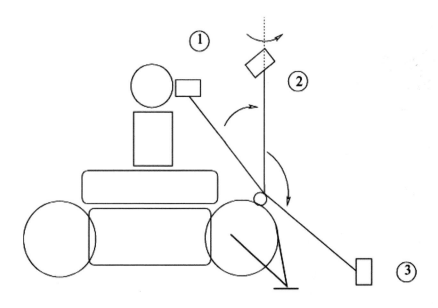

The deployment sequence of the E-8-5 robotic arm: 1, stowed position; 2, azimuth rotation (180°); 3, deployed position. Operating the arm between positions 1 and 2 took 5.5 minutes, and between 2 and 3 took 7 minutes more.

5.3   THE DECADE OF THE ROBOTS BEGINS

After another E-8-5 on 6 February 1970 was lost when the Proton's first stage was erroneously shut down, the Soviet third generation finally proved its worth.

Luna 16 was launched on 12 September 1970, and four days later entered a 110-km 71° circular orbit around the Moon. After two days, it lowered its perigee to 15 km, and on 20 September the four auxiliary tanks were jettisoned, and the main KTDU-417 engine was fired. About 600 metres above the lunar surface, the radar altimeter detected the Moon approaching at some 700 km/h and commanded one last engine firing. At a height of about 20 metres the engine stopped. The smaller manoeuvring rockets continued firing until the probe was within 2 metres of the surface – these being mounted so as to minimise contamination of the regolith around the probe. (Another possible source of contamination had been eliminated by sterilising the whole probe before launch.) At 05.18 GMT, the probe, decending in darkness at 2.4 m/s, came to rest in northern Mare Foecunditatis, at 0°.68 S, 56°.3 E, some 1.5 km from the target point. TASS announced the third successful lunar landing of a Soviet probe, without adding any further details.

Luna 16 was identical to the other E-8-5 probes, consisting of a descent stage and an ascent stage. On the descent stage was mounted a toroidal pressurised module

An artist's depiction of Luna 16's ascent stage lifting off from the Moon, with its lunar sample in the spherical capsule at the top.

containing the mission instruments, to which was attached a 90-cm-long robotic manipulator. This had two degrees of freedom, enabling it to rotate up to 120° in elevation and 180° in azimuth. The manoeuvre from the stowed position to the drilling position consisted of three movements: rotation up to a vertical posture, azimuth rotation, and then rotation down to contact with the surface. In addition, the azimuth joint could be used to clear the sampling area of debris by slightly moving the arm left and right. On the free end of the manipulator was a 90-

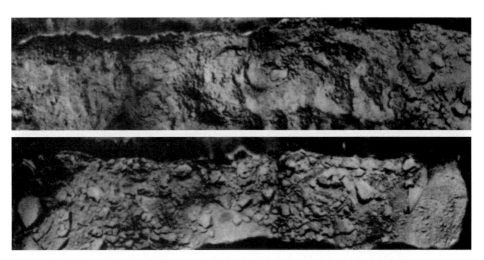

The majority of the 30-cm core sample retrieved by Luna 16, deepest at lower right.

mm-diameter, 290-mm-long cylinder containing the drill auger and two electric motors – one for drill rotation, and the other for drill translation – drawing up to 140 W of power. The hollow drill-bit was 417 mm long, had a diameter of 26 mm, and could collect samples from a maximum depth of 380 mm. It could rotate at up to 508 rpm, advance at a speed of 56 mm/min, and penetrate compacted regolith by percussion at 1,500 impacts per minute. The exterior of the drill was lubricated with a special synthetic oil designed to sublimate in vacuum. The interior was sculpted with a particular helical shape designed to retain finer dust during sample removal. The instrument had a total mass of 13.6 kg, and could take samples from material as soft as dust or as hard as granite.

On top of the descent stage was mounted the 520-kg ascent stage, which brought the total height of the probe to 390 cm. This comprised a KRD-61 (Kosmicheskii Raketnii Dvigatiel – Space Rocket Engine) 18.8-kN-thrust engine unit with a pair of spherical propellant tanks, and a cylindrical pressurised instrument module, and was topped by a 39-kg hermetical capsule with a diameter of 50 cm, designed to survive atmospheric re-entry at more than 10 km/s.

The scientific payload of the E-8-5 probes usually included a scanning camera similar to that of the E-6 probes. However, as Luna 16 had landed in darkness and was not equipped with floodlights, the camera could not be used. As a result, the manipulator's azimuth was selected without any knowledge of the landing site. Nevertheless, camera telemetry is said to have been received by the radio astronomy observatory at Bochum, in what was then West Germany.

The robotic arm was meanwhile carrying out its tasks. At a depth of 35 cm the drill ground to a halt, and sample acquisition had to be stopped for fear of damaging the auger. The final few centimetres fell out as the tube was extracted. Once the 101-g sample had been stored inside the return capsule, ground stations determined the precise position of the probe, in order to compute the optimal return trajectory.

The Luna 16 sample capsule after Earth return.

Finally, at 07.43 on 21 September, 26 hrs 35 min after landing, the ascent stage separated and left the Moon, accelerating to the escape speed of 2.7 km/s. The return flight, which took three days, was essentially ballistic, as no correction manoeuvre was made. About 48,000 km out, the ascent stage was jettisoned, and four hours later the re-entry capsule hit the atmosphere. The re-entry could not have been endured by a cosmonaut, because the probe was subjected to a 350-g deceleration, and its thermal shield was heated to a temperature of 10,000° C! About 14.5 km from the ground, a drogue parachute opened, followed immediately by the main parachute and four beacon aerials. In fact, the capsule was already being tracked by a helicopter recovery team. At 05.26 on 24 September, what was left of a probe with a launch-mass of more than 5 tonnes landed about 80 km south-east of Dzhezkazgan, in Kazakhstan. The following day, *Pravda* triumphantly announced: 'The decade of space robots begins!'. With the Americans grounded after the drama of Apollo 13, the Soviet success could not have come at a better time.

Chemical analyses were carried out in a purpose-built laboratory at Moscow's Vernadsky Institute of Geochemistry and Analytical Chemistry. This included an admission chamber filled with neutral helium, where the samples could be heated for sterilisation, a vacuum chamber (of the order of 10^{-12} hPa) equipped with mechanical manipulators, and a very-low-background radiation chamber built underground and protected by 1-metre-thick concrete walls for radioactivity analyses.

The samples were found to have a slightly different composition to those that were recovered by Apollo 11 and Apollo 12 – in particular, being poorer in titanium. Three grammes of Luna 16 regolith were was given to America in exchange for 3 grammes each from Mare Tranquillitatis and Oceanus Procellarum – and a few milligrammes were given to British scientists. Three tiny fragments of rock, two grains of regolith and a tiny basalt chip were presented to Korolyov's widow, and on 11 December 1993 were auctioned by Sotheby's in New York, in the first and so far only legal sale of lunar material. The winning bid was an astonishing $442,500. At the same price-rate, the 380 kg returned by the Apollo missions could be worth an unbelievable $850 billion! Although all Apollo material is the property of the US government, some dust grains reportedly taken from the space-suit of Apollo 15 astronaut David Scott were later sold for $42,500 – but in this case no proof of authenticity was available.

Luna 17 lifted off from Tyuratam on 10 November 1970, and entered lunar orbit four days later. The orbit was circular at 85 km altitude, and had an inclination of 141°. The next day, the perigee was lowered to 19 km, and on 17 November the probe landed in a previously unexplored region of the Moon, near the northern rim of Mare Imbrium, and close to Sinus Iridum (Gulf of Rainbows) and Promontorium Heraclides,* at 38°.28 N, 35° W. Its payload was the first E-8 rover.

The idea of a mobile laboratory on the Moon was first published in the Soviet Union in 1954, in a popular science article which described a tracked 'tankette-laboratory' weighing about 100 kg that would explore the landing site by remote control from Earth.

Korolyov's team began studies of a lunar rover, called L-2, in 1960. It was decided quite early that it was to have a mass of at least 600 kg, but this was far beyond the lifting capability of the Molniya. Korolyov therefore designed the E-8 probe and its rover for an N-series rocket – in particular, for the N-2, which was to be capable of placing 16 tonnes in Earth orbit. To design a lightweight chassis, Korolyov turned to the State Committee of Tractor and Agricultural Machine Building, but in May 1963 the proposal was turned down because it was deemed impossible to design a lunar chassis weighing only 100 kg. Korolyov therefore turned to the Mobile Vehicle Engineering Institute – better known as VNII 'TransMash'. The rover project was given the formal go-ahead on 3 August 1964, as part of the edict that ordered human lunar missions. Soon thereafter, work on the N-2 was halted, so the rover was transferred to the Proton variant that would employ Korolyov's stage D for translunar injection. This switch further constrained the E-8's mass. The first TransMash-designed prototype was demonstrated in 1965. Referred to somewhat mundanely as Sh-1 (Shassi – Chassis), it had four tractor wheels, of which only the rear pair could be steered. Other track-mounted vehicles, as well as six- and eight-wheeled vehicles, were also studied. The definitive 150-kg, eight-wheeled prototype was built in 1967.

---

*   Heraclides Ponticus (*c.* 390–310 BC), a student of Plato.

Sh-1 (Shassi – Chassis), built in 1965, was the first Soviet lunar rover prototype. (Courtesy VNII TransMash.)

When the management of unmanned lunar missions was transferred to Lavochkin in 1965, VNII TransMash and Lavochkin had to coordinate their activities. This led to some pathfinding experiences such as the soil-mechanics analyses undertaken by Luna 13 and the transmission-gear tests onboard Luna 11 and Luna 12. Motors, transmissions and wheels were tested on simulated lunar regolith and simulated one-sixth gravity on a Tu-104 flying laboratory.

The end result of these studies was Lunokhod 1. This cylindrical vehicle had a diameter of 215 cm, a height of 135 cm, and a mass of 756 kg. Under the cylinder were four articulated pairs of 51-cm-diameter wheels (built by the Kharkov state bicycle plant) made of a metallic mesh designed to be both light and flexible. Each wheel was equipped with an electric motor and a small explosive charge to sever its axle and so set it free to rotate in case the motor failed. Two tractor wheels on each side were enough for a successful mission. Steering was accomplished by rotating the wheels on each side at different speeds, the minimum steering radius being 80 cm. Each wheel had a disk brake to immobilise the vehicle during stops, particularly at night. An automatic system stopped the vehicle each time its tilt was high enough to risk tipping over. The vehicle's track was 2.21 by 1.6 m. A ninth non-tractor wheel with spikes on its rim to prevent it from slipping was mounted on the back to act as an odometer to measure the distance travelled. By comparing the measurements of the odometer with those of the main wheels, further information on the lunar regolith could be inferred. The design speed was 0.8–2 km/h. The hermetically sealed magnesium cylindrical 'bathtub' body contained the computer, the cameras, the scientific instruments, and the environmental control system. The internal temperature was kept constant by the forced circulation of air that was heated by the radioactive decay of a small mass of polonium-210.

On top of the cylindrical module was a rear-hinged cover with solar cells on its interior surface. During the day, the cover was kept open to generate electricity, and at sunset it was closed to minimise internal heat dissipation. From the probe protruded the telemetry antennae and a 3.7-kg TL laser retro-reflector analogous to those carried on the Apollo missions. The TL (Télémetrie Laser – Laser Telemetry; and Terre–Lune – Earth–Moon) consisted of fourteen glass prisms, and was built in France by Sud Aviation Cannes after an agreement between the Soviet Union and the French space agency CNES (Centre National d'Etudes Spatiales).* Because of the secrecy which cloaked Soviet space projects, the French designers had been shown only the instrument mounting interfaces and informed that it would be mounted on a type of lunar vehicle.

A Lunokhod mock-up on show at the Paris Air Show in 1971. The French-built retro-reflector is visible under the conical antenna at left. The 'boxy' object under the twin front cameras is the RIFMA spectrometer. The 'lid' with solar cells is open. (Courtesy Patrick Roger.)

*   In 1966, President De Gaulle had signed a cooperation agreement with the Soviet Union in the field of space exploration, which was to prove very fruitful. Other than Lunokhod, early examples of French–Soviet cooperation included instruments onboard the Soviet Mars probes and the unflown Rozo (Reed) scientific satellite. Rozo (Roseau in French) would be launched on a highly elliptical Earth orbit with an apogee of 250,000 km, although studies were also carried out into the possibility of placing it in lunar orbit. It was eventually flown, as a Soviet-only mission, as Prognoz (Expectation).

The Lunokhod cameras were grouped according to their task. To the first group belonged a single camera, mounted on the front, which took a photograph every 20 seconds and relayed it to Earth for guidance purposes; and to the second group belonged four cameras, which were, between them, able to photograph a 360-degree panorama when required. The two front cameras, the guidance camera and one of the panoramic cameras were housed in a pair of protruding ports which gave the vehicle a vaguely 'biological' appearance.

The scientific instrument suite consisted of an X-ray telescope, a cosmic-ray telescope, an X-ray fluorescence spectrometer (often referred to by the acronym RIFMA – Roentgen Isotopic Fluorescence Method of Analysis) to measure the composition of the surficial lunar material, and the PrOP (Pribori Ochenki Prokhodimosti – Cross-Country Capability Evaluation Instrument) penetrometer,

One of the most important instruments carried by the Lunokhods was PrOP, which combined an odometer (the wheel at left) and a pantograph-mounted penetrometer (at right). (Courtesy VNII TransMash.)

As Lunokhod's camera built up a panoramic view of its landing site, it caught the ramps on its landing stage.

which measured the force required to drive a 4-cm-long metallic wedge into the regolith and the torque required to rotate it against the grip of four projecting vanes.

The probe was controlled from the Deep Space Communication Center near Moscow, where the Lunokhod's 'crew' consisted of five people: a commander, a driver, a navigator, a system engineer, and a radio operator, all of whom had trained on simulated lunar regolith produced in Simferopol.

Two hours after landing, and after ejecting the dust covers, the front and back cameras imaged the landing site. After confirming that no boulder was obstructing them, the four ramps were unfolded, and the vehicle slowly crawled to the surface. (The second pair of ramps were a contingency in case the first were blocked.) After travelling 20 metres, the rover checked the status of the landing module. On the second day, Lunokhod stood still to recharge its batteries; on the third day it travelled 90 metres; and on the fourth day it travelled another 100 metres, climbing up a small hill with a slope of 10°. Meanwhile, a problem was discovered that would greatly reduce the distance that could be travelled during the mission: the braking system had failed in the 'on' position, and the robot had to drive against this friction. On the fifth day, some 197 metres from the landing module, Lunokhod closed its cover and waited for the encroaching lunar night. On 5 and 6 December, the first and last distance measurements were taken using laser beams shot from Pic du Midi, in the French Pyrenees and Simeis in the Crimea. These measurements proved to be barely distinguishable from the background 'noise' level, so the retro-reflector was abandoned. It may have been mounted incorrectly, its cover may have failed to open fully, or possibly it was contaminated with lunar dust.

On 9 December the Sun rose over Lunokhod, and the ground crew could see the lunar panorama crossed by the long shadows of dawn. The next day was dedicated to battery charging, and on the following days the robot travelled further, to a total of 1,522 metres. On one day it crossed 300 metres, on another it climbed a hill with a slope of 23°, and on another the automatic control system had to intervene when the slope reached 27°. Finally, a few days before Christmas, and a new sunset, the mountains of Promontorium Heraclides appeared on the horizon ahead. On 12 December the RIFMA observed a solar flare that, according to TASS, would have been fatal for astronauts on the Moon. The third lunar day was mostly devoted to approaching a boulder which appeared huge in the guidance images but was only a small rock, and Lunokhod ran over it repeatedly – highlighting the difficulty of remotely-controlling the vehicle. Moreover, the cameras were mounted very low on the body, and this altered the perception of the environment. Craters into which it would be quite dangerous to step, appeared as simple strips of dark shadow until the very last moment. Other problems surfaced at lunar noon, when the lack of shadows and contrast rendered steering impossible. On 18 January, the navigation system was tested by returning to the landing site. On 10 February 1971, the probe experienced a lunar eclipse (or, from the probe's vantage point, an eclipse of the Sun by Earth), during which the temperature dropped by almost 250° C, although it recovered within a comparatively short time. It spent the rest of its fourth lunar day exploring some small craters, and was then parked 1 km from the lander.

During the fifth day, the rover explored most of the perimeter of a 500-metre-diameter crater. During the sixth day it had to climb the rim of a crater. Because this required the activation of all the motors at their maximum torque, it had then to halt to recharge its batteries. During the seventh day the rover explored a plain covered by ejecta from a nearby crater, although it travelled only 197 metres. On the next day it travelled 1,559 metres, on the ninth day 220 metres, and on the tenth day 215 metres – when the sunlight glinting on it enabled it to be spotted by the Apollo 15 astronauts flying overhead. During the eleventh and last day only 88 metres were travelled. Finally, on 14 September the internal pressure suddenly dipped, after which telemetry was lost. On 4 October – the anniversary of the launch of Sputnik – the end of the mission was announced.

Lunokhod 1 was designed to work for 90 days, but continued its mission for eleven months. Despite its brake being stuck 'on', it travelled a total of 10.54 km, took 20,000 guidance pictures and 206 panoramas, and mapped 80,000 square metres of the lunar surface. It carried out 500 soil-mechanics measurements and twenty-five analyses of the composition of the regolith, which proved to be similar to the Apollo 12 site in Oceanus Procellarum and quite unlike Mare Tranquillitatis, being much poorer in aluminium and potassium, and slightly richer in calcium. It was the most successful Soviet lunar mission ever undertaken.

5.4   MORE SOVIET MISSIONS

In May 1971, while Lunokhod 1 was active, the Soviets took advantage of a very favourable launch window to dispatch three sophisticated orbiter/lander probes to Mars. Although the first became stranded in parking orbit, and was written off as Cosmos 419, the other two were successfully dispatched as Mars 2 and Mars 3. Luna 18 was launched on 2 September, on a sample-return mission. On 6 September it entered lunar orbit (100 km circular, at 35° inclination) and four-and-a-half days later (almost twice the time required by Luna 16 and Luna 17) it began to descend. The landing site was in a mountainous region dividing Mare Crisium from Mare Foecunditatis at 3.57° N, 56.5° E. Unfortunately, contact was lost about 100 metres

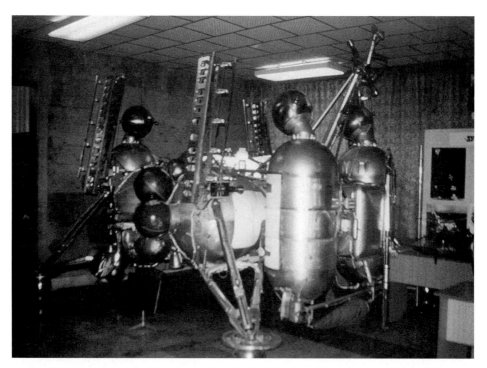

The KT lander stage of the unflown Lunokhod 3, showing the twin sets of slides. (Image copyright Wolfgang Hausmann.)

above the surface, as the probe was orienting itself to the correct attitude, and it crashed. Moscow acknowledged the failure of the mission, which was caused, it said, by the unfavourable conditions of an automatic landing in a highland region.

Luna 19 – the first third-generation orbiter of the E-8LS type – was launched shortly afterwards, on 28 September 1971. On 3 October, it entered a 140-km-high orbit with an inclination of 40° and a period of two hours. This was later lowered to a 127-km-high circular orbit, from which the first part of the scientific mission was carried out, during which it mapped the lunar gravity field of the near-side, measured the micrometeoroid flux and the magnetic field of both the Moon and of interplanetary space, in cooperation with Mars 2 and Mars 3, at that time *en route* to Mars, Venera 8 in solar orbit, and Prognoz 1 and Prognoz 2 in high Earth orbit. In November, the perigee was lowered to 77 km in order to carry out a photoreconnaissance of some areas of particular interest. Among the few photographs taken by the probe and published by the Soviets were images of the highland area between 30° and 60° S and 20° and 30° E, and pictures of Eratosthenes,* Sinus Aestuum (Seething Bay) and Mare Vaporum.

---

*   Erathostenes (276–194 BC), Greek geographer.

Luna 19 also studied the Sun, observing ten solar flares, and exploited fifteen radio-occultations to study near-lunar plasma. However, only seven attempts yielded useful data. During the first three attempts on three consecutive orbits on 8 May 1972, no trace of plasma was found; but on 11 June a slight electron concentration was noted, having a maximum some 10 km above the lunar surface. The difference between the two attempts resided in the different Earth–Moon–Sun angle, which on 11 June was close to zero (when the Moon was 'new'). Nothing was detected over the shadowed hemisphere on other occasions. Finally, gamma-ray spectrometric measurements were taken of the composition of the lunar surface, the electrical characteristics of the regolith were studied, an altimetric reconnaissance was carried out using the Vega radar altimeter, and the orbital perturbations of the mascons were mapped.

The Luna 19 mission lasted more than a year, up to October 1972, during which time the probe completed 4,000 orbits of the Moon.

Meanwhile, on 14 February 1972, Luna 20 was launched with the aim, as stated by TASS, of further exploration of the Moon and near-lunar space; in fact, it was the replacement for Luna 18. After four days of cruising – during which time optical telescopes in the Crimea and the Caucasus were used to track the spacecraft in order to compute a very accurate trajectory – the main engine fired to brake the E-8-5 probe to a 100-km circular orbit at an inclination of 65°. Twenty-four hours later, the perigee was lowered to 21 km in preparation for a landing. On 21 February the engine was fired for 4 min 27 sec and, after following a slightly different descent profile, Luna 20 set down at 3°.53 N, 56°.55 E, near the small crater Ameghino* in the highlands between Mare Foecunditatis and Mare Crisium, about 150 km north of Luna 16. The landing site was at an elevation of about 1,000 metres, and was a plain sloping at 8°. The navigation had been so good that Luna 20 set down only about 1,800 metres from the wreck of Luna 18.

As the probe had landed in daylight, with the Sun around 60° above the horizon, it was possible to use the camera to image the ground, which was littered with regolith-free rocks, and to take photographs of the sky to accurately identify Earth's position. After a suitable sampling point had been identified, a modified drill was brought into action. However, after about 10 cm it encountered very hard material and had to repeatedly pause to prevent the drill bit from overheating, with the result that after forty minutes it had gained only another 5 cm. The sampling operation was therefore terminated 2 hrs 14 min after landing, and the sample of only 30 grammes (or according to another source, 50 grammes) was recovered.

On 23 February, 27 hrs 39 min after landing, the ascent stage lifted off. The greatest peril came at the end of the mission, on 25 February, when the capsule descended into a severe snow storm and, to the alarm of the recovery team when they spotted its high-visibility orange parachute, drifted towards the Karakingir river, about 40 km north of Dzhezkazgan. The possibility of a splash-down on water had been contemplated, and the probe was equipped with two inflatable floatation bags,

* Florentino Ameghino (1854–1911), Italian historian and naturalist.

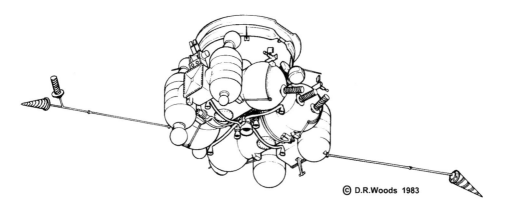

© D.R.Woods 1983

A line drawing of the Soviet E-8LS probe. Its family links with the Lunokhods are evident.

but the recovery operation would have been quite difficult because the river was partially frozen. Fortunately, the capsule landed on a small island and was recovered the following day, as by then the storm had settled, and it was possible to reach the island with Mil Mi-6 helicopters. (There was also an attempt to reach the island with off-road vehicles, but this proved impossible.)

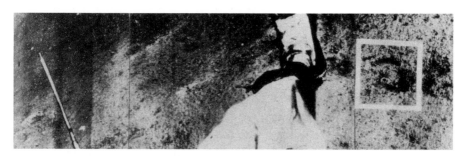

Luna 20 took this image immediately after withdrawing its robotic arm (centre) from the sampling area (in the white square). Luna 20 was the only Soviet sample-return spacecraft to land in daylight and, apparently, the only one to return images. (Courtesy ITAR-TASS.)

Despite the very low mass of the sample it was considered priceless, as it had been collected in the lunar highlands. The very high percentages of aluminium and calcium oxides in the sample were particularly interesting – a composition that was indicative of anorthosite. Indeed, most of the fragments were anorthositic, whereas on the mare sites sampled such material had comprised only a few percent. This reinforced the Apollo 15 conclusion that the highlands are the intensely bombarded remnant of the original lunar crust, formed as 'scum' on a global 'magma ocean'. As after Luna 16, 2 grammes of the Luna 20 sample were exchanged for 1 gramme from

Hadley–Apennine. This exchange included the two largest grains – weighing 40 milligrammes and 30 milligrammes – for which American scientists determined a precise date, which the Soviets were unable to do on such tiny samples: 3 billion years.

Luna 21 was launched on 8 January 1973, with Lunokhod 2. The translunar cruise was problematic, as telemetry was lost at least once, and the Lunokhod cover had to be opened to recharge the batteries of the descent stage. Nevertheless, it achieved lunar orbit (90×100 km and $60°$ inclination) and, on 16 January, touched down at a speed of about 1.9 m/s – the slowest lunar landing ever – inside the breached crater Le Monnier* at $25°.85$ N, $30°.45$ E.

Lunokhod 2 was an improved version of the original. Its mass had been increased to 840 kg, its top speed had almost doubled, and more instruments were carried, including an ultraviolet telescope, a magnetometer, and a third front camera at an adult human's height above the ground for a better sense of perspective. The guidance images were now relayed to Earth at four different rates (every 3.2, 5.7, 10.9 or 21.1 seconds) – the previous fixed rate of an image every 20 seconds, combined with a 2.6-second round trip, had created some driving problems. The French TL laser reflector was still in place, and the Soviets had added a photocell that, with a radio signal, registered strikes by the laser beams. This time the performance of the reflector proved faultless. A new and improved version of RIFMA was also carried, using a tritium and zirconium alloy on a tungsten bead as a source of X-rays instead of a tritium and titanium alloy on a molybdenum bead.

A few hours later Lunokhod 2 was on the surface. After pausing to recharge its batteries, the vehicle scrutinised its landing stage – risking a bump with it, as it was only 4 metres distant when it was stopped. A serious malfunction in the navigation system could have complicated the mission had not the Soviets been unofficially provided by NASA with some very-high-resolution Apollo 17 photographs of the landing area.

During the first lunar day, Lunokhod 2 set a record when it crossed 1,148 metres in six hours, as it travelled south to the crater rim, 6 km away. During the following night the thermometers measured a temperature of $-125°$ C at the wheels, and $-183°$ C on top of the antennae. As it slowly climbed up an $18°$ slope to the summit of a 400-metre-high hill during the second day, its wheels sank into the regolith up to their axles. From the summit, it returned a panorama which included the rim of Le Monnier and the Taurus Mountains, in a valley of which some 150 km to the south, Apollo 17 had landed a few months earlier. Looking upwards, it also took images of the Earth, of which only a thin slice was visible. The third and fourth lunar days were dedicated to the exploration of a small canyon called Fossa Recta (Straight Crevasse), 50–60 metres deep, 400–500 metres wide, and 6 km long, of which the robot explored both rims. Alongside the crevasse, the signature of a

---

*   Pierre Charles le Monnier (1715–1799), French astronomer.

A model of Lunokhod 2, exhibited at the 1973 Paris Air Show. Note the different front camera arrangement. (Courtesy Patrick Roger.)

The French-built TL2 laser retro-reflector was mounted on the Soviet Lunokhod 2 rover. It is the only Soviet deep-space experiment that is still in service. (Courtesy Philippe Jung, Alcatel Space.)

weak fossil magnetic field was detected. On 9 May, Lunokhod left the canyon on a heading to the north-east. The same day, it rolled into a small crater and some dust covered the solar panels, which dramatically reduced the available power. Shortly after dawn on the fifth day, on 3 June, TASS announced that it had malfunctioned.

The most important discovery made by Lunokhod 2 was the variability of the composition of the regolith across the mare-filled crater. Near the landing site, 6 km from the crater rim, the percentage of aluminium and iron was intermediate between the mare and the highlands. As it approached the rim, the composition became more like the highlands. As it travelled eastward, parallel to the rim, no measurable variations were recorded. Near Fossa Recta, the percentage of iron was higher than elsewhere. It was also discovered that a good correlation existed between the 'tone' of the regolith, measured by a 39-point grey-scale, and its composition – its iron content, in particular.

Lunokhod 2 took 80,000 guidance images and 86 panoramas, and travelled about 37 km (although this figure is approximate, as the navigation system had failed). Interestingly, photometric scans of the sky indicated that optical astronomy from the Moon in daylight would be impaired by suspended dust that diffuses sunlight to create a sky background fifteen times brighter than the darkest terrestrial night. Of course, a telescope operating during the lunar night would have a magnificent view!

Learning from previous experiences, a third Lunokhod was prepared with two stereoscopic and orientable cameras on the front, and without the laser reflector, but it never flew, and is now on show in the small museum of the Lavochkin Association.

A Lunokhod 2 panorama of the south wall of the crater Le Monnier. (Courtesy the Vernadsky Institute, Moscow.)

The grey-scale calibration target mounted on Lunokhod 2. Rover tracks can be seen in the background. (Courtesy Dr. Vladislav Shevchenko.)

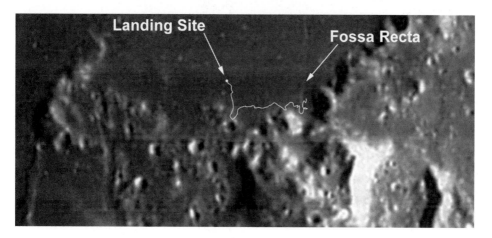

Lunokhod 2's route along the southern rim of the breached crater Le Monnier.

5.5 A LUNAR RADIO TELESCOPE

A few months after Apollo 17 returned to Earth, NASA launched Explorer 49 as RAE-2, the second spacecraft in the Radio Astronomy Explorer series, which was to be stationed in lunar orbit. Although no-one realised it, no further American probes would be sent to the Moon for twenty-one years!

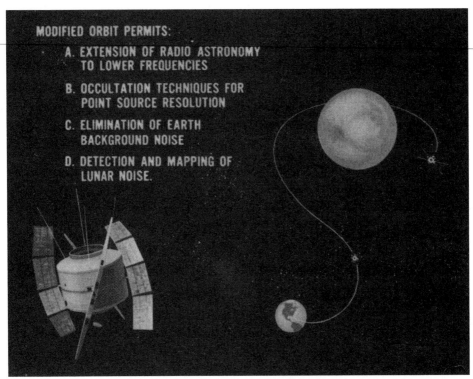

MODIFIED ORBIT PERMITS:

A. EXTENSION OF RADIO ASTRONOMY
   TO LOWER FREQUENCIES

B. OCCULTATION TECHNIQUES FOR
   POINT SOURCE RESOLUTION

C. ELIMINATION OF EARTH
   BACKGROUND NOISE

D. DETECTION AND MAPPING OF
   LUNAR NOISE.

A NASA drawing of the Radio Astronomy Explorer spacecraft and its mission.

The RAE project was conceived at NASA Goddard Space Flight Center in 1962, and began with an investigation to identify the observations most suited to this kind of spacecraft. The study culminated in the preliminary design of a radio telescope intended to observe the sky at wavelengths unobservable from below the Earth's ionosphere. The scientific objectives were to map the spatial and spectral distribution of cosmic radio noise from the Milky Way in the chosen frequencies, and to collect information on galactic magnetic fields; solar coronal radio emissions; the Jovian magnetosphere and possibly the magnetospheres of the other giant planets of the outer Solar System, and, hopefully, to discover new cosmic radio sources and new astronomical phenomena.

Construction of the RAE-1 satellite began in 1965, and on 4 July 1968 it was launched on a Delta rocket as Explorer 38 from Vandenberg AFB in California into an unusual circular orbit 5,800 km high, with an inclination of 120° with respect to the equator (that is, a retrograde orbit). For several months, the small satellite observed the radio sky at frequencies of 0.2–9.2 MHz, but it was also subjected to the continuous radio interference emanating from Earth – both natural (aurorae and thunderstorms) and artificial – which surpassed extraterrestrial sources by as much as 40 dB (a factor of 100). The mission was a success, but because it was considered to be experimental no catalogue of observed radio sources was published.

Encouraged by the results obtained by the first satellite, NASA decided to modify the flight plan of the second RAE and send it to orbit the Moon, where it would be safely removed from terrestrial interferences and would benefit from periods of complete occultation of our planet by the lunar disk, lasting as much as 20% of its orbital period. In addition, as the Moon does not possess an ionosphere, it was also possible to stretch the spectrum of detectable frequencies.

At 328 kg, RAE-2 was considerably heavier than its predecessor (200 kg), in part due to the need to include a solid-rocket motor. It was a small cylindrical satellite, with a diameter of 92 cm and a height of 79 cm. Power was generated by four solar panels, and the radio telescopes used two dipole antennae 18 metres long, and four wire antennae each 229 metres long with a diameter of 0.005 cm, which, once the satellite was in lunar orbit, were unfurled to form a huge 'X' configuration. The probe was also equipped with a 129-metre deployable boom for control of libration motions (pendular movements around the yaw and roll axes). The radiometers could observe the sky at nine frequencies in the range 25 kHz to 13.1 MHz – a part of the electromagnetic spectrum rarely observed from Earth, extending from VLF to LF, MF and HF (Very Low, Low, Medium and High Frequencies). Three receivers, each with thirty-two channels, recorded transitory events, and a known-impedance probe was used for calibration of the instruments. For reference, two small vidicon cameras took images of the area of sky that the radio telescope was observing.

RAE-2 was launched on a Delta-1913 rocket from Cape Canaveral on 10 June 1973. The navigation to the Moon was quite problematic, because a new panoramic sensor designed to measure the attitude of the probe with respect to the Sun, the Moon and the Earth proved sensitive to stray light, and mission controllers had to discriminate by hand between real data and spurious data. On 15 June the probe entered a lunar orbit with a perigee at 1,123 km, an apogee at 1,334 km, and an inclination of 61°.3. This was changed to 1,053 × 1,063 km at 38°.7, with a period of 221 minutes. The spacecraft then jettisoned its solid-rocket motor. For the first phase of the scientific investigation, the dipole antennae were extended and the spacecraft set spinning at 4 rpm. After three weeks, the dipoles were retracted and the wire antennae were unfurled, which had the effect of halting the rotation and orienting the spacecraft in the gravity gradient so that it maintained a fixed axis with respect to the Moon as it pursued its orbit. Although the Moon-pointing antennae extended only to 183 metres, observations were feasible.

Observing our planet was particularly easy at the time of 'full' Moon, when the Moon is inside the tail of the magnetosphere. The real advantage of being in lunar orbit, however, was that there were lengthy periods of time during which the Earth was occulted by the lunar disk and thus did not interfere with observations. In addition, the sharp limb of the Moon – which from the vantage point of RAE-2 had an apparent angular diameter of 76°, some 150 times the apparent diameter as seen from Earth – provided for occultation-observations of several radio sources both inside and outside our Solar System.

Between August 1973 and June 1974, RAE-2 observed occultations of the Sun and Jupiter, the most powerful radio sources in the Solar System, but Saturn and at least seven deep-space objects proved too faint to be detected. (The deep-space radio

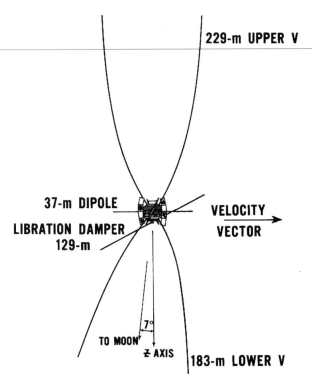

The lunar orbital configuration of the Radio Astronomy Explorer spacecraft. RAE-2 remains the largest spacecraft (in terms of geometric dimensions) to have ever orbited the Moon. (From Alexander, J. K., 'Scientific Instrumentation of the Radio Astronomy Explorer 2 Satellite', *Astronomy & Astrophysics*, **40**, 365–371; copyright European Southern Observatory.)

sources that RAE-2 tried to detect were Fornax A, Pictor A, Taurus A, Hydra A, Virgo A, Hercules A and Cygnus A, several of which are galaxies.) The probe mapped the Milky Way at low and very low frequencies, to provide a map of galactic magnetic fields and their sources. At shorter wavelengths, the centre of the galaxy was clearly defined, and with increasing wavelengths, large structures of our galactic neighborhood (including supernova remnants only a few hundred light years away) became more and more evident.

In November 1974 the two jammed antennae finally unfurled to their full length. The mission was completed in June 1975, and contact with the craft was maintained sporadically until August 1977.

5.6   THE LAST MISSIONS

After Luna 21, there was a hiatus of more than a year before the Soviets returned to the Moon, during which time their piloted lunar programme was permanently

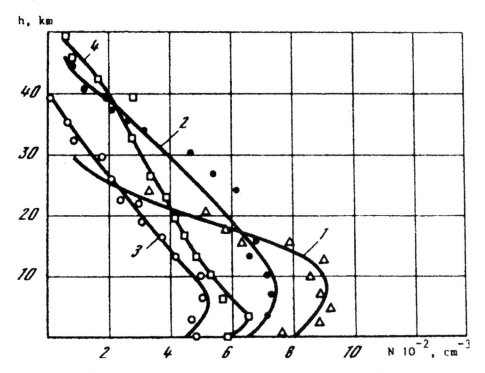

Altitude profiles of electron concentration near the Moon, as measured by Soviet lunar orbiters using radio occultation techniques: 1, Luna 19, 11 June 1972; 2, Luna 22, 18 August 1974; 3, Luna 22, 19 August 1974; 4, Luna 22, 21 August 1974. The variability of electron concentration with time (and Moon phase) is evident, as is the maximum concentration some 8 km above the surface. (From Kotenikov, M. A., Savich, N. A., Yakovlev, O. I., 'Spacecraft Radiophysical Investigations of the Sun and Planets', in Kotelnikov, V. A. (*editor*), *Problems of Modern Radio Engineering and Electronics*, Moscow, Nauka.)

cancelled, and the focus shifted to the exploration of Mars; unfortunately, however, only one of the four probes launched toward the Red Planet in 1973 achieved any sort of success.

Luna 22 lifted off from Tyuratam on 29 May 1974, and on 2 June entered a 220-km-high lunar orbit with an inclination of 19° and a period of 130 minutes. It was a follow-up to Luna 19, the mission of which had ended a little more than a year earlier. After temporarily lowering its perigee to about 25 km, it conducted a four-day photoreconnaissance, surveyed the topography using a laser altimeter and mapped the composition of the regolith using a gamma-ray spectrometer. The perigee was then raised. It spent the next five months in an almost circular orbit to study the mascons, and to help design the last three sample-collection missions, of which only two were eventually carried out. During this phase, twenty-three micrometeroid impacts were recorded – three times less than during the

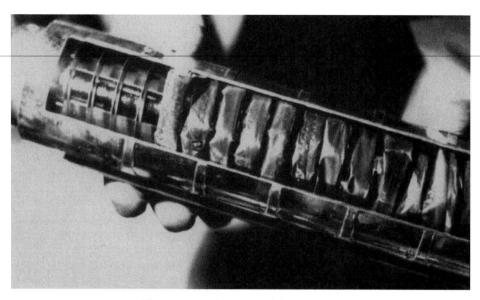

The core sample recovered by Luna 24.

initial, lower photographic phase. In November the probe manoeuvred into an eccentric orbit of 171 km perigee and 1,437 km apogee. This orbit enabled non-focused observations of the entire lunar environment. On 24 August 1975, the perigee was once again lowered to 30 km, and the cameras – dormant for fourteen months – were switched on again, after which a single image, of rather good quality, was obtained. On 2 September the probe, by then in a $100 \times 1,286$ km orbit at $21°$, ran out of attitude-control gas. Contact was maintained sporadically for another two months. For some reason, no image taken by (or of) the probe has ever been published.

Like Luna 19, Luna 22 carried out twenty radio occultation experiments to detect cislunar plasma. The occultations of 18, 19 and 21 August 1974 proved particularly useful, as the Moon was then between 'new' and 'first quarter' phases. The model inferred from the occultation experiments revealed that a layer of ionised gas, tens of kilometres deep, forms above the sunlit hemisphere, and has a maximum electron concentration at an altitude of about 8 km.

Luna 23 was launched on 28 October 1974, and entered lunar orbit (94×104 km, at $130°$) on 2 November. On 5 November the perigee was lowered to 17 km, and the following day the probe landed inside Mare Crisium at $13°$ N, $62°$ E. This was the first E-8-5M – an uprated version of the E-8-5 – which was equipped with a new drill, mounted on a rail rather than an arm and capable of penetrating to a depth of 2 metres and preserving the sample's stratification, but did not provide the choice of sampling site. Unfortunately, the drill was damaged in a very rough landing, and no samples were collected. The useless probe returned diagnostic and engineering data for three days, until its battery power was depleted.

A model of an E-8-5M probe, showing the modified, rail-mounted drill.

A replacement mission was launched one year later, on 16 October 1975, but stage D did not ignite, and so it failed to reach parking orbit. The second replacement, however, was successful. Luna 24 was launched on 9 August 1976, and on 14 August it entered a 115-km circular lunar orbit at an inclination of 120°. By 17 August the perigee had been lowered to 12 km. On 18 August, it landed in darkness south-east of Mare Crisium, at 12°.75 N, 62°.2 E, a few kilometres from Luna 23 and where Luna 15 was supposed to have landed seven years earlier. To the north-west, 17 km away, was the 6-km-diameter crater Fahrenheit*, and it was hoped that ejecta from that impact might be collected.

*   Daniel G. Fahrenheit (1686–1736), British physicist.

This time, the drill survived the landing. The drills of Luna 16 and 20 had lost part of their loads when they had been extracted. The new rotary percussion drill was not only designed to drill deeper, its configuration would retain the sample. In order to store a 2-metre-long core sample on the small return capsule, the drill fed its material into an 8-mm-diameter flexible plastic tube, which was coiled by a helical screw into a compact package. The drill managed to penetrate to its full depth and recover a 170-gramme sample of the regolith. The ascent stage lifted off 22 hrs 49 min after the probe landed. As the return capsule flew back to Earth, the descent stage conducted engineering tests on the drill. The capsule was recovered on 23 August some 200 km from the small town of Surgut in Siberia – a recovery site that had not previously been used. The sample was extremely interesting, as, mixed with the regolith, there were some centimetre-sized pebbles. Three grammes were given to America in exchange for 0.5 gramme of Apollo samples, and 0.91 gramme was given to British scientists. The samples from Mare Crisium were quite rich in iron, but depleted in titanium.

As Lunokhod 3 was being prepared to fly as Luna 25 in 1977, its launch vehicle was requisitioned for a communication satellite. A third E-8-5M was also ready, but the politicians switched the effort to the ambitious 5M programme, which was to return a sample of Martian soil by the early 1980s, and even when this was cancelled in 1978 the waiting lunar probes remained grounded.

In total, the three successful Soviet sample-return missions returned only about 300 grammes of lunar soil, compared with the 380 kg returned by the six Apollo missions.

Due to the reliability of the E-8 probes, it is not surprising that some components were reused. The KTDU-417 engine and part of the descent-stage tanking were used

The Fregat upper stage of the Russian Soyuz launcher reuses several components of the E-8 lunar probes.

at the end of the 1980s in the new Fobos class interplanetary probes, and are now marketed as an upper stage of the Soyuz launcher, with the name Fregat (Frigate). Its first operational flight was in early 2000, and the payloads have included ESA's Mars Express probe in 2003.

5.7 LAST AMERICAN GLIMPSES

Some sixteen-and-a-half hours after launch on 3 November 1973, Mariner 10 took six mosaics of the Moon as it headed towards Venus and to three historic fly-bys of Mercury, the innermost planet. By a curious coincidence, it was the last probe built according to an architecture and a philosophy invented at JPL for the Ranger lunar probes, almost fifteen years earlier. The Mariner 10 images showed, for the first time, vertical perspectives of some areas of the lunar north polar zone, which during the Lunar Orbiter missions had been either in shadow or had been viewed only obliquely.

On 18 September 1977, two weeks after its launch, Voyager 1 took the first images showing both our planet and the Moon from a distance of 11.66 million kilometres, as it set off toward Jupiter and Saturn on an epic voyage that will take it to other stars in around 100,000 years.

On 12 August 1978, NASA launched ISEE-3, the third spacecraft in the International Sun–Earth Explorer series. It took up station in a 'halo orbit' around the Lagrangian point some 1.5 million km sunward of the Earth to provide early warning of gusts in the solar wind. Its base was a sixteen-sided 1.58-metre-tall polygon with a diameter of 1.77 metres, and it was spin stabilised. From the base protruded two long antennae, taking the total length to 14 metres. From the sides protruded four wire antennae with a total span of 92 metres, and four 3-metre-long booms carrying a magnetometer and plasma sensors. The launch mass was 479 kg.

It was discovered that, despite the lack of funds to build an American probe to study comet Halley – which was expected to reach perihelion in 1986 – it would be possible to modify the orbit of ISEE-3 to fly by the lesser-known periodic comet Giacobini–Zinner, and later, at a far greater distance, comet Halley. The project was authorised and, in order to accelerate the probe to the correct heliocentric orbit, it was decided to use the Moon's gravitational field for a series of fly-bys.

At the end of 1982 the probe was returned from the halo orbit and over the next year fly-bys of the Moon were made on 30 March at 19,570 km, on 23 April at 21,137 km, on 27 September at 22,790 km, on 21 October at 17,440 km, and on 22 December at a distance of only 120 km. This increased the heliocentric speed of the probe from 4,679 to 9,278 km/h, and allowed it to break free of the Earth's gravity. The final manoeuvre was complicated by power requirements, as the battery had failed at the beginning of the year and it proved necessary to completely shut down the probe for 28 minutes – the time it took to cross the lunar shadow. The operation was successful, and at this point the spacecraft was renamed ICE (International Comet Explorer). It made the first fly-by of a comet when it passed within about

7,000 km of Giacobini–Zinner on 11 September 1985. On 25 March 1986 it flew by Halley at a distance of about 20 million km.

The orbit of ICE is such that in August 2014 it will again approach the Earth. If the spacecraft systems are found to be in good condition, it will be possible to send it on a second fly-by of comet Giacobini-Zinner in 2018; or, as an alternative, the perturbations of the Earth–Moon system will be able to recapture it in Earth orbit, in which case it may then be possible to recover the probe to examine its coating of cometary particles.

5.8   PROJECTS OF THE 1980S

From 1978, the Soviets repeatedly announced that they were about to restart lunar exploration. As a first move, they were to launch an orbiter (based on Luna 19 and Luna 22) into a 100-km-high polar orbit, from where it would carry out a complete reconnaissance of the surface. The expected scientific payload was overwhelming: a camera, gamma-ray, X-ray and neutron spectrometers, a spectrophotometer, an infrared photometer, a laser altimeter, a mass spectrometer, a plasma sensor, a micrometeoroid detector, a magnetometer, and several other instruments. The data collected were then to be used to prepare a new updated E-8-5 probe, able to return samples from the far-side. But the years passed without a launch.

Among many unflown Soviet projects presented at the International Astronautic Federation's meeting in Brighton in England in 1987 were Vesta to the eponymous asteroid and Korona to fly by the Sun at a distance of 3 million km from its photosphere 'surface', and several lunar projects, including a polar orbiter in 1993 based on the new UMVL (Universalnyi Mars, Venera, Luna – Universal for Mars, Venus and the Moon) bus which first flew in 1989 as the Fobos probe, and a far-side mission scheduled for 1996, plus a fleet of newly designed and long-duration Lunokhods.

And much the same was happening in America. The first post-Apollo NASA lunar probe project was LPO (Lunar Polar Orbiter). Born as a Goddard Space Flight Center study, it was transferred to JPL in 1975, with its launch – via Delta rocket or Space Shuttle – expected in 1979 or 1980. The 482-kg probe was to have 64.5 kg of instruments, including a gamma-ray spectrometer to map the surface composition and investigate the possibility of water ice in permanent shadows at the poles, a multispectral camera, a magnetometer, an electron reflectometer, an X-ray spectrometer, two infrared radiometers, a microwave radiometer, a radar altimeter, and a 28-kg relay satellite to be placed in a 3,424-km-apogee orbit to study the gravity field of the far-side; the main probe was to operate in a 100-km-high orbit. A few years later this project was superseded.

Around the mid-1980s, NASA decided to reduce the cost of interplanetary missions by designing two standardised spacecraft buses on which mission-specific instruments could be mounted. The first bus – Mariner Mark II – was designed for the exploration of the outer planets, and the second – called Observer, and based on NOAA and Tiros satellites – was designed for exploration of the inner Solar System.

The proposed American Lunar Polar Orbiter probe orbiting the Moon, as the sub-satellite travels on a higher orbit. (Courtesy L5 News and National Space Society.)

The first two Mariner Mark IIs were to be the Comet Rendezvous and Asteroid Flyby (CRAF) and the Cassini Saturnian orbiter, but in 1992 Congress cancelled CRAF and Cassini was redesigned as a one-off configuration, and after several delays was launched in 1997 on a trajectory leading to Saturn in 2004. The first inner Solar System probe to fly was Mars Observer, which was launched in 1992, but it fell silent as it prepared for Mars orbit insertion.

One of the many proposed Observer missions was, of course, a lunar mission. Lunar Observer would be the second probe of the series, and would reuse as many components of the Martian mission as possible. The launch by an Atlas–Centaur rocket was tentatively scheduled for 1993, after which the probe would enter a polar orbit in order to carry out a complete cartographic, mineralogical and gravimetric reconnaissance of the Moon. However, after the loss of Mars Observer the series was cancelled.

As regards the prospect of human flights to the Moon resuming, on 20 July 1989 – the twentieth anniversary of the Apollo 11 landing – President George Bush (senior) approved a project elaborated by the National Space Council, which was headed by Vice President Dan Quayle. Flanked by the Apollo astronauts, Bush announced: 'First ... Space Station Freedom, our critical next step in all our space endeavours. And next, for the new century, back to the Moon; back to the future; and this time, back to stay. And then a journey into tomorrow, a journey to another planet – a manned mission to Mars.' This bold vision was called the Space Exploration Initiative (SEI).

On 2 November, details were presented of a study which involved 160 NASA managers. Five possible schedules were envisaged, depending on the funding and timing. The first human lunar mission would be carried out sometime between 2001 and 2004; a permanent base – the First Lunar Outpost – would be built between 2005 and 2011; and the first self-sufficient experiments (such as oxygen extraction from lunar rocks) would begin between 2005 and 2018. The first human mission to Mars would be carried out between 2011 and 2018. The fundamental cornerstone of these projects was the completion, by 1997, of the space station, to be used as a stepping stone for deep-space exploration. The cost of human lunar return was estimated at $100 billion, and the mission to Mars was another $158 billion. The operation of the lunar base through to 2025 would cost $208 billion, and the project as a whole was estimated to cost a staggering $541 billion between 1991 and 2025! The National Research Council recommended that, for economic reasons, the development of the requisite new technologies proceed at a pace less frantic than had been the case during the Apollo programme. In June 1990 – less than a year after Bush's speech – Congress deleted all SEI funding.

Timetable for the Space Exploration Initiative (c. 1989)

Milestone	Approach				
	A	B	C	D	E
Lunar emplacement	1999–2004	1999–2004	1999–2004	2002–2007	2002–2007
Lunar consolidation	2004–2009	2004–2007	2004–2008	2007–2012	2008–2013
Lunar operations	2010	2005	2005	2013	2014
Humans on the Moon	2001	2001	2001	2004	2004
Permanent habitation	2002	2002	2002	2005	–
Construct a habitat	2005	2006	2007	2008	2011
Eight crew	2006	2007	2007	2009	–
Oxygen production	2010	2005	2005	2018	–
Far-side sortie	2012	2008	2008	2015	2022
Lunar steady-state mode	2012	2008	2012	2015	–
Mars emplacement	2015–2019	2010–2015	2015–2019	2017–2022	from 2024
Mars consolidation	2020–2022	2015–2018	2020–2022	2022	–
Mars operations	2022	2018	2022	–	–
Humans on Mars	2016	2011	2016	2018	2016
Extended stay on Mars	2018	2014	2018	2023	2027

Two lunar probe projects were born out of SEI. The first was Artemis (the Greek goddess of hunting, and sister of Apollo) – a series of landers using a common bus on which would be mounted one of four possible 200-kg payloads. The most important payload was the Lunar Ultraviolet Telescope Experiment (LUTE). This 1-metre-diameter telescope, powered and heated by an RTG, would produce a catalogue of ultraviolet sources in a thin strip of the sky down to magnitude 27. Another very interesting payload included a pair of 60-kg rovers, each of which was designed to last for ten days. They were to carry a suite of instruments consisting of an alpha-ray spectrometer (similar to the old Surveyor spectrometer), a Mossbauer spectrometer to carry out high-quality mineralogical analysis, and a stereo camera. The first pair were to land near the eastern rim of Mare Tranquillitatis, and the second would visit the Apollo 15 site and carry out a dramatic descent into Rima Hadley to investigate the stratified rocks discovered by the astronauts.

The second project was Lunar Scout, consisting of two orbiters. Equipped with a stereo camera with 4-m maximum resolution, an X-ray spectrometer, and a neutron spectrometer, Lunar Scout 1 would enter a low polar orbit in 1996. Lunar Scout 2 – to be launched twelve months later – would carry a gamma-ray spectrometer and other instruments to study the composition of the surface. It would fly an elliptical orbit so that it could be used as a relay satellite for Lunar Scout 1 as that mapped the gravity field on the lunar far-side.

The two missions, coordinated by the Johnson Space Center, were designed in 1991, and were to cost $100–150 million each, but after receiving no funding in either 1993 and 1994 they were quickly forgotten.

6

A small invasion

After Explorer 49, launched in 1973, more than twenty years passed before another American probe was sent to the Moon. In contrast, Luna 24 is destined to remain the final Soviet lunar mission.

A similar situation occurred in the broader arena of interplanetary exploration, when, after the amazingly successful missions launched in the 1970s (Viking and Voyager), America took a long pause through the 1980s, during which time other actors stole the show: the Soviets with their missions to Venus, Mars and its satellite Phobos, the international project to study Halley's comet involving the European Space Agency's Giotto probe, Japan's Sagigake (Pioneer) and Suisei (Comet), and the two Soviet VeGa probes.

6.1   GALILEO

On 18 October 1989, Space Shuttle *Atlantis* was launched from Cape Canaveral, carrying a precious load: a 2,550-kg probe which had cost around $900 million, and more than $2 billion dollars including launch and operational costs. The probe was named Galileo, in honour of the Italian scientist. It was to have been launched in May 1986 by the Shuttle mission STS-61G using a powerful Centaur upper stage to reach Jupiter. In January of the same year, however, *Challenger* was lost high above Florida during launch. This was followed by a two-and-a-half-year suspension of American piloted spaceflight, and the introduction of new safety regulations which included a ban on the use of the cryogenic Centaur. The Galileo mission had to be completely redesigned. The new route was much longer, and had to rely on the less powerful Boeing IUS (Inertial Upper Stage) two-stage solid-propellant upper stage, augmented by gravity assists in the inner Solar System. The new mission profile included a launch toward Venus, a fly-by of that planet, a fly-by of Earth, the first encounter with an asteroid (951 Gaspra), a second very close fly-by of Earth, an encounter with another asteroid (243 Ida), and the release of an atmospheric probe shortly before the spacecraft became the first human artefact to enter orbit around the gas giant, in December 1995.

Two images of the Moon's north polar region taken by Mariner 10 (left) and Galileo (right). The Mariner 10 image shows half of the near-side and part of the far-side. The bright area towards the upper right is the young crater Giordano Bruno. The Galileo image shows near-side regions. Mare Crisium is visible in both images.

The launch was perfect, as was the Venus fly-by – which was even more scientifically important, as another NASA probe, Magellan, was heading for the planet. During the first Earth encounter, it was decided to test Galileo's instruments on the Moon. The fly-by date was 8 December 1990, and Galileo took images of the Moon on both the inbound and outbound legs of its trajectory, with a surface resolution as good as 400 metres, and also produced the first infrared map of the lunar far-side. The maria appeared rich in basalt and thus rich of iron, and the highlands were metal-poor. The ejecta blanket surrounding the Orientale basin was found to be primarily plagioclase-rich (anorthositic) 'highland' rock. It had been thought that such a major impact would have excavated pyroxene-rich rock from deep within the crust.

On 11 April 1991, with Galileo finally in the relatively benign environment beyond the Earth's orbit for which the spacecraft had been designed, the command was sent to deploy its umbrella-shaped high-gain antenna, but this did so only partially because several of its 18 graphite-epoxy ribs had become welded to the catches which held their tips tight against the central support column – ironically, it turned out, due to inadequate lubrication during the three years that the probe had been in storage awaiting launch. This seriously jeopardised the scientific potential of the mission. To make the best possible use of the low-gain antenna, the engineers reprogrammed the spacecraft with image-compression algorithms. Meanwhile, the images and other data from the historic asteroid Gaspra fly-by on 29 October 1991 took weeks to relay to Earth.

The second fly-by of Earth took place on 8 December 1992, when Galileo skimmed just 304 km over the Earth's surface and finally attained a speed sufficient to reach Jupiter. During this occasion, it once again observed the Moon with a surface resolution of 1.1 km. The same work as was carried out on the far-side two years earlier was this time carried out on the near-side. The infrared imagery highlighted the compositional differences between the various maria. This time, Galileo's trajectory took it over the Moon's north pole, which was observed by the ultraviolet spectrometer to search for signs of water ice inside shadowed craters, and to test a controversial theory by Iowa University's Louis Frank – a former assistant of James Van Allen, working with the Pioneer lunar probes during the 1950s – that the Earth and the Moon are each day struck by hundreds of thousands of small comets, which release water. Frank predicted that the Moon should be hit by about 300 of these 'mini-comets' per day. The Apollo seismometers, however, had not detected signals from such impacts. Galileo's spectrometer did not see any evidence of the water that such impacts would leave. This seriously questioned the theory. Nor did the instrument shed any light on the putative ice in the shadowed north polar region.

At last leaving its mother planet for good, Galileo took a dramatic sequence of the Earth turning on its axis, with the Sun reflecting on the oceans and the Moon slowly orbiting around it.

6.2   RISING SUN OVER THE MOON

As early as 1982, NASDA (National Space Development Agency), the Japanese space agency dealing mostly with space technology applications, announced that it was considering a lunar mission, not unlike the US Lunar Observer. The 650-kg probe would be launched by an H-1 rocket (a Japanese version of the American Delta) and enter a 100-km-high lunar polar orbit. Its 85 kg of scientific instruments were to characterise the lunar surface and fossil magnetic field for no less than a year, after which the orbit was to be progressively lowered to an eventual impact. However, this project was cancelled, in favour of a proposal by Japan's other space agency, ISAS (Institute of Space and Astronautical Sciences).

The ISAS mission involved two spacecraft. The first was an 84-cm-tall, 1.4-metre-diameter drum with a mass of 185 kg. Referred to as MUSES-A (MU Rocket Space Engineering Satellite) during development, it was renamed Hiten (a Buddhist angel playing music in sky) once safely in space. The second was a tiny twenty-six-faced polyhedral satellite with a maximum diameter of 36 cm and a mass of 12 kg, including a 4-kg solid-propellant rocket, called Hagoromo (the angel's robe). Hiten carried some technological experiments and a single scientific instrument: a dust detector built by the University of Munich. Hagoromo did not carry scientific instruments, it was to relay telemetric and diagnostic data to the ground. Hiten's role was to test navigation and communication techniques to be used by a new generation of Japanese deep-space probes. In particular, the mothership had to demonstrate the use of the WSB (Weak Stability Boundary) transfer orbits, using

The Japanese Hiten–Hagoromo stack was the first lunar probe launched neither by the United States nor by the Soviet Union. Hagoromo is the polyhedral object on top. The upper surface of the Hiten spacecraft is covered with thermal protection used during the aerobraking manoeuvres.

solar gravity perturbations at great distances from Earth to modify the parameters of its orbit without using fuel.

The two probes were launched on 24 January 1990 by a small Mu-3SII rocket – the same type as had launched the twin probes Sagigake and Suisei towards comet Halley. They were injected into an orbit that intersected that of the Moon. An anomalous burning of the last stage, however, produced a total speed increment 50 m/s less than required. This was solved by starting Hiten's engine and consuming 20 kg of hydrazine, and by delaying the encounter with the Moon from the fifth to the sixth apogee. Some days before the rendezvous, however, the transmitter on Hagoromo failed. Nevertheless, it was decided to try to insert the silent probe in lunar orbit.

On 18 March, when Hiten was 16,472 km from the Moon, Hagoromo was released and its solid rocket was fired. The engine firing was photographed by the 105-cm Schmidt camera of the Kiso Observatory, and it is therefore certain that the tiny probe entered lunar orbit. The estimated orbit had a perigee of 7,400 km, an apogee of 20,000 km, and a period of a little more than 48 hours. Japan thus became the third country – after America and the Soviet Union – to launch a lunar probe.

Hiten, meanwhile, was still working, and accomplished a second lunar fly-by, at a distance of about 76,050 km, on 10 July, another on 4 August, and a fourth on 7 September. After a fifth fly-by on 2 October, the apogee rose to 1.35 million km, and the spacecraft was thus able to visit the terrestrial magnetotail for three months, to conduct a navigation experiment to assist the international Geotail mission that would be devoted to the exploration of this region of the magnetosphere. The fly-bys on 3 January, 28 January and 4 March 1991 lowered the perigee for an interesting and little-known experiment carried out on 19 March and repeated on 30 March. On each occasion, the probe skimmed the outer fringes of the Earth's atmosphere at 10 km/s. The first pass, at a height of 125 km, reduced its speed by 1.7 m/s and apogee distance by 8,600 km, and the second by 2.8 m/s and 14,000 km. This technique – called aerobraking – has since been used operationally by the Magellan orbiter at Venus and by several Mars orbiters.

At this point, the main mission of the probe was over, although a quarter of the original fuel load was still on board. With support from JPL, a mission extension was designed which included more fly-bys of the Moon and the first reconnaissance of the L4 and L5 Lagrangian points of the Earth–Moon system.* During the 1950s and 1960s, the Polish astronomer K. Kordilevsky published observations of small 'clumps' of dust near the two Lagrangian points – effectively, microscopic natural satellites of our planet. But this is not at all unusual, as asteroids are found near the Lagrangian points of the Sun–Jupiter, Sun–Neptune and Sun–Mars systems, and there are small moons near the Lagrangian points of some of the large moons of Saturn. As the only scientific instrument onboard Hiten was the dust detector, this investigation had scientific as well as navigational benefit.

After a ninth fly-by of the Moon in April which raised the apogee from 405,700 km to 1,532,100 km, on 2 October the probe once again flew by the Moon to begin its journey to the Lagrangian points: L4 was visited during the second half of October, and L5 five orbits later, at the end of January 1992. Although the dust detector did not record any increase in dust particle counts, the results are deemed inconclusive because Hiten never approached closer than about 10,000 km from the Lagrangian points.

---

*   These two points – which precede and follow the Moon on its orbit by 60° – were discovered theoretically by the French–Italian mathematician Joseph Luis de Lagrange in 1772, and have the characteristics of being the only stable equilibrium points for any body having a negligible mass with respect to the Earth's mass and the Moon's mass.

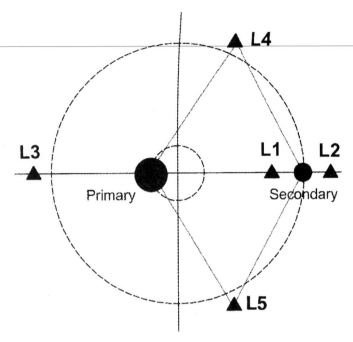

A diagram of the five Lagrangian points – a simplified solution of the equilibrium problem in a three-body gravitational system. L1 and L2 are called 'colinear points', and can fairly easily be calculated. L4 and L5 are the only stable equilibrium points. The L4 and L5 points of the Earth–Moon system were visited for the first time by the Japanese Hiten spacecraft. (Courtesy ESA.)

On 15 February 1992, Hiten encountered the Moon for the eleventh time, and its engine was fired for ten minutes, after which it entered lunar orbit with a perigee of 422 km and an apogee of 49,400 km.

With time, gravitational perturbations lowered the orbit more and more, and the last drops of fuel were used to move the computed impact point from the lunar far-side to the near-side, and at 18.03 GMT on 10 April 1993 it hit the surface at 55°.3 E, 34° S, near the large crater Furnerius.* It was the first artificial object to reach the lunar surface after a respite of seventeen years. On this occasion it proved possible, for the first time, to use the 3.6-m Anglo–Australian telescope to image the 5-km diameter cloud of hot plasma produced by the impact. In contrast with previous claimed visible-light observations of the same kind, which are probably spurious, the impact was observed in the infrared, and against the chill of the darkened Moon; and of course this time there was photographic proof.

As previously mentioned, Hiten acted as a pathfinder for the Geotail scientific mission, which must be included in this history because it accomplished fourteen lunar fly-bys and used lunar gravity to keep the orientation of its orbits fixed in order

---

* Georges Furner (*f*.1643), French mathematician.

to fly in the terrestrial magnetotail during each apogee. Geotail – a joint NASA–ISAS project – was a drum-shaped, 790-kg satellite built by NEC of Japan. It was launched, on a Delta-6925 rocket from Cape Canaveral on 24 July 1992. During its fifth orbit, on 8 September, it flew by the Moon at 12,647 km and raised its apogee from 426,756 km to 869,170 km. The fly-bys then continued at an almost monthly rate, with passes at distances between 22,498 km and 32,637 km raising the apogee to 1,400,000 km.

An interesting incident on this mission occurred during the ninth fly-by, at a distance of 34,076 km, on 26 September 1993. The computer of one of the most important scientific instruments (dedicated to the analysis of low-energy particles) had locked up and could not be rebooted or shut down. The encounter trajectory was then modified to allow the satellite to pass, for several minutes, inside the lunar shadow. Once in darkness, the batteries were disconnected, and the instrument resumed normal operations when the spacecraft emerged from the shadow and the solar panels again began to generate power.

Four more encounters then followed, at distances between 30,432 km and 74,300 km. The fourteenth and last fly-by, at a distance of 22,445 km, on 25 October 1994, placed Geotail in an orbit with a perigee of 50,000 km and an apogee of 180,000 km – the 'Near Tail' orbit – to explore the magnetotail close to Earth.

Geotail also contributed to lunar science by discovering new components of the tenuous lunar atmosphere: oxygen, silicon, sodium and aluminium – all of which were present in traces.

6.3   HUCKLEBERRY HOUND'S SPACECRAFT

When an American spacecraft was at last sent to the Moon after a hiatus of more than twenty years, it seemed like a return to the 1950s. The sponsor of the project – as for the early American lunar probes – was not NASA, but the Department of Defense.

In 1983 President Ronald Reagan created the SDIO (Strategic Defense Initiative Organisation) to pursue what became popularly known as the 'Star Wars' project. By 1993 this controversial project had spent several tens of billions of dollars on research without producing any operational systems, and had acquired a focus and changed its name to the BMDO (Ballistic Missile Defense Organisation). Several spacecraft were developed in order to assess technologies for detecting and tracking objects above the atmosphere, most notably MSX (Midcourse Space eXperiment), which carried infrared telescopes designed to detect warheads against the cold background of space, and DSPSE (Deep Space Program Science Experiment) to test a suite of optical sensors.

When DSPSE began in 1990 the plan was to develop a specific target vehicle for it to track, but when funding for ballistic missile defence was cut back, DSPSE found itself without an artificial target. Undeterred, the project leader, Stewart Nozette of the Lawrence Livermore Laboratory, had the idea of modifying the mission to track natural targets. As such a mission might well make a contribution to space science, in

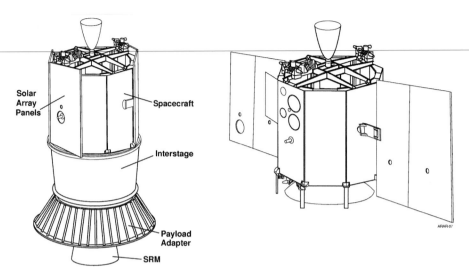

Solar Array Panels

Spacecraft

Interstage

Payload Adapter

SRM

ARAR-07

Two views of the lunar orbiter Clementine in both its (left) launch and (right) orbital configuration.

January 1992 he contacted NASA, which assured its support and made available its Deep Space Network. Eugene Shoemaker, a lunar geologist of considerable repute since the time of the Apollo flights, led the thirteen-member team of scientists assigned by NASA to the project.

With this redesigned mission, the probe was renamed Clementine, because – like the girl in the song about the 1849 Californian gold rush made famous by the Huckleberry Hound cartoons – after testing its suite of optical sensors on the Moon and one or two asteroids it would be 'lost and gone forever'.*

Clementine was built by the Naval Research Laboratory (NRL) in less than two years. Overall, the project cost $75 million, including launch and mission support. It was one of the cheapest interplanetary missions ever carried out. It should not be forgotten, however, that it benefited from several billion dollars of ballistic missile defence research.

The original programme included a launch to a low Earth parking orbit, from where Clementine would first be accelerated to a 169,000-km-apogee orbit, and then to a 386,000-km-apogee orbit. A month after its launch, the probe would enter a lunar polar orbit, where it would remain for two months to carry out the first global photoreconnaissance since the Lunar Orbiter missions. It would then exit lunar orbit and make a 19,134-km fly-by of the Earth to enter a 555,000-km-apogee orbit where, one month after leaving the Moon, it would again encounter it at a distance of 7,342 km and be inserted into solar orbit. Three months later, on 31 August 1994, it would fly by 1620 Geographos – one of the first of the near-Earth

---

* To further complicate the nomenclature, the Clementine probe was dedicated to the late Gerard K. O'Neill, an early promoter of space colonies, who died in 1992.

asteroids to be discovered – and image it in detail. After this, the probe would be really 'lost and gone forever'.

But since this flight plan would have left some unused fuel on board, two possible mission extensions were considered: one to lower the last lunar orbits to a few tens of kilometres, from where detailed pictures of all the Apollo landing sites could be obtained, and the other to perform a second asteroid fly-by. It was decided to fly by the unnamed asteroid 3551 (also known as 1983 RD) in October 1995. The probe was also expected to turn its cameras to Jupiter in July 1994 in order to observe the impact of comet Shoemaker–Levy 9, which was discovered in March 1993 by Shoemaker, his wife Carolyn, and amateur astronomer David Levy.

Clementine's payload included four CCD (charge-coupled device) cameras, each of which was mated to a Cassegrain telescope. The first camera covered the near ultraviolet and visible part of the spectrum with a six-filter carousel, five of which were chosen by the scientific team. Its resolution at the planned 430-km lunar perigee was 79–106 metres per pixel. The second operated in the infrared, had a six-filter carousel, and yielded a maximum resolution 112–128 metres per pixel. The third was a mechanically cooled infrared camera without filters which yielded a resolution of 43 metres per pixel. And the fourth was the high-resolution camera. It had six filters (military, opaque, and four scientific). Its maximum resolution – subject to motion blur – was 13–30 metres per pixel. By combining the different wavelengths, it was possible to assemble multispectral images which coded the surficial composition using 'false' colours. The probe also had a laser altimeter to monitor the spacecraft's height above the ground to an accuracy of 40 metres. In contrast with every other NASA probe, Clementine's instruments were not designed to be calibrated during the mission.

But Clementine was, first of all, a spacecraft designed to test new technologies for the military. These included very lightweight gallium arsenide solar panels, 350-gramme stellar cameras for attitude determination (a conventional star camera is up to ten times this mass), lightweight laser gyroscopes, a 2-kg reaction-wheel system, nickel–hydrogen batteries yielding an energy-to-mass ratio twice that of traditional space batteries, a solid-state 1.6-Gbyte data recorder, and a CNES-built chip that would compress data in JPEG format in real time.

To reduce the mass of the probe, its structure was made of carbon composites. It had an octagonal shape with a maximum width of 1 metre. One base carried the 1-metre-diameter high-gain antenna, and the other carried the main engine, stellar cameras and attitude control system nozzles. One of the sides carried the optics for the cameras, protected by a side-hinged light shield, and other sides carried the two solar panels. The total launch mass was 482 kg, divided between the structure and instruments (232 kg) and the fuel (195 kg of monomethylhydrazine and nitrogen tetroxide for the main engine, and 55.5 kg of monomethylhydrazine for the attitude control engines).

The launcher – the Titan IIG – also helped to minimise mission costs. During the 1960s, Martin–Marietta built 140 Titan II intercontinental ballistic missiles for the US Air Force, and twelve more for NASA, which used them to launch its Gemini spacecraft. The military Titan IIs – tipped with 9 MT nuclear warheads – spent

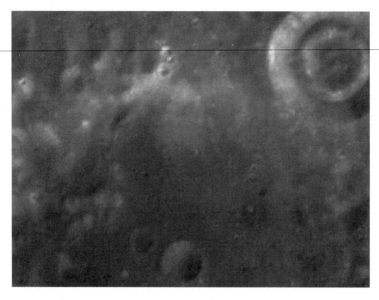

A raw Clementine ultraviolet/visual image of one of the Moon's most beautiful craters, Hesiodus A ($\sim$15 km), with its characteristic double wall. Clementine's orbit was selected so that it took its pictures looking straight down at terrain under full solar illumination, for optimal spectroscopic observation. (Courtesy US Naval Research Laboratory, Washington.)

twenty years in underground silos at Davis Monthan AFB, near Tucson, Arizona, at Little Rock AFB, Arkansas, and at McConnell AFB, near Wichita, Kansas. When they were decommissioned in 1987, thirteen missiles were converted to space launchers, and one (previously based at Little Rock AFB) was chosen to launch the first American lunar probe in twenty-one years. The refurbished missile lifted off from Vandenberg AFB in California at 16.34 GMT on 25 January 1994. The second stage placed its 1,150-kg payload of Clementine and a Star-37FM solid-propellant rocket stage into the desired 259 × 296-km parking orbit inclined at 67°.2 to the equator. On 3 February the Star engine was fired, and it faultlessly accelerated Clementine to its planned trajectory. Immediately after separation, the probe turned to take some pictures of its third stage. This was, in fact, a satellite in its own right, with a telemetric system and an experiment on the degradation of materials in deep space, which operated until mid-May before eventually re-entering the Earth's atmosphere.

On 19 February, after firing its retro-rocket for several minutes, Clementine entered lunar orbit. The initial orbit ranged from 430 km to 2,950 km, and had an inclination of 90° and a period of 5 hours. The instruments were tested, and mapping started less than a week later. The perigee of the initial orbit was in the southern hemisphere, between 27° and 30° of latitude, which facilitated mapping of the south polar region from a near-vertical perspective from low altitude, as Lunar Observer would have done. During each orbit, Clementine took pictures around perigee, and

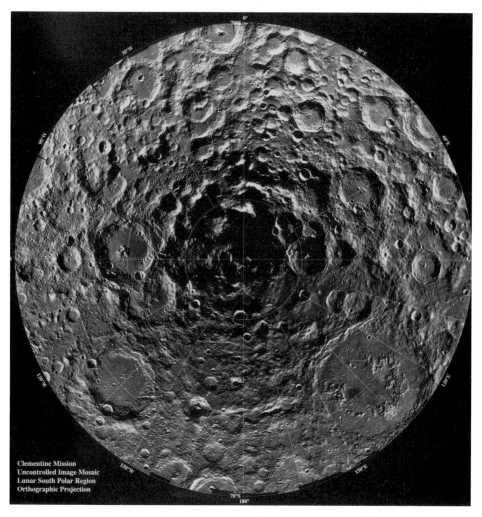

Clementine Mission
Uncontrolled Image Mosaic
Lunar South Polar Region
Orthographic Projection

A mosaic of the lunar south pole, taken by Clementine, including several previously unmapped areas. (Courtesy US Naval Research Laboratory, Washington.)

then, while approaching apogee, pointed its antenna towards Earth and relayed between 4,000 and 6,000 images (per orbit!), altimeter data, telemetric data, and data collected by a three-axis accelerometer that was to be combined with tracking data to yield a detailed map of the lunar near-side gravitational field and its anomalies. Over the period of a month, during which the Moon rotated on its axis beneath the spacecraft's plane, a longstanding cartographic hole near the south pole was filled in. On 25 and 26 March, the probe manoeuvred to flip its perigee into the northern hemisphere for a month of mapping the north polar region. Some small cartographic holes remained in certain wavelength. As a bonus, very-high-resolution images of some interesting sites were taken. The most impressive of these is an oblique panorama of the Apollo 16 landing site which, although not showing the LM's descent stage directly, clearly shows a bright patch of lunar dust disturbed by its exhaust. Clementine took almost 1,800,000 images of the Moon – more than all previous missions combined.

During Clementine's last days in lunar orbit, the most interesting observations were made. It was decided to use the probe as a bistatic radar (a radar with separate transmitter and receiver) to shed some light on a longstanding mystery. In 1961, Bruce Murray, Harrison Brown and K. Watson at Caltech had noted that because the Sun never strays more than 1°.6 from the lunar equator (compared with 23° on Earth) there might be some craters near the poles where the Sun never rises. These craters could preserve water ice deposited on the Moon by comets. The Apollo samples were completely arid, and seemed never to have encountered water, but as the Apollo missions did not venture into the polar regions it was impossible for them to test this theory. The observations by the Galileo spacecraft had not been encouraging. But Clementine, in a polar orbit, was ideally located to investigate this possibility. Water ice reflects radio waves much better when the angle of incidence of the 'illumination' is close to zero – a phenomenon called coherent backscatter – and in contrast to the reflection of radio waves by rock, reflection by ice preserves the polarisation of the wave. In March and April 1994 the spacecraft's radio signal was beamed into the darkened craters at each pole and the reflection was monitored by antennae on Earth. The results in March were inconclusive, but those in April were intriguing: two passes over the north pole were unremarkable but one of two passes over the south pole produced a reflection suggestive of ice.

During orbit 234 Clementine directly illuminated the south pole, and the receiving antenna on Earth recorded both an increase of reflection and an increase of the radiation percentage preserving the original polarisation. During the following orbit an area near 83° S was illuminated, but exhibited none of these effects. Clementine images also revealed a small crater exactly at the pole, and these factors led mission scientists to announce the possible discovery of ice on the Moon. Following this announcement, some results made by the Arecibo radio telescope a few years prior to the Clementine mission were published which called into question this conclusion. The Arecibo observations had been made using the big dish as a radar, sending out a pulse and then analysing the reflection from the Moon. But these results, too, were ambiguous. On the one hand, strong reflections were received from locations in the south polar zone that were not permanently shadowed, but on the other hand the strongest reflections were from sites that were. The Arecibo scientists preferred to explain their results in terms of particularly rough and radio-reflective terrain.

Once the mapping phase and other observations were over, on 3 May 1995 Clementine fired its engine for 4.5 minutes, exited lunar orbit, and headed for a fly-by with its home planet. Four days later, however, the probe experienced a major problem when a software bug commanded a firing of the attitude control engines, which lasted for 11 minutes and depleted the attitude fuel, setting the spacecraft into a fast spin at 80 rpm. When the mishap was discovered, consideration was given to using the main engine to slow the rotation rate, but this would not permit clear imaging, and so on 17 May, the Bat Cave (Clementine's mission control centre) announced that the asteroid fly-bys were cancelled, and that the spacecraft would remain in Earth orbit to test sensor resistance during repeated passes through the Van Allen belts. To set this up, it would be necessary to prevent the probe from returning to the vicinity of the Moon, for that perturbation would deflect the craft into solar orbit. A manoeuvre proved impracticable because the spin prevented the solar panels from generating sufficient power, and so on 20 July 1994 it encountered the Moon and slipped into heliocentric orbit, and operations were terminated on 8 August.

Clementine, however, proved to be much tougher than expected. The following February it was noticed that the probe's spin had temporarily placed it in the correct attitude to recharge its batteries. NASA's Deep Space Network antennae were again

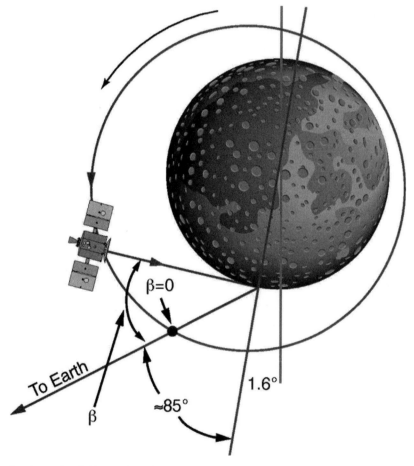

A schematic of Clementine's bistatic radar experiment that may have provided the first evidence of ice on the Moon. The lunar tilt relative to the ecliptic ($1°.6$), the lunar tilt toward Earth ($\sim 5°$), and the bistatic angle β between the spacecraft, the lunar surface, and the Earth receiver are all shown. (Courtesy Dr Stewart Nozette, DARPA.)

aimed at the probe on 8 February 1995, and contact was re-established twelve days later. After regaining control of the spacecraft several commands were executed, including firing the main engine to slow the spin, and taking engineering images using the science cameras. In consideration of the slowly increasing distance between the Earth and the spacecraft (by now travelling in an orbit with a 153-million-km perihelion, a 159-million-km aphelion, and a period of 387 days) operations were again terminated, this time permanently, on 10 May 1995.

A Clementine 2 mission was also studied. The first proposal was to send the probe to fly by the near-Earth asteroids 433 Eros and 4179 Toutatis, at which it would fire small kinetic projectiles originally designed to destroy enemy missiles. Consideration was also briefly given to converting Clementine 2 into a probe that would enter lunar orbit and fire small insect-rovers – called Huey, Dewey and Louie – to the surface. Louie would inspect the remains of Apollo 15, while Huey and Dewey were to carry out astronomical experiments. In the end, the first mission design was preferred. Clementine 2 would be launched in 1998 or 1999, and would fire projectiles at three near-Earth asteroids. In 1997, however, President Clinton vetoed this overtly military mission.

The satellite Wind served a similar mission to Geotail, in that it collected data on solar wind velocity, density, temperature and composition, but in this case from a vantage point upstream of Earth. The drum-shaped, 1,240-kg spacecraft had eight instruments, including the first Russian experiment – a gamma-ray detector – ever to be flown on an American satellite. It was launched on a Delta-7925 rocket on 1 November 1994. Its mission design included using lunar fly-bys to maintain the orientation of its orbit fixed with respect to the Sun.* The first Wind lunar fly-by, at a distance of 11,833 km, was accomplished during the fifth orbit, on 27 December 1994. More than thirty fly-bys have since followed, at distances from the surface between 2,809 km and 80,834 km. In 2001 the probe had its brief moment of fame when it was 'discovered' by one of the telescopes searching for near-Earth asteroids. After being classed as a 'space rock' with the designation 2001 DO_{47}, its true nature was revealed by orbital determination and by the observation of a course correction manoeuvre.

Like Geotail, Wind made several studies related to the Moon, particularly of the cavity produced in the solar wind plasma using a space-and-time resolution higher than ever before. A detailed map of ion and electron densities and energies, of magnetic fields, and of electromagnetic and electrostatic waves within the cavity downstream of the Moon, was obtained during the first fly-by. And on one fly-by, its plasma radio astronomy experiment served as the receiving antenna for a bistatic lunar radar at 8.075 MHz, a frequency never investigated before, and this, too, provided evidence of the existence of highly reflective locations.

The initial Wind mission design was to include several months in lunar orbit in order to relay data collected over the far-side by the polar orbiter which the Soviets were planning during the 1980s, but that mission never flew.

---

*   The same had been done for Geotail, but in this case the apogee was to remain downstream of Earth, whereas with Wind it was to remain upstream.

6.4   EUROPEAN FALSE STARTS

As briefly noted in the second chapter, Europe studied a lunar probe programme in the early 1960s; but it was never carried out, due mostly to the lack of a suitable launcher. The same fate was shared by several other programmes proposed during the 1980s and 1990s. The first of these was the Polar Orbiting Lunar Observatory (POLO – *Ital.* Pole), proposed in 1980.* It was to have been deployed from the Space Shuttle and used a PAM-A solid-propellant upper stage. Once inserted into a 300-km-high polar lunar orbit, it would have released a sub-satellite (their combined masses being 1,050 kg). Several years later this project was updated and proposed on three occasions by a team of Italian scientists under the name Moon Orbiting Observatory (MORO – *Ital.* Moor), initially in response to the tender for a 'mid-class' ESA scientific mission. MORO was to hitch a ride into space with a communications satellite launched by an Ariane rocket, and be released during the cruise up to the apogee of the geostationary transfer orbit, whereupon it would fire its own engine to head for the Moon. This 'piggyback' method of launch offered cost-savings, but the constraints on the launch window limited the number of launch opportunities because few payloads were that flexible.

Like POLO, MORO was to enter a 100-km-high polar lunar orbit and deploy a sub-satellite. Structurally, MORO was similar to the design adopted for ESA's Cluster scientific satellites. Its dry mass was 636 kg, including the 9-kg, 40-cm-diameter sub-satellite, and it would start with 570 kg of propellant. Its mission was to conduct very-high-resolution photoreconnaissance and studies in geology, gravimetry, altimetry and radiometry. Unlike previous probes, which had mapped the gravitational field on the far-side using indirect methods, MORO would have been able to conduct a more accurate study using the ultrastable quartz oscillator on its sub-satellite. By measuring the Doppler shift in the transmissions between the two spacecraft, orbiting about 100 km apart, it would be possible to determine the relative speed to an accuracy of 0.1 mm/s, and thus to map the gravitational field of the far-side by using the same technique as used for the near-side. Although spin-stabilised, MORO would carry a 16-metre-resolution camera, an ultraviolet, infrared and visual spectrometer, a gamma-ray spectrometer, a radar altimeter and a radiometer. However, a satellite to study anisotropy in the cosmic microwave background won the funding for the 'mid-class' mission, and MORO was redesigned as a 'small' mission, with a corresponding reduction in budget. In its second version, MORO became a three-axis spacecraft with a reduced payload of a camera capable of a resolution of four metres per pixel in selected situations, a radiometer, an X-ray fluorescence spectrometer and the sub-satellite. When this was rejected, MORO was redesigned once again and proposed to the Italian Space Agency as a 'small' scientific mission as the Lunar Polar Orbiter (LUPO – *Ital.* Wolf). In this form, its mass would have been a mere 250 kg. Nevertheless, it would have been able to carry

* ESA's current member states are Austria, Belgium, Denmark, Finland, France, Germany, Ireland, Italy, The Netherlands, Norway, Portugal, Spain, Sweden, Switzerland, the United Kingdom, and, as an external member, Canada.

An image of the first spin-stabilised version of ESA's MORO probe. The tiny sub-satellite is shown flying nearby. (From Chicarro, A., Racca, G., Coradini, M., 'MORO: a European Moon-Orbiting Observatory for Global Lunar Characterisation' *ESA Journal*, **18**, No. 3.)

a camera that could both take pictures and record spectra, a radiometer, a radar altimeter and the sub-satellite. To keep costs down, it would have been launched by a Ukrainian Tsiklon (Cyclone) rocket fitted with an additional upper stage, possibly the Italian Research Interim Stage (IRIS). Unfortunately, even this reduced mission failed to secure finding.*

In 1994 another ambitious technological mission was proposed: the Lunar European Demonstration Approach (LEDA). After hitching a ride on an Ariane V in late 2002, the 1,007-kg probe would first enter a lunar polar orbit and then land near the south pole between 83° and 85° latitude, where it would collect samples with an experimental robotic arm, and release a French rover powered by fuel cells and derived from Russian martian rover studies. Other instruments might have included an instrumented drill to measure the thermal flux and thermal gradients of the

---

* The Italian scientific community has also proposed an interesting mission to monitor the thin lunar atmosphere, using of an ultraviolet spectrometer mounted outside the International Space Station.

regolith, geochemical instruments, prototype synthesizers to release oxygen from lunar material, and an experiment for the detection of ice.

One of the biggest problems was the choice of landing site, which had to be free of large boulders that could jeopardise the landing. As the available imagery of the selected area had insufficient resolution (of a few metres), the probe would carry out a detailed reconnaissance of the target area prior to attempting to land – in the same manner as the Vikings did before they landed on Mars in 1976. Still greater problems were posed by the necessity of generating solar power while at the same time communicating with Earth, which, due to lunar libration, is sometimes below the horizon from such a latitude. A detailed analysis showed that these two conditions are met during just four consecutive months each year. For a launch in 2002, the landing could take place between January and April 2003. LEDA was submitted to ESA in October 1995, re-using existing technologies as much as possible in order to keep costs down, but it was soon superseded by a different project.

In 1996, the former Dutch astronaut Wubbo Okels proposed the Euromoon-2000 probe – a fusion of LEDA and MORO. The mission was born out of a request by the ESA committee for long-term space politics for a mission to introduce the new millennium. This high-profile mission would be managed and financed by unconventional methods: through sponsorship via the European Union, through fund-raising campaigns, and with a contribution of 50 million Euros (25%) from ESA. The total cost was estimated to be 1 Euro for each citizen of the EU. The 2,900-kg probe would be launched on a dedicated Ariane 4 towards the end of 2000, and would enter a 200-km-high lunar orbit, where a 400-kg MORO-based orbiter with 50 kg of scientific instruments would be released. After several months of photoreconnaissance by the main spacecraft of the target area – an unnamed 20-km-diameter crater precisely at the south pole – a specific landing site would be selected and the 1,000-kg lander would relay descent imagery to Earth, which would be shown live on television. Once on the surface, it would raise one of its solar panels for power, and release a rover (possibly even a group of mini-rovers) with a payload designed to search for water ice in the regolith. Meanwhile, the orbiter would spend a full year working with its sub-satellite to map the Moon's gravitational field. As an alternative to the sub-satellite, it might have been possible for the orbiter to work with an ultrastable quartz oscillator on the lander. As a revision of the project, it was proposed to replace the 200-km circular orbit spacecraft and sub-satellite with a 100-kg probe that would hitch a ride with a communications satellite on an Ariane V, be released in geostationary transfer orbit, and then establish itself in a highly elliptical lunar polar orbit with its perigee over the south pole. Once it had surveyed the target area, the lander would be launched. Unfortunately, in March 1998 ESA's administrators declared the 'anomalous' funding strategy to be unsatisfactory, and cancelled the project.

The European lunar lander Euromoon in its landed configuration. (Courtesy ESA.)

6.5   THE SEARCH FOR ICE

In a period of a little more than a year, NASA's image was seriously compromised by the technical problems of the multi-billion-dollar Hubble Space Telescope, by the loss of Mars Observer, and by the serious problems with Galileo's antenna. At the end of 1993, therefore, NASA administrator Dan Goldin, in an attempt to improve the agency's public image, initiated a new series of low-cost, rapid-development interplanetary missions called Discovery.

The first two missions were announced at the same time: Discovery 1 would be NEAR (Near-Earth Asteroid Rendezvous), designed to fly by asteroid 253 Mathilde and then rendezvous with and enter orbit around 433 Eros; and Discovery 2 would be Mars Pathfinder, designed to demonstrate that probes could make 'hard' landings on Mars using airbags to absorb the shock, and it was to carry a meteorological station together with a mobile robot equipped with an instrument for a mineralogical study of the rocks. On 17 February 1996, NEAR lifted off from Cape Canaveral on a Delta-7925 rocket. A few days later, its camera was turned on for the first time

Lunar Prospector, about to be mated to its launcher's third stage (in the background). On the left is the cylindrical gamma-ray spectrometer, on the right is the electron reflectometer, and at the tip of the folded boom is the magnetometer.

to take forty calibration pictures of the Moon, 1.5 million km away. A few weeks later, it took twenty-five pictures of the bright comet Hyakutake, which at that time was prominent in the skies of the northern hemisphere. After flying by asteroid Mathilde, NEAR flew by the Earth on 23 January 1998, took several photographs of the Moon's southern hemisphere, and finally headed for Eros.

Discovery 3 was chosen in February 1995. Lockheed–Martin was awarded a $63-million contract (a fortieth of the total cost of the Cassini probe, and a quarter of the cost of a film such as *Titanic*) to build and launch a NASA lunar probe, 25 years after Apollo 17 and RAE-2. This project – Lunar Prospector – was conceived some six years earlier as a private initiative, when the possibility was studied of launching a probe with the objective of detecting water ice at the lunar poles.

To minimise costs, the 300-kg probe was extremely simple, with a triangular carbon-fibre body based on the structure of the Iridium mobile telephone satellites housing a tank of hydrazine, the communications system, and a sequencer – there was no real onboard computer, as once every orbit the ground control centre would relay a list of operations to be undertaken. To this triangular structure were attached three curved solar panels, so that the probe was cylindrical, 129 cm tall and 137 cm wide. The attitude and course correction engines projected from the base, which mated with the interface to the launch vehicle. On the other end were two antennae – a directional medium-gain antenna and an omnidirectional antenna, mounted one on top of the other on a 1.6-metre-long conical structure. The omnidirectional antenna was used only for the translunar part of the mission, and all scientific data were relayed by the medium-gain antenna. The scientific instruments were mounted at the tips of three folding booms, 2.5 metres long. The act of unfolding the booms took five minutes, and slowed the rotation of the spacecraft from 48 rpm to 12 rpm.

Lunar Prospector had five instruments: an alpha-particle spectrometer, a neutron spectrometer, a gamma-ray spectrometer, a magnetometer, and an electron reflectometer. The gamma-ray and neutron spectrometers were to be used to study the composition of the regolith and, in the case of the neutron spectrometer, also to investigate the possible existence of water ice at the poles. The alpha-particle spectrometer (mounted on the same boom as the neutron spectrometer) would search for outgassing. The magnetometer and reflectometer, which were mounted together on the same boom, with the magnetometer on the tip of a 14.4-cm-long extension in order to position it 3.5 metres from the probe for magnetic isolation, were to chart the Moon's very patchy magnetic field. In addition, the Doppler shift of the radio signal would be used to map the gravity field of the near-side.

Besides building and managing Lunar Prospector together with NASA Ames, Lockheed was also awarded the launch of the probe on its three-stage solid-fuel Athena 2 (previously known as LLV-2 and LMLV-2). The launch was scheduled for June 1997, but as the date approached NASA became concerned that the Athena 2 had never flown, and that its two-stage version, Athena 1, had failed on its maiden flight. The launch was therefore slipped to the end of September. In August, Athena 1 launched a small scientific satellite, which failed shortly after entering orbit; and the investigation caused further delays. Then, to ensure that there was no conflict with the launch of the Saturn-bound Cassini mission in October, Lunar Prospector was rescheduled to 5 January, but on that day one of the trajectory control radars malfunctioned. Finally, on 7 January 1998 – thirty years to the day after the previous NASA unmanned lunar mission – at 02.28 GMT the Athena 2 lifted off from Cape Canaveral's Pad 46 (the pad last used to test Trident submarine-based ballistic missiles) and, after functioning perfectly, placed Lunar Prospector in a lunar transfer orbit. (This launch also marked another significant 'first', because the spacecraft carried a small capsule containing the ashes of the planetary geologist Eugene M.

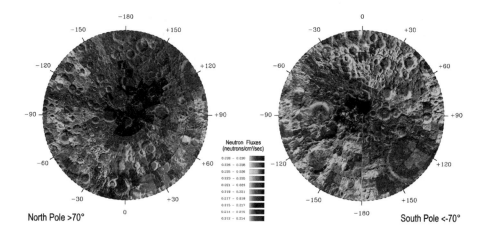

Maps of the counts of epithermal neutrons taken by Lunar Prospector over the lunar poles. Darker areas correspond to lower epithermal neutrons fluxes and, possibly, to larger ice deposits. (Courtesy NASA/Ames.)

Shoemaker, who had died in a car crash in Australia on 18 July 1997). During the first hours of the mission, the booms were extended and the scientific instruments were activated. On 11 January, after two course correction manoeuvres, Lunar Prospector entered a polar orbit ranging between 71 km and about 8,500 km with a period of 11.8 hours. Over the following two days, three engine firings decreased the orbital period to 3.5 hours and the apogee to 1,870 km, and then to 153 km, to enter the operational circular orbit (99×100 km), and data collection began on 16 January. The first rumour that something had been discovered leaked during the second half of February, and was confirmed by a science team press conference on 5 March: the neutron spectrometer had detected a widespread signature of water ice at *both* poles.

The preliminary results of the mission were published in *Science* on 4 September. In the search for ice, the neutron spectrometer had remotely sensed neutron counts in three energy classes: low (*thermal* neutrons, with energies of < 0.3 eV), medium (*epithermal* neutrons), and high (*fast* neutrons, with energies of $> 300,000$ eV). Calculations had established that the hydrogen released by the water ice should significantly reduce the flux of epithermal and fast neutrons, slowing them by collisions, and yet leave the flux of thermal neutrons unaltered. The observations revealed a decrease in particle counts of 4.6% within 30° of the north pole and 3.0% near the south pole for epithermal neutrons, and of 0.35% and 0.5% respectively for fast neutrons. Assuming ice deposits to be 2 metres deep, buried under a 40-cm-thick layer of regolith, and covering 1,850 square kilometres at each pole, the total mass of water ice on the Moon was estimated at 3,000 billion tonnes. This mass could have been deposited by cometary impacts, formed from the reduction of iron oxides by hydrogen in the solar wind, or erupted by volcanoes shortly after the Moon formed.

Although there are alternative explanations for the anomalous neutron counts, the water ice hypothesis is considered to be the most probable.

The neutron spectrometer was also used in concert with the gamma-ray spectrometer to study the composition of the surface globally. In particular, the gamma-ray spectrometer searched for radioactive elements such as thorium and potassium (which, as previously mentioned, are important in the study of lunar geology). A total of 356,691 gamma-ray spectra were used to produce a global map of the composition for every 150×150-km 'tile', of which a mean of 74 minutes of data were collected – a figure to compare well with the results obtained with the spectrometer used on the Apollo 15 and Apollo 16 missions, which mapped only the equatorial zone and collected a mean of 12.6 minutes of data for a similarly sized tile. It was thus discovered that thorium and potassium are usually found near the rims of maria in the western hemisphere, with a region of high concentration running from the southern arc of the Imbrium basin, near the recently added crater Copernicus, out across the basin ejecta that is the Fra Mauro Formation. Concentrations were discovered on the far-side near Mare Ingenii and close to the Aitken basin. The distribution of iron was found to be similar to that inferred from Clementine multispectral images.

Another important result was obtained in gravity field studies, as the near-side was mapped from an optimal low and polar orbit to a very high precision by measuring, every ten seconds, the relative speed between the probe and ground-tracking antennae to an accuracy of just 0.2 mm/s. New mascons were found on the near-side, corresponding to Mare Humboltianum and the craters Mendel and Schiller.* In fact, the shapes of single craters could be distinguished in the data. It was also possible to map the gravity field of the far-side by noting how the probe's orbit drifted between disappearance and reappearance from behind the Moon. Four previously unknown mascons were thereby discovered, of which only one – Mare Moscoviense – is filled with lava, the other three – Hertzsprung, Coulomb and Freundlich – being craters.**

The magnetometer and electron reflectometer analysed the fossil lunar magnetic field in detail, and confirmed the magcons identified by the Apollo P&F satellites and discovered several more. The data, collected in concert with the Wind satellite, have yielded the important discovery of a large magcon antipodal to the Imbrium basin with such an intense magnetic field that a small magnetosphere is created – which is even accompanied by a minuscule bow shock in the solar wind, the smallest such structure ever detected in the Solar System.

The alpha-particle spectrometer data suffered from the increased solar activity leading to the solar maximum in 2000, and preliminary results were published only in 2001. Despite covering only a small swath of the lunar surface within 15° of the equator, they reveal that a difference also exists between the near-side and far-side in

---

*   Gregor Johann Mendel (1822–1884), Austrian biologist; Julius Schiller (*d*.1627), German monk, and
    author of a celestial atlas.
** Ejnar Hertzsprung (1873–1967), Danish astronomer; Charles-Augustin de Coulomb (1736–1806),
    French physicist; Erwin Freundlich (1885-1964), German-British astronomer.

the production of alpha-rays. In particular, alpha-rays have been detected over some uranium-poor and thorium-poor areas, but it is unclear whether these originate from the release of deep gases triggered by lunar quakes.

As a remarkable cooperation between scientists and engineers, new gravimetric data were used in almost real-time for mission management, enabling the time between correction manoeuvres designed to keep the height above the surface fixed at 100 km to be extended to two months, in order to make the best use of the propellant remaining.

The mission proceeded smoothly until December, the only difficulty being the particularly arduous eclipse period during the autumn, which often required turning off some spacecraft systems to preserve power. On 17 November, Lunar Prospector was just one of the many near-Earth spacecraft facing the dangers of the Leonid meteor shower, which was expected to be particularly active in 1998 following the recent perihelion of its parent comet. For four days, the probe oriented its lower face toward the constellation Leo – the part of the sky from which the Leonids appear to radiate. However, not a single satellite of the many hundreds orbiting Earth was hit, and Lunar Prospector was unscathed. Meanwhile, many lucky Europeans (including the author) watching the skies during the night hours of 17 November saw a fireball show that they are unlikely to forget.

By 19 December the primary mission was over, and a one-month transition period was started before the beginning of the extended mission: the collection of equivalent data from a 25-km-high orbit, the maintenance of which would require much more fuel. The probe remained in this low orbit until the end of July 1999, when funding was due to end, but because it was still operational (with the single exception of the alpha-particle spectrometer) it was decided to end the mission in a dramatic way.

On 30 July the apogee was raised to 234 km, and the next day the engine was fired for the final time over the far-side, without any possibility of intervention from the ground. At 09.52 GMT on 31 July, Lunar Prospector hit the Moon at a speed of 1.69 km/s, impacting inside a permanently shadowed unnamed crater at 87°.7 S, 42° E. At that precise moment, at least twenty astronomical observatories, hundreds of amateur astronomers and the Hubble Space Telescope were observing the lunar south pole in the hope of detecting either the dust or, better still, the OH radicals produced by the dissociation of some 20 kg of ice that was expected to be thrown up by the impact to catch the light of the Sun. Unfortunately, at that time NASA's Submillimetre-Wave Astronomy Satellite (SWAS), designed to detect the OH radical, was above the side of the Earth opposite the Moon, and could observe the impact site only half an hour later. Although it is possible that the 64-m Kaliazin radio telescope near Moscow detected a change in lunar radio emission associated with a minor quake caused by the impact, there was no positive observation of Lunar Prospector's final experiment.

The mystery of the lunar ice deepens ... Perhaps the only way to be certain will be to land a probe to make an *in situ* study. It is unfortunate that ESA cancelled the Euromoon polar lander.

6.6   THE MOON IS BUSY AGAIN

A very unusual lunar mission began with a launch which predated Lunar Prospector by just a few days. On 24 December 1997 a Proton launcher lifted off from Tyuratam carrying the Asiasat-3 geostationary communication satellite for a Hong Kong-based consortium. The $70-million Proton put the satellite and its now well-known stage D, in its 'export' DM3 version, into parking orbit. Later, the upper stage fired and the complex was placed into geostationary transfer orbit. Upon reaching apogee, stage DM3 was to fire again for 110 seconds to circularise the orbit and allow Asiasat-3 to slowly drift to its operational position at a longitude of 122° E. But instead, the engine fired for just 1 second, stranding the satellite in the transfer orbit at an inclination of 51°.4, ranging between 365 km and 36,000 km.

Asiasat-3 was built by Hughes (the builder of the Surveyor landers) using the HS-601HP bus – one of the most successful communication satellite buses ever, of which dozens are in orbit. It was constructed around a 4-m-wide cubic central body carrying the attitude control engines, onboard systems, twenty-six C-band and sixteen Ku-band transponders. To the sides were attached two folding solar panel wings of 26.2 metres total span, generating a maximum power of 9.9 kW. The transponders used a pair of 2.72-metre-diameter antennae which increased the width of the satellite to some 10 metres. The dry mass of the satellite was 2,534 kg. It carried 1,680 kg of propellant.

With the satellite stranded in a useless orbit, the insurance companies repaid the satellite's owner, and Hughes bought it back and renamed it HGS-1 (Hughes Global Services). Then, starting in April 1998, a series of six manoeuvres managed to raise the apogee from 36,000 km to 413,600 km. On 13 May the spacecraft – spin-stabilised, and with its solar panels still folded – passed some 6,248 km from the lunar surface, the perturbation from which placed it in an orbit with its perigee inside the geostationary belt, its apogee beyond the Moon's orbit, at 488,000 km, a period of 15 days, and – significantly – with a reduced inclination. A month later, on 6 June, HGS-1 once again flew by the Moon at the much higher distance of 34,300 km. In mid-June, one last series of manoeuvres was carried out. On 14 June the engine was fired to lower the period to 46 hours, and was again fired two days later. On 17 June the orbit was finally circularised to become geosynchronous, with a slightly higher inclination than is usual. In this orbit, the satellite could be useful for services such as maritime communications, which do not require such precise pointing of the antennae. At the end of the operation – undoubtedly the most complex spacecraft rescue ever attempted, and the first lunar fly-by of a commercial vehicle – HGS-1 was leased to the American communication company PanAmSat, which operated it as PAS-22 at a longitude of 60°W over Brazil. The $4-million operation was entirely repaid by the end of 2001, and in mid-July 2002 the satellite was boosted off its station to await a new customer. Meanwhile, in March 1999 a replacement satellite, Asiasat-3S, had been launched on a Proton for the original customer.

The rescue of HGS-1 proved the feasibility of using the Moon as an intermediate step in the transfer of satellites to geostationary orbit. Although deliberate use of this approach would require some redesign of several spacecraft systems,

An artist's depiction of an HS-601 satellite not unlike HGS-1, the first private spacecraft to fly by the Moon. (Courtesy US Department of Defense.)

it would be possible to save some hundred of kilogrammes of fuel on each satellite.

A similar rescue mission has been proposed for Orion-3, an HS-601HP that was launched from Cape Canaveral on the second Delta III rocket – a new and more powerful version of Delta – on 5 May 1999. Once in parking orbit, the cryogenic second stage was to fire to move the spacecraft to geostationary transfer orbit, but the engine shut off after only a few seconds, stranding Orion-3 in an orbit at 162 km perigee and 1,378 km apogee. In 1999 a rescue mission was proposed using the Space Shuttle, as such missions had been carried out previously. In August 2000 an internal NASA study on the rescue of Orion-3 was released. The proposal was to grab the satellite with the Shuttle, mate it to a new solid-fuel upper stage and to a scientific instrument suite, and then to send it to a fly-by of the Moon before heading for geostationary orbit. But there would be no opportunity for such a mission before 2002, because the Shuttle was busy assembling the International Space Station, and in the end the project was put aside as overly expensive.

All of the other spacecraft that have observed the Moon in recent years have been on their way to targets in deep space. A few days after HGS-1 reached its final orbit, another spacecraft was heading for the Moon. On 3 July 1998, a Japanese Mu-5 rocket lifted off from the Kagoshima missile range carrying the small Planet-B probe – later renamed Nozomi (Hope) – which was designed to study the planet Mars, and in particular the interaction of its atmosphere with the solar wind. The strategy was to gradually increase the probe's energy by a series of lunar fly-bys such that when the window for Mars opened in December the final fly-by would accelerate it to escape speed and put it on course for Mars. It was first placed on a very eccentric 15-day Earth orbit, and on 24 September flew by the Moon at a distance of about 5,000 km, raising its apogee to 1.5 million km. On descending from the apogee on 18 December, the probe once again flew by the Moon, this time at a distance of 2,809 km. On returning to Earth on 20 December, it finally reached escape speed. The arrival at Mars was scheduled for October 1999, but an engine problem prompted this to be postponed to December 2003.

Nozomi has a 1.6-m-wide, 58-cm-tall body, to which are attached a parabolic antenna, some whip antennae, the magnetometer boom, and two 6.22-metre solar panel wings. The launch mass is 536 kg. One of the scientific instruments is a colour camera, and this was used to take some dramatic pictures of the Moon during its fly-bys. Its extreme ultraviolet camera also collected data on the composition of the lunar surface in an area between Mare Crisium and Mare Smythii.

Cassini–Huygens, a mission to Saturn staged by NASA, ESA and ASI (Agenzia Spaziale Italiana, the Italian Space Agency), lifted off from Cape Canaveral on 15 October 1997 on a Titan IV – Centaur launch vehicle. The huge spacecraft made two fly-bys of Venus and then returned to fly by Earth at a distance of 1,171 km on 18 August 1999. At this time, the probe closed to 377,000 km from the Moon, close to the mean Earth–Moon distance. For 15 minutes, calibration images of the Moon were taken using both the narrow-angle camera with a resolution of 2.3 km per pixel (comparable with the resolution of small amateur telescopes) and the wide-angle camera with a resolution of 23 km per pixel. Calibration spectra were taken by the combined infrared and visual spectrometer. In contrast with the images obtained by Galileo, the Cassini images have no particular scientific interest, as they are of the near-side, and at low surface resolution, which has been mapped in detail. After the fly-by, Cassini headed towards Jupiter, which it passed in December 2000, using the gravity-assist to head for Saturn, which it will reach on 1 July 2004.

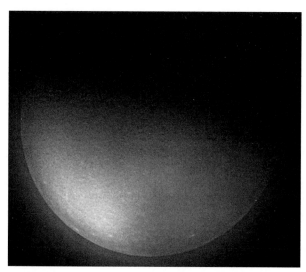

The Moon's north polar region as seen by the Stardust spacecraft from a distance of 108,000 km during its fly-by in January 2001. The bright spot is the crater Aristarchus. The haziness is due to heavy contamination of the camera mirror. (Courtesy JPL/ NASA/Caltech.)

Stardust (Discovery 4), launched on a Delta II rocket on 7 February 1999, is to fly by the periodic comet Wild 2 (pronounced Vilt) on 2 January 2004. When it arrives there it will image the nucleus and collect samples of the coma for return to Earth in January 2006. To reach the comet, it was necessary to perform an Earth fly-by at a distance of 6,000 km on 15 January 2001. Seventeen hours later, twenty-three calibration images of the Moon were taken from a distance of some 100,000 km. Unfortunately, the main camera mirror was contaminated by an unidentified oil-like substance, and in the images the Moon appeared as if surrounded by a halo. Attempts to sublimate the substance have since been successful, and the camera is now fully functional – as evidenced by the images of asteroid 5335 Annefrank, which the probe flew by in November 2002. In addition, the probe carries two pairs of metallic chips on which are etched the names of many hundred thousands of people (including the author) willing to express their support for the mission; one pair of chips will remain on the probe, and the second will be returned to Earth with the cometary coma samples.

Another Discovery probe – Comet Nucleus Tour (CONTOUR) – was to observe the Moon during several Earth fly-bys beginning in 2003, but was lost when its solid-propellant motor exploded at the time of solar orbit injection on 15 August 2002.

On 7 April 2001 a Delta II rocket launched the 2001 Mars Odyssey spacecraft. Twelve days later, the probe's visual and infrared camera snapped some images of the Earth–Moon system from a distance of 3.5 million km. On 8 May 2003, an even more distant image of the Earth–Moon system was taken by its sister probe Mars Global Surveyor, which had been orbiting Mars since September 1997. At that time, the Earth was 139 million kilometres away and in conjunction with Jupiter in the martian sky. This image shows both planets, three moons of Jupiter and the Earth's Moon, the latter having a bright spot, probably the large ray system associated with the crater Tycho. Pictures of the Earth–Moon system were also taken on 3 July 2003 by the European Mars Express spacecraft.

The most recent probe to visit the Moon was the Wilkinson Microwave Anisotropy Probe (WMAP) which was launched on a Delta II on 30 June 2001, on a mission to map the fine detail of the cosmic microwave background left over from the Big Bang. In order to reach its vantage point at the L2 Lagrangian point of the Sun–Earth system, 1.5 million km from Earth, the 840-kg probe made three highly elliptical Earth orbits, and flew by the Moon at a distance of 5,200 km on 31 July.

From a scientific point of view, an event likely to have a major impact on the next decade of lunar exploration was the Leonid meteor shower of 17 November 1999. As European observers were experiencing the most active meteor shower since the 1966 Leonids, some American observers, experiencing a less impressive local show, were instead concentrating their attention on the shadowed side of the Moon, using visual techniques or video cameras on small telescopes. It was thus possible to observe, for the first time, the impact of six large Leonid meteors with an estimated diameter of 20 cm, and to record their approximate impact sites; a feat which was repeated during the Leonid meteor shower in 2001.

A target for the next missions – and in particular for the Japanese SELENE probe (discussed in the next chapter) – will be to locate these fresh impact craters, which are estimated to a have diameters of several dozen metres.

7

The future

7.1 AMERICAN PLANS

With the end of the Lunar Prospector mission, America has no continuing or firmly scheduled lunar missions, with the possible exception of some fly-bys. None of the presently approved Discovery missions will visit the Moon, although some might carry out observations for instrument calibration. Deep Impact may observe the Moon after its launch in late December 2004, and Messenger could carry out some observations during its 2 August 2005 fly-by as it sets off for Mercury. But Genesis, launched in August 2001 to orbit the L1 Lagrangian point of the Sun–Earth system, will never approach the Moon closer than 100,000 km. And, of course, the Origins science project – intended to search for life in the Solar System and elsewhere – quite obviously has no need of the Moon.

However, a recent study undertaken by the National Research Council, on American policy in Solar System exploration in the decade 2003–2013, may revive interest in the exploration of our natural satellite. Signed by sixty top American planetary scientists, this study proposes a new programme to overlap with the Discovery programme, which is cost-capped at $325 million. The new programme – called New Frontiers – will fund a $650-million mission every two years. In addition, a single Flagship mission costing more than $650 million will fly during the decade.

Several lunar probe designs have been proposed for the Discovery programme, but apart from Lunar Prospector they were not approved. Mention is briefly made of the orbiters Icy Moon, Lunar Discovery, PELE and Diana – the latter being a mission using an ion-engine-propelled probe. Like Lunar Prospector, Diana would have mapped the Moon, but would have made use of a sub-satellite deployed in a 400-km-high orbit to study the gravitational field of the far-side, and after one year in lunar orbit would have headed for asteroid 4015 Wilson–Harrington, known to be a dormant comet.

There have also been unsuccessful proposals for landers and rovers to explore the polar regions in search of ice, or to study the Aitken impact basin, which is probably deep enough to have excavated the lunar mantle, in addition to scattering crustal material.

In 1997, Carnegie–Mellon University proposed the Lunar Ice Explorer to explore craters at the south pole. This was derived from a joint NASA–Carnegie–Mellon prototype called Nomad, a large four-wheeled 800-kg rover powered by fuel cells that tested in the Atacama desert in Chile in 1997. It travelled 215 km in 15 days, at an average speed of 1.5 km/h. In 2000 the mission was proposed again under the name Victoria, having as its target a four-month exploration. Another mission was Interlune One, proposed by former Apollo astronaut Harrison Schmitt. It would have used a large rover and a micro-rover to explore a geologically favourable area. The larger rover would also have carried a small telescope to study the lunar skyglow, the outer planets of the Solar System, and stars and galaxies.

The Johnson Space Center proposed the Jules Verne mission, which would have landed two rovers inside the Aitken basin to study the composition of rocks in order to find samples from the mantle.

One New Frontiers mission proposed by the National Research Council study is a probe to return samples from the Aitken basin. A preliminary mission could return 4.6 kg of material, including a 2-metre-deep sample, several selected small rocks, and some regolith. The probe will include four components: a lander, an ascent vehicle, a rover, and a radio-relay orbiter. The total mass of the vehicle will be close to 2,585 kg, and the launcher will probably be one of the Atlas rockets. However, such a mission is unlikely before the end of the decade. Similar probes may return samples from some of the most geologically interesting spots. Other than the Aitken basin, these include the craters Giordano Bruno and Copernicus, the floor of Tsiolkovsky, and the shadowed regions of one of the poles. Another sample-return mission, Moonraker, consisting only of a lander equipped with a sampling arm, was proposed as a Discovery mission in 2000 and rejected due to the cost.

A recent Discovery proposal is Polar Night, which would determine the chemical composition and deuterium-to-hydrogen ratio (an indicator of the origin of ice) in the polar ices by landing six penetrators in the polar regions. To perform this type of analysis, each penetrator would have to carry a mass spectrometer and a neutron spectrometer.

Apart from the Discovery and New Frontiers programmes, other lunar missions have also been proposed.

Blue Moon was proposed by the US Air Force Academy. It would have been a very small orbiter carrying a magnetometer, and operated in concert with Lunar Prospector but it was not funded. The American military is also studying the feasibility of very small probes costing around $10 million, which would hitch rides into geostationary transfer orbit, each of which would carry a couple of scientific instruments to the Moon.

A mission that will surely reach the Moon is Magnetospheric MultiScale (MMS), a NASA project using a number of small satellites to be built by Surrey Satellite Technology Ltd in England. After being launched into highly elliptical orbits in the summer of 2008, they will use two lunar fly-bys both to increase the inclination of their orbits to 90° and to reduce the eccentricity of their orbits, so as to explore the polar regions of the terrestrial magnetosphere.

Finally, one of several advanced launchers under NASA study is Lunar Tug – a reusable upper stage to carry payloads from low Earth orbit to geostationary or lunar orbit, and to return to Earth on a free-return trajectory.

7.2   JAPANESE PROBES

The Hiten and Nozomi missions (discussed in the previous chapter) prepared the ground for a new Japanese lunar probe: Lunar-A. This is a mission which will carry out some very detailed lunar seismological studies. The 520-kg probe is a short cylinder, 1.11 metres tall and 1.2 metres wide. On one end is the Mu-5 launcher interface and an antenna, and on the other is the orbital manoeuvering engine and two antennae. On the sides are hinged four square solar panels and two penetrators

A model of the ISAS Lunar-A probe in its current configuration, with two penetrators and four solar panels. (Courtesy Brian Harvey.)

designed to dig into the lunar regolith at very high speed carrying a suite of scientific instruments. The only instrument mounted on the probe is a monochromatic camera protruding from under a solar panel, to take 30-metre-resolution pictures of the surface.

The 84-cm-long, 16-cm-diameter, 13.6-kg penetrator spikes are provided with two independent propulsion systems. The first – which increases the total length of the craft to 138 cm – is a spherical solid-fuel rocket with a total burning time of 20 seconds, to brake the penetrator from orbital speed in order to ensure its fall to the surface. The second is a cold-gas system, of which two are mounted on each side of the centre of mass, to ensure the correct impact angle within 8° of the vertical. Each penetrator carries a two-axis seismometer (far more sensitive than those deployed on the Apollo missions), an array of twenty-three sensors of three different types to measure the thermal flux of the regolith, an accelerometer, an inclinometer, batteries with a life of one year, and a miniature radio system for relaying data to the mother ship. Each is designed to withstand an impact at speeds of 250–330 m/s and a maximum deceleration of 10,000 g as they bury themselves to a depth of up to 3 metres.

The penetrators are the newest part of the mission, as this is the first time that they will have been used on the Moon, despite having been proposed as early the 1950s.* Penetrator probes have also been designed for Mars, although as yet without success. Two were to have been deployed by Mars 8, but this spacecraft was lost within hours of launch in 1996. Shortly before it attempted to land in 1999, the Mars Polar Lander spacecraft deployed a pair of penetrator probes (named Scott and Amundsen) for the Deep Space 2 project, but they disappeared without trace. Due to their originality, it is not surprising that the penetrators for Lunar-A have suffered engineering problems: the electronic boards were deemed too delicate; the attitude control system malfunctioned; the separation system was insufficiently rigid and would have released the penetrators in an incorrect attitude; and during impact tests to certify the batteries – carried out at the Sandia laboratories in New Mexico during the winter of 1998 – it was discovered that the impact shock had cracked the instrument electronics.

The orbiter, too, had some development problems. In particular it was discovered that, depending on the launch window, the probe could experience long eclipse periods six months after lunar orbit insertion, during which time the batteries would not be able to provide enough power. However, due to the mass of the required additional batteries, the three original penetrators had to be reduced to two, with a consequent reduction in the scientific potential of the mission.

The launch of Lunar-A was originally scheduled for August 1997, but it has been repeatedly postposed, and is not now expected to fly until 2004. It will use a WSB transfer orbit, first entering a very elliptical Earth orbit with an apogee near the Moon, and after four orbits using a lunar fly-by to raise its apogee to 1,185,000 km.

---

*   According to one source, penetrators equipped with miniature seismometers were dropped over China by American U-2 spyplanes in 1960 in order to detect nuclear tests. The Chinese, however, did not carry out their first such tests until 1964.

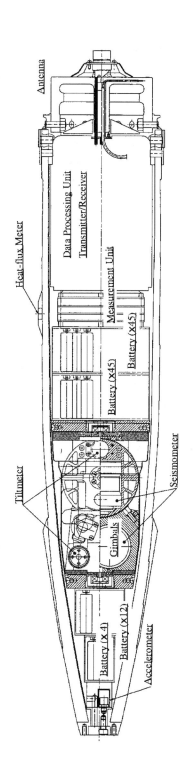

A cutaway of a Lunar-A penetrator. (Courtesy Takahashi Nakajima (ISAS).)

After another fly-by, around six months after launch, it will be inserted into a lunar orbit with an inclination of 30° and a perigee as low as 40 km.

During the next month, the penetrators will be deployed close to the equator, the first being aimed at the centre of the crater Mendeleyev on the far-side, and the second on the near-side, not far from Apollo 12.* For the third penetrator of the original mission, a high-latitude landing site inside Mare Crisium was chosen. It is deemed particularly important for calibration of the instruments that at least one of the penetrators lands on an area of the lunar surface for which samples have been returned to Earth, and for which the mechanical characteristics of the surface are known. Lunar-A will then raise its orbit to some 200 km, where the rest of the mission will be carried out. It will take photographs during terminator crossings, and act as a radio relay for the seismometers by dowloading data to Earth every 15 days. The mission is designed to last for one year after orbit insertion.

In 2005, Japan plans to launch SELENE (SELenological and ENgineering Explorer) as a collaboration between its two space agencies.** The project has a long and troublesome history. It was first proposed by NASDA in the 1980s as LOOM (Lunar Orbiting Observatory Mission), and afterwards became a joint project including a lander and two orbiters: a high-altitude orbiter working as a relay satellite to map the gravity field of the far-side, and a low-altitude orbiter with a large scientific payload. Due to financial problems, the lander was later modified to act as a propulsion module for the low-altitude orbiter and would land only when the latter's mission ended after one year. Although it would have a camera to inspect its landing site its primary role would be to act as an antenna on the near-side for a two-month very-long-baseline interferometry radio navigation experiment. Due to cost, however, the mission has recently been divided into two separate probes.

The 1,700-kg SELENE-A will be launched on Japan's H-IIA rocket, which has a capability similar to that of the European Ariane 4. The main spacecraft, which will enter low lunar orbit, will deploy a pair of 39-kg relay satellites, one of which (called RStar) will have an apogee of 2,400 km and the other (VStar) will have an apogee of 800 km; both will have a perigee coincident with the main spacecraft's 100-km circular orbit. RStar will be used as a relay to measure the position of the main spacecraft over the far-side. VStar will carry a transmitter for the very-long-baseline radio interferometory experiment which, by operating in concert with an identical source on RStar, will provide the positions of the three satellites with a mean accuracy of 20 cm – essentially a thousand times better than on than previous missions – to produce an extremely detailed global map of the Moon's gravitational field. The main spacecraft, in low polar orbit, will have a comprehensive scientific payload: an X-ray spectrometer, a gamma-ray spectrometer, an alpha-particle spectrometer, a multispectral camera capable of working in the ultraviolet, visual and infrared and with a maximum resolution of 20 metres, a spectral camera, a 10-

---

*   Dimitri Ivanovich Mendeleyev (1834–1907), Russian chemist.
** On 1 October 2003 NASDA and ISAS merged to form the Japanese Aerospace Exploration Agency (JAXA).

metre-resolution stereo camera, a laser altimeter with 5-metre accuracy, a sounding radar to determine the structure of the regolith to a maximum depth of 5 km, a magnetometer, plasma sensors and analysers, a radio astronomy experiment to study the tenuous lunar ionosphere and to observe the giant planets of the Solar System, and the very-long-baseline radio interferometry experiment. At the end of the one-year mission, the low-altitude orbiter may be lowered to 40 km in order to carry out more detailed observations.

In 2006, the 2,000-kg SELENE-B will be launched, carrying a second orbital module and a fully instrumented lander. However, studies have only just begun, and no details are yet known concerning either the scientific payload or the mission itself, except that it is to last six months.

In addition, a 3,700-kg SELENE-2 is being planned. Once in orbit around the Moon, this is to deploy a 50-kg relay satellite and a flotilla of seven Lunar-A penetrators, four of which will land on the far-side and three on the near-side. The sub-satellite will collect penetrator data for relay to Earth. The main vehicle, meanwhile, will land near one of the poles. Its most important instrument will be a telescope with a lightweight liquid-metal mirror that will maintain its parabolic shape by spinning very fast. This will be used to track stars near the lunar celestial pole to detect lunar librations in order to infer the internal structure of the Moon. This lander will also deploy a small 60-kg four-wheeled rover carrying a camera, a mass spectrometer and a gamma-ray spectrometer on the tip of a robotic manipulator to analyse the composition of the most interesting rocks within 25 km of the landing site, possibly including a permanently shadowed site in which the regolith may contain water.

After this, Japan may launch a 500-kg Nissan six-wheeled rover equipped with a robotic arm for sample collection. After landing inside Mare Serenitatis, it will travel up to 1,000 km to reach the crater Copernicus. Future missions may return up to 50 kg of samples to Earth.

7.3 EUROPEAN PROJECTS

After the false starts of POLO, MORO, LEDA and Euromoon, the first European lunar project is SMART-1 (Small Missions for Advanced Research in Technology) – the first of a series of small scientific spacecraft which put the emphasis on the introduction of new space technologies such as ion engines, nanorobots, and space interferometry, in an effort modelled on NASA's New Millennium programme.

The SMART-1 spacecraft was built for ESA by the Swedish Space Corporation. It is a boxy structure with a maximum dimension of 1.57 metres and a minimum dimension of 1.04 metres, with a pair of solar panels generating 1,400–1,850 W of power. It has four reaction wheels and eight hydrazine thrusters for attitude control. The primary propulsion system is an experimental electrostatic Hall-effect ion engine. This PPS-1350 engine was supplied by SNECMA of France. Although it

SMART-1, the first European lunar probe (without solar panels) is put through a series of thermal tests at ESA's solar simulator in Noordwijk, in The Netherlands. (Courtesy Swedish Space Corporation.)

A model of the PPS-1350 plasma thruster mounted on the SMART-1 European probe. Although it provides a thrust equivalent to the weight of a sheet of paper, it requires just 70 kg of xenon to boost the spacecraft from Earth orbit to lunar orbit.

provides a thrust of only 70 mN, equivalent to the weight of a sheet of paper on Earth, it will be able to boost the spacecraft to the Moon using just 70 kg of xenon 'propellant'.

Of the 370-kg launch mass, 19 kg is allocated to instruments – both scientific and engineering – including a 1.5-kg experimental infrared spectrometer, a camera, also to be used for optical communication experiments using lasers (a field in which ESA is at the forefront), two ion engine diagnostics instruments, a radio communication experiment, a Doppler tracking experiment, and a 2.5-kg X-ray spectrometer to map the composition of lunar rocks.

After hitching a ride into geostationary orbit with two communications satellites on an Ariane V launched on 27 September 2003, the probe switched on its ion engine to start to spiral out towards the Moon's orbit. In March 2005, following several fly-bys, it will enter an orbit with a perigee between 300 km and 2,000 km, and an apogee of around 10,000 km. The last six months of the mission will be dedicated to lunar studies, using the instrument suite. Many of the technologies tested during the SMART-1 mission will be reused by future ESA scientific missions – in particular, the BepiColombo Mercury orbiter.

Unfortunately, European plans involving the Moon are tentative, at best. One idea is to explore a crater near one of the poles using a revised form of the 7-kg mini-lander that was designed for the BepiColombo mission, using airbags to cushion the 'hard' landing. Another proposal – MoonShine – has been put forward by SSTL, a company that specialises in small satellites with masses of some tens of kilogrammes to several hundreds of kilogrammes. This would be a lunar orbiter based on the UoSAT-12 satellite, to be launched at a total cost of no more than $15–25 million. After hitching a ride into geostationary transfer orbit, MoonShine could carry up to 70 kg of payload to a low lunar orbit. Based on this proposal, in 1998 ESA awarded SSTL a one-year preliminary study of LunARSat (Lunar Academic and Research Satellite), a small probe to be built by a team consisting of university students, engineers and scientists from fifteen European countries. The main purpose of the mission was to be educational, as the images taken by the probe were to be used to stimulate an interest in space exploration and in science generally. It would carry a back-up SMART-1 camera, a wide-angle camera weighing 35 grammes, and a regolith-sounding radar. In order to reduce the mass of fuel required, the probe would utilise a WSB transfer orbit, which lowers the magnitude of the braking manoeuvre for entering lunar orbit.

ESA has also begun a long-term programme called Aurora, the ultimate aim of which is the human exploration of the Moon, Mars, and near-Earth asteroids. This will study the requisite enabling technologies and, in this context, two classes of unmanned mission are foreseen: infrequent, large-scale costly Flagship missions, and smaller, less costly Arrow missions that can be rapidly developed and launched. The four missions currently under study are a long-duration Mars rover, a Mars sample-return mission, a Mars aerobraking demonstrator, and an escape-speed Earth re-entry demonstrator. A future prospect involving the Moon is a probe to produce fuel and oxygen from lunar rocks.

7.4   THE RETURN OF RUSSIA

After the collapse of the Soviet Union in 1991, Russian cosmonautics entered a very deep crisis. Its launch rate plummeted from more than a hundred satellites launched per year during the 1980s, to as low as forty-seven satellites in twenty-five launches in 1999. Military satellites are still being launched at a much-reduced rate compared with in the past, but for civilian spaceflight, matters are much worse. Russian space agency annual funding currently amounts to about $200 million (lower than the Indian space agency's funding), and very few scientific missions are planned or are being launched. With the exception of spacecraft to support Mir, and after its demise, the International Space Station, the civilian programme has effectively ceased.

One early lunar probe proposal to emerge from Russia – and one of the most unconventional ever proposed – was called Dva Orla (Double Eagle). It envisaged a large orbiter, designed in cooperation with the US Pentagon, together with several Russian institutes, and built by McDonnell–Douglas in America. After launch on a Proton, Dva Orla would enter lunar orbit with a smaller target satellite to be used for calibrating the main instrument on the probe: a powerful laser. After laser calibration, the probe would lower its orbit to 200 km, to carry out topographical mapping with a camera, and compositional mapping with the laser and with an X-ray spectrometer. The laser beam would be fired to the surface, and the spectrometer would analyse the vaporised rock to determine its composition. The orbit could later be lowered to about 40 km in order to study in detail some fifty sites of particular interest. As the applications of these technologies would have been mainly military, it is not too regrettable that the mission was 'postponed' (cancelled) during the mid-1990s. However, the spectrometer will reportedly be flown on a Russian Foton satellite to study the Sun.

After the total loss of the Mars 8 mission in 1996, the section of the Russian Academy of Sciences dealing with Solar System research proposed a new exploration programme with three primary targets: the Moon, minor bodies (asteroids and comets), and Mars. Three missions were to be launched between 2000 and 2005: Fobos-Grunt, to return samples from the larger of Mars' two satellites; Mars-Aster, to explore Mars, and several asteroids and comets; and Luna Glob, to explore the Moon. Fobos-Grunt would be the first Russian mission to use an ion engine. Unfortunately, at the end of 2000 the whole plan was shelved due to lack of funds.

The extremely ambitions Luna Glob mission was to have included an orbiter, a lander, two large penetrometers, and ten small high-speed penetrometers. The two large penetrometers, similar to those on Mars 8, were designed to withstand a deceleration of 1,000 g upon striking the lunar surface. They were to have been placed close to the Apollo 12 and Apollo 14 landing sites and been capable of monitoring seismic events across a wide spectrum of frequencies. The smaller penetrometers were designed to withstand an impact at 2.5 km/s and 10,000-g deceleration, burying themselves some 10 metres into the regolith. They were to be emplaced in a 10-km-diameter circle to establish a small-aperture seismic array in

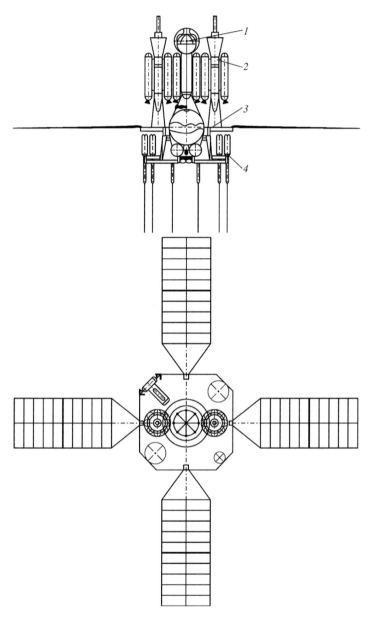

Two views of the ambitious Russian Luna Glob probe: 1, polar lander; 2, large penetrator; 3, orbiter; 4, high-speed penetrators. (Courtesy MAIK Nauka/Interperiodica Publishing, Moscow.)

the southern part of Mare Foecunditatis. The penetrometers were to operate for a year. As the lander, which was not unlike the E-6 'egg' of the 1960s Luna missions, was to set down in one of the permanently shadowed polar craters, it would have its own illumination system, a panoramic camera with a maximum resolution of 2 mm at a range of about 70 cm, a neutron- and a gamma-ray spectrometer to detect water ice, a mass spectrometer to analyse the regolith samples collected by a miniature sampling system, an alpha-ray spectrometer, a magnetometer, a thermograph, and an accelerometer. Although the primary task of the orbiter was to relay to Earth the data from the various probes, it would undoubtedly have had instruments of its own.

It was to have been launched by a Molniya rocket or a Soyuz Fregat. Twenty-eight hours before encountering the Moon, the ring on which the ten high-speed penetrators were mounted would be jettisoned to continue its flight in full autonomy. About 700 km from the ground, it would release the first salvo of five penetrators, followed by the second salvo at a distance of about 350 km. A ring spin speed of approximately three revolutions per second would ensure the correct dispersion of the penetrators over two areas of 10 km and 5 km in diameter. The two large penetrators would then separate. These were to be braked by a solid-fuel retro-rocket system that would halt them some 2 km from the surface, after which they would fall and impact at only 60–120 m/s. Shortly after deploying the last penetrator, the probe would enter a 500-km-perigee polar orbit, from where the lander would descend to the surface, be braked by a retro-rocket package similar to that on the E-6 probes, and then roll to a stop. It was an ambitious plan, but nothing came of it.

After the strategic weapons-reduction treaties, several ballistic missiles had to be eliminated, and they can now be 'destroyed' by using them as space launch vehicles – and some may even be used for lunar missions.

The Dnepr launcher, for one, is jointly marketed by Russia and the Ukraine, the other heir of Soviet space technology. It is based on the Yangel R-36M2 Voivoda ICBM, known in the West as SS-18 Satan, the heavyweight of the stable, which was first tested in October 1971. With a fourth solid-fuel stage added, it can send almost 1,000 kg to the Moon.* A similar launcher, with a similar payload, is the Rockot, based on the Chelomei RS-18 ICBM, known in the West as SS-19 Stiletto. The Shtil launcher is a derivative of the Mekeyev R-29RM submarine-based ballistic missile, known in the West as SS-N-23 Skif. The Babakin space centre (the former Lavochkin design bureau) has recently proposed the use of Shtil, mated to a Kaplya upper stage (a derivative of Fregat, which itself is a derivative of the E-8 lunar probe descent stage) to launch a 100-kg lunar orbiter.

A lunar project has also recently been proposed by the Ukrainian space agency, which inherited the production of the Tsiklon and Zenith launchers.

---

* The firing sequence of the Dnepr third stage is quite unusual. This stage was designed to provide the ICBM with a very high targeting accuracy, with its engines firing backward to slow the warhead. In its satellite launcher role, the third stage needs to carry out a 180-degree pitch manoeuvre before firing, in order to accelerate the payload.

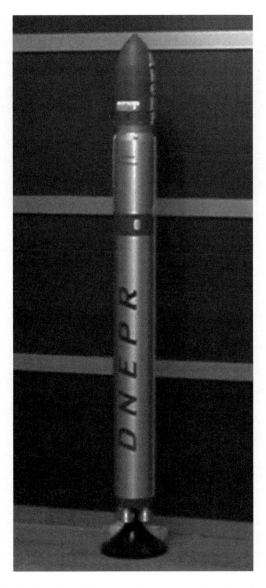

A scale model of the Russian–Ukrainian Dnepr launcher, based on the decommissioned SS-18 'Satan' ICBM. The Dnepr may launch its first lunar probe as early as
2004.

This 300-kg lunar polar orbiter would be extremely interesting, as its emphasis lies in
the study of the lunar regolith and polar areas. It would have a synthetic-aperture
radar for detailed mapping of the shadowed polar areas and for searching for
deposits of ice, a ground-penetrating radar to study the stratigraphy of the regolith,

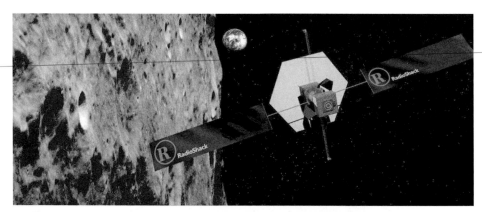

The LunaCorp SuperSat probe in lunar orbit. SuperSat might be the first spacecraft assembled in orbit onboard the International Space Station. (Courtesy Mark Maxwell of LunaCorp.)

and a polarimetric camera to determine the physical characteristics of the regolith material. Time will tell whether this plan is pursued.

7.5   PRIVATE LUNAR MISSIONS

As mentioned in the previous chapter, HGS-1 was the first commercial lunar mission – but much more is in store for the future.

One of the first companies to work on a private lunar mission was International Space Enterprises (ISE), based in San Diego, California. ISE was a pioneer of the commercial use of space (and had even filmed advertisements onboard Mir), and in 1993 announced its agreement with the Russian Lavochkin Association to build and launch privately funded lunar probes. The American company would manage the marketing side of the venture, and the Russians would build and launch the probes, and operate them throughout their flight. Two models were offered: ISELA-600, derived from the Fobos bus, to carry up to 1,500 kg into lunar orbit or 600 kg to the surface; and the completely new ISELA-1500, carrying 3,000 kg into orbit or 1,500 kg to the surface. Funds for the missions were expected to be raised from the sale of fifteen or so payload spaces – at a cost of \$125,000 per kilogramme – to research centres, film producers, advertising agencies and amusement parks. The first launch was expected in 1997, and the payload for the first landing mission could have been NASA's ultraviolet telescope, LUTE. With time, the enthusiasm dissipated, and the repeatedly delayed project was cancelled in mid-1999, due to a lack of funds. One of the first partners in ISE was LunaCorp, based in Arlington, Virginia, whose consultants include former astronauts Buzz Aldrin and Scott Carpenter, and Lunar Prospector principal scientific investigator Alan Binder. ISE is now producing electric vehicles, while LunaCorp has continued to study lunar missions on its own.

LunaCorp's first probe will be SuperSat, which may become the first spacecraft

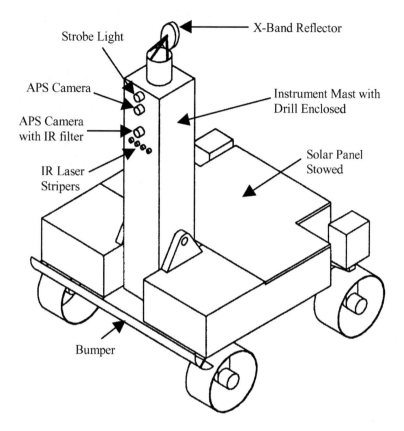

Strobe Light

APS Camera

APS Camera
with IR filter

IR Laser
Stripers

X-Band Reflector

Instrument Mast with
Drill Enclosed

Solar Panel
Stowed

Bumper

A schematic of LunaCorp's Lunar IceBreaker rover. The solar panel is shown in its undeployed position. (Courtesy Richard Blomquist.)

ever assembled in space. It is designed be carried disassembled to the International Space Station, and later, space station astronauts will assemble it, check that it is working correctly, and release it into space. Using an ion engine, SuperSat will reach lunar orbit, where it will undertake mapping at very high resolution (about 1 metre per pixel). Other than its camera, the probe will carry a radar altimeter to track lunar topography. Images and altimetric data will then be combined to yield a three-dimensional model of the lunar surface, to be used by scientists and by the planners of future lunar missions, and to produce virtual-reality video games. Both NASA and the Department of Defense have shown interest in the SuperSat concept, as they see it as an interesting test of the assembly of spacecraft at a space station. It had been hoped to deliver SuperSat in late 2004, but following the loss of the Space Shuttle *Columbia* in February 2003, consideration is being given to the possibility of launching it on a European Ariane 5.

LunaCorp also plans to launch Lunar IceBreaker, which will make a landing in the 75-km-diameter crater Peary, near the north pole. It will deploy a four-wheeled rover built by Carnegie–Mellon University (which also built some of the robots used in operations after the Three Mile Island and Chernobyl nuclear accidents, and NASA's Dante rover). Powered by solar panels and batteries, the rover will travel at a speed of 4 km/h, following the Sun on a path that will return it to its starting place after a lunar day. From time to time, it will venture into one of the permanently shadowed areas to take samples, using a drill to reach to a depth of 120 cm. After a sample is warmed with a laser it will be possible to measure its water content, if any. The rover will also carry a bistatic radar, the receiving antenna of which will be on Earth, to investigate the deep structure of both the regolith and ice. The cost of the mission – estimated at no more than $200 million – will be covered by private funding, but it is possible that NASA will provide some scientific instruments under the Discovery project. The first mission sponsor, announced in June 2000, is the RadioShack Corporation, an American chain of domestic electrical appliance stores. Before launch, LunaCorp is expected to secure sponsorship from three or four more companies. In July 2001 a Lunar Ice Breaker rover demonstrator called Hyperion – funded in part by NASA – was tested in northern Canada to prove its ability to follow the Sun and to avoid obstacles.

Based on its experience with Lunar IceBreaker, LunaCorp plans to stage a Grand Tour consisting of two rovers on a 'lunar industrial archaeology' mission. After landing in southern Mare Tranquillitatis, the two remotely controlled rovers will pay a 'respectful' visit to the Apollo 11 site, staying at a safe distance from the descent stage of the LM Eagle and the other mission artefacts, after which they will make a short trip of a few dozen kilometres to the north, to visit Surveyor 5 and then the crater produced by the crash of Ranger 8. The next stop will be some 500 km to the north, beyond the shore of Mare Tranquillitatis, at the Apollo 17 site in the Taurus Mountains. Then they will venture into Mare Serenitatis, to reach Le Monnier crater to locate Luna 21 and the Lunokhod 2 rover. After one year, during which 1,000 km will be travelled, the mission will be over. But who would be interested in such a mission? One propect is to sell rover movies to amusement parks and television networks, which could even stage 'lunar guiding' sessions in virtual reality. A fee could be charged to carry personal messages or small items, and the first lunar Webcam site could be installed, with which Internet users could see real-time images. It will also be interesting to see how artificial structures have reacted to such long exposure in space.

In the summer of 1999 another competitor appeared – and could be the first to launch a private lunar probe. The company is TransOrbital, and its first mission, TrailBlazer, will be launched on a Russian–Ukrainian Dnepr rocket. The eight-sided orbiter will be 90 cm in diameter and 85 cm tall, with a dry mass of around 41 kg and a fuelled mass of 95 kg. The base will carry the main engine, and the top a parabolic antenna and two imaging systems: a medium-angle camera to take short movies of Earthrise or similar dramatic views, and a high-resolution colour camera. It will also carry out an interesting engineering test by attempting to receive Navstar-GPS (Global Positioning System) satellite signals in lunar orbit. These satellites orbit the Earth at a height of 24,000 km, and their antennae face Earth. If it proves possible to receive the lateral lobes of the antennae at lunar distance, this would greatly simplify navigation on later missions. TrailBlazer

(Left) the TrailBlazer mock-up being integrated on the Dnepr launcher in late 2002. Several small satellites hitching a ride on the same launch are visible on the platform. (Right) the TrailBlazer mock-up being enclosed in the Dnepr launcher fairing. (Courtesy Yuzhnoye Design Office, Ukraine, and TransOrbital.)

might also have the TIR (TrailBlazer Italian Radiometer) experiment designed to measure the attenuation of artificial electromagnetic noise over the far-side to pave the way for lunar radio telescopes (discussed in some depth, later in this chapter).

Four days after being launched from a Tyuratam missile silo, the probe will enter an 18-hour lunar orbit. From its 50-km-high perigee, it will take high-resolution pictures of the lunar surface to compile an atlas. About 30 days later – one lunation – the perigee will be lowered to just 10 km, after which the pictures should reveal details as small as the tracks left by the Apollo rovers. The solar-panel-powered probe should last no more than three months before impacting on the surface, where it will release personal objects and messages which will be carried on board at a cost of about $100 each. Further funds to pay for the probe may be made by selling the probe's high-resolution pictures, and by the sale of the footage of the trip from the Earth to the Moon.

In August 2002, TransOrbital became the first private company involved in lunar exploration to receive the US government's approval to export its probe to Russia for launch in late 2003 or early 2004. Moreover, at the end of December 2002 a

Ukrainian-built TrailBlazer mock-up was launched to Earth orbit to demonstrate its compatibility with the Dnepr launcher.

TransOrbital is also studying a family of landers called Electra. The first lander has a dry mass of 45 kg and a fuelled mass of 200 kg, and will be launched by Dnepr. After entering lunar orbit, it will descend to Mare Anguis (Sea of the Snake) – a narrow, sinuous plain near the eastern limb of the near-side, north of Mare Crisium, at 22°.6 N, 67°.7 E. On the surface, it will deploy two cameras: a stereoscopic system based on the camera used by Mars Pathfinder, and a boom-mounted camera to take photographs of the probe. After sunset, it will be put into hibernation in the hope of surviving the cold lunar night. Sample-return missions carrying up to 15 kg of regolith, or missions for *in situ* analysis of the polar ice, could also be carried out using the same spacecraft bus.

Another company working on private lunar probes is SpaceDev, which is currently developing NEAP (Near-Earth Asteroid Prospector) to be the first private asteroid mission. Based on this experience and NASA-funded studies, SpaceDev is developing a lander to carry a small dish antenna for multi-wavelength radio astronomy to the lunar south pole.

A company that had to withdraw due to its failure to raise sufficient funding is Applied Space Resources. It proposed the Lunar Retriever 1 probe. This would be launched by an Athena 2 rocket and land one week later inside Mare Nectaris, a few dozen kilometres from the large crater Theophilus,* in order to return 10 kg of regolith using a patented capsule. Re-entry was expected to take place over the Utah desert. The descent stage was to continue relaying pictures for at least one full lunar day. The company hoped to cover the cost of the mission – about $60 million – by selling samples to laboratories, collectors or jewellers.

Later on, the company envisaged a real lunar assault. Lunar Pointer 1 would have taken very-high-resolution movies from a 10-km-perigee orbit, to be shown in IMAX-equipped theatres. Lunar Retriever 2 would have landed in a southern polar crater, would have illuminated it with floodlights, and would have returned a 500-gramme core sample to Earth. Small explosive charges and a geophone net would have investigated the deep structure of the icy regolith. Lunar Pointer 2 would have carried out another very-high-resolution movie mission over the landing sites of the historical missions of the 1960s and 1970s. Finally, Lunar Retriever 3 would have inaugurated a family of mini-probes, launched in clusters of three each, able to return samples from any part of the Moon. Other landers would have carried small rovers (Lunar Huskies) to explore interesting sites such as Alphonsus or the bright 'swirl' near Reiner-γ, and yet others would have explored the far-side, deploying antennae for a simple radio astronomy observatory. These missions would have required relay satellites at the L2 Lagrangian point – the Lunar Watchdogs.

This family of probes, however, threatened to revive the heated debate concerning the ownership of lunar material. The 1979 United Nations *Moon Treaty* declared deep space, the Moon and the planets to be the property of the whole of mankind,

---

*   Theophilus (*d.* 412), patriarch of Alexandria.

and no-one is allowed to claim any part of them. The treaty has not been ratified by America, but it is quite certain that the thought of priceless samples being sold by a jeweller for $200 per gramme would raise some eyebrows – including both of the author's.*

7.6   THE SECOND MOON RACE?

Today, the country with the most 'aggressive' space programme is undoubtedly China. It launched its first Dong Fang Hong (East is Red) satellite in 1970, and following a successful but low-profile programme of meteorological, telecommunication, Earth observation and microgravity research missions it decided to enter two high-profile and high-prestige fields: human spaceflight and deep-space probes. It is ironic that the father of Chinese astronautics, Tsien Hsue-shen, was one of the founders of JPL, and that during the McCarthy 'witch-hunt' era he was accused of being a Communist and was expelled from the United States.

The piloted spaceflight programme is already known. After prototype spaceships, called Shenzhou (Heavenly Boat), were flown in space, Yang Liwei became the first yuhangyuan, or taikonaut (taikong – Chinese 'space') when he was launched on 15 October 2003 for a 14-orbit mission. Project 921 (as the programme is called) reuses Russian-designed electronics and docking systems and the entire Shenzhou re-entry capsule is modelled upon that of the Soyuz. The programme could lead, in fifteen years time, to a reusable space shuttle and to a modular space station similar to Mir.

Shenzhou is currently launched by the CZ-2F (Chang Zheng – Long March), a powerful two-stage rocket using four liquid-fuel boosters, able to place 10 tonnes into low Earth orbit, but a more powerful CZ-2EA is under study for future flights. Human spaceflights start from the Jiuquan space centre (Shuang Cheng Tzu) in the Gobi desert, not far from the Mongolian border, at a latitude of 41° N. The Chinese have two more launch centres: Taiyuan, 450 km from Beijing, and Xichang, at a latitude of 28° N, in the mountainous Sechouan region, from which geostationary satellites are launched. There are also plans for the construction of a launching range on Hainan island, which will launch out over the ocean.

China began to study deep-space probes as long ago as 1963, when a small group of engineers had to collect information on American space projects in order to help define a national space programme. The interest of this group was focused mostly on applications satellites such as Tiros (meteorology), Discoverer CORONA (a recoverable spy satellite), Transit (navigation), and Echo, Telstar and Syncom (telecommunications). Also studied were scientific spacecraft such as the Canadian Alouette and the British Ariel satellites, piloted spacecraft such as Mercury and Gemini, and JPL's

---

* The 1979 Moon Treaty has been ratified only by Australia, Austria, Chile, Mexico, Morocco, the Netherlands, Pakistan, the Philippines and Uruguay. Its Article 11 forbids exploitation of lunar resources.

Chinese Long March launchers. From left to right: the CZ-2C/CTS (up to 800 kg to the Moon), the CZ-3A (1,700 kg), the CZ-2E and the powerful CZ-3E (3,400 kg).

Ranger probes. Probably as a result of these studies, in 1964 two papers were published in Chinese technical journals concerning the design of 'Moon-crasher' probes.

It is possible that the first, ambitious Chinese space programme, developed during the difficult years of the Cultural Revolution under Defence Minister Marshal Lin Biao, also included the development of lunar probes, besides other projects such as the tiny Shuguang (Dawn) piloted spacecraft. However, after the death of Lin Biao in a mysterious air accident in the early 1970s, the programme was redirected toward more useful applications. Party Secretary Deng Xiaoping even publicly declared: 'China has no need to go to the Moon in order to modernise itself.'

In March 1986, the Chinese Ministry of Science and Technology began its Project 863, to develop technological innovation in several fields from biotechnologies to robotics. One of the most important programmes developed under Project 863 dealt with lunar exploration. In 1993 and 1995, the Chinese Academy of Launch Vehicle Technology (CALT) carried out two studies of a lunar mission, but neither was implemented.

The existence of one such programme was revealed in only January 1995, when Jiang Jingshan, responsible for space planning of the Chinese Academy of Sciences, announced that the recently proposed Five-Year Plan included the development of non-defined planetary missions, including a lunar probe to be launched 'around 2000' – a plan later confirmed by Ma Xingrui, Vice President of the Chinese Academy of Space Technology, who stated, in March 1998: 'We will launch a small lunar probe when possible.' In November 2000 an official white paper on the next ten years of Chinese space activities was made public, calling for the development of missions ranging beyond Earth orbit.

In 1997 a workshop on lunar probe design was organised by CALT, and its topics possibly covered navigation and orbital design. In May 2000 and January 2001, two more symposia were held at Tsinghua University, and the resultant plan, called the Chang'e Programme, after the character who flies to the Moon in the fairy tale *Huai Nan Zi*, was presented for approval after the sixteenth congress of the Communist Party in 2002.

During the first phase, around 2005, China will launch a simple fly-by probe or a more complex orbiter. This mission should cost no more than $120 million – a figure which can be compared with the $2.3 billion which China has reportedly spent on human spaceflight up to 2003. At the same time, Chinese scientists may cooperate with Japanese colleagues on the SELENE lunar mission.

After this first phase – before 2010, and possibly timed to coincide with the Beijing Olympics in 2008 – China is to launch its own lunar lander. The third step will see landing missions equipped with small rovers, and during the fourth phase, after 2015, sample-return missions will be launched. Finally, after 2030, taikonauts are to reach the Moon to build a long-duration base. According to some Chinese space officials, the central government may approve this plan (or part of it) 'soon'.

The role and motivations of the political power in this phase of the Chinese space programme does, however, appear unclear. Some observers of China state that national prestige constitutes a strong motivation, while others point out that the programme's projects, being entirely controlled by the military, are most likely evaluated on and judged by their military value. In the case of piloted flights, the propaganda value cannot be doubted, but the military value remains unclear. In the case of automatic lunar probes, the scientific value of such missions is high, but they can hardly be justified from a 'prestige only' point of view, and even less from a military point of view. In keeping with the tradition inaugurated by the former Soviet Union, details of the missions remain sketchy. What will be the planned

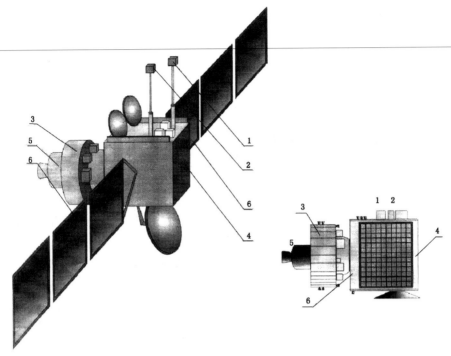

The DFH-3-based Chinese lunar orbiter in both its deployed and undeployed configurations. 1, scientific instrument; 2, TV camera; 3, propulsive module, to impact the Moon after lunar orbit insertion; 4, lunar orbiter; 5, solid-fuelled motor; 6, solar panels. Image reprinted with permission from CALT. (From Li Dong, et al.; A Tentative Idea about Lunar Exploration, Missiles and Space Vehicles, 2002 No. 5, pp. 20–28.)

launcher? And what are the objectives and characteristics of these probes? These questions, and others, remain to be answered.

To obtain a clearer picture, the author has sifted through Chinese technical and refereed publications available on the Internet, in order to answer, at least partially, the above questions. This search has located more than a hundred papers, abstracts and citations, and has helped to clarify Chinese objectives and capabilities.

China has a formidable arsenal of space launchers, several of which can be used for lunar missions. The three-stage CZ-2C/CTS uses the first two stages of the piloted CZ-2F mated to a third solid-fuel stage. Depending on the launch site – Xichang or Jiuquan – this rocket can launch 700–800 kg to escape speed. Its first use will probably be to launch the two scientific Xinguan (Double Star) satellites, built in cooperation with ESA. Of these, the first will be launched from Xichang, and the second from Taiyuan.

The most powerful Chinese launchers are the CZ-3 family, launched from Xichang. The basic CZ-3, having a liquid oxygen/liquid hydrogen cryogenic third stage, can launch around 1,000 kg to the Moon – a payload equivalent to the American Surveyor probes of the 1960s. CZ-3A – a stretched CZ-3 – can launch up to 1,700 kg to the Moon. CZ-3B – currently the most powerful Chinese launcher – is a CZ-3A with four liquid-fuel boosters, and can launch up to 3,400 kg to the Moon. The unflown CZ-3C – a CZ-3B with only two boosters – can launch up to 2,400 kg of payload to the Moon. There are clear indications that the CZ-3 family will be used for the first lunar probes.

Even the CZ-3, however, will not be able to launch sample-return missions. For these, the 2000-2005 Five-Year Space Plan reportedly includes the development of the new CZ-5 launcher – a family of modular rockets developed with Russian help, using stages with diameters of 5 metres, 3.25 metres, and 2.25 metres. The heaviest version will use a hydrocarbon first stage and cryogenic upper stages to carry 25 tonnes into low Earth orbit or 13 tonnes into geostationary transfer orbit. CZ-5 will be able to launch 4.4, 8.1 or up to 10.6 tonnes to the Moon, depending on the version, and could also provide the basis for a Chinese version of the Saturn V, if and when China decides to put humans on the Moon. Indian scientists have even suggested that China may launch a piloted circumlunar mission similar to Zond within only a few years, and one such mission would surely be feasible using the heaviest CZ-5 version. Certainly, Chinese sources say that the CZ-5 will facilitate 'deep-space exploration, including flights to the Moon and planets'. Very little information has been published concerning the early Chinese lunar probes, their projects, and their missions.

The first lunar orbiter will be based on the DFH-3 communication satellite already tested in geostationary orbit, and may reuse components and instruments developed for other spacecraft such as Feng Yun-1 (meteorological), Zi Yuan-1 (Earth observation), and Shenzhou. In particular, it may reuse the 2.5-kN-thrust nitrogen tetroxide and UDMH-fuelled main engine of the latter. The probe will also use a propulsion module probably based on the 'smart dispenser' used in the past to insert Iridium mobile communication satellites into their individual orbits. This 500-kg module would carry out the lunar orbit insertion manoeuvre and would then deorbit itself to head for a lunar impact. Although the module itself would not carry any scientific instruments other than a radio beacon, the impact would be monitored by the orbiter, yielding interesting results on the composition of the regolith. A scientific mission will be conducted from a 200-km-high polar orbit using a CCD camera, an imaging spectroscope, a laser altimeter, a microwave radiometer, X-ray and gamma-ray spectrometers, and instruments to study the Sun, cosmic rays, and low-energy particles. A study indicates that pointing accuracy for these instruments will be at least as good as expected for the European SMART-1 probe.

Other proposals have, however, been put forward. One, elaborated under the aegis of Project 863 by researchers of the prestigious Tsinghua University, is called LunarNet, which envisages a polar orbiter equipped with sixteen 28-kg landers, to be released in equally spaced areas on two mutually perpendicular orbital planes. The landing system – probably utilising air-bags – must ensure survival after an impact at speeds of 12–22 m/s. Each lander will carry a camera, temperature sensors, cosmic ray detectors, a penetrometer, an instrument for the measurement of regolith magnetic properties, and various other instruments. The researchers have also

A computer graphic rendering of the small Chinese Moon Rabbit probe, consisting of a spin-stabilised orbiter and a small lander for engineering tests. (Courtesy Le Stelle.)

proposed an ingenious system to compensate for the lack of a Chinese deep space network. A relay satellite is to orbit the Earth in an orbit with apogee near the Moon and perigee at some 6,000 km, collect data from the orbiter and landers during its frequent lunar fly-bys, and relay them to Earth at perigee. This system, however, does not simplify the uplink of commands from Earth to the probes.

Another proposal is designated Moon Rabbit, after a traditional Chinese story. This 330-kg probe will cost only $30 million, and will be launched on a geostationary transfer orbit from the Xichang space centre. Insertion into a lunar transfer orbit will be carried out using the onboard bipropellant engine. At the time of its third apogee, the probe will be inserted in a 100×200-km-high lunar orbit, where it will split into two components. The first of these – apparently based on the Xinguan scientific satellites – will carry out an orbital mission using a CCD camera, an infrared camera, a radar altimeter and a radiometer. The second will head for a lunar landing. This lander, braked by a solid-propellant engine, will carry only a camera, and will test the optimal control algorithms which have been discussed at length in the Chinese literature. Once on the surface, the lander will release a 60 m^2 Chinese flag.

Several other studies of lunar missions have apparently been recently carried out in China: a small spin-stabilised orbiter; an ion-propelled 300-kg probe designed under the aegis of Project 863; a 600-kg lander; and vertical landers using a similar technique to the Soviet E-6 probes. Moreover, studies have been carried out of solar sails and planetary penetrators – experiments for the latter have been carried out using projectiles of steel, titanium and tungsten fired at very high speed against concrete targets. Such experiments are probably rooted in the military, as 'bunker buster' bombs and planetary penetrators have several points in common. A 4-tonne sample-return lander, launched by the CZ-5 heavy launcher, is also being studied.

As previously mentioned, one of the problems which China will have to face – if and when it decides on a programme of lunar exploration – is that of communicating with deep-space probes. One possibility is FAST (Five-Hundred-Metre-Aperture Spherical Telescope), a radio telescope, similar to, but larger than, that at Arecibo, that will be built in a natural recess in the Guizhou region. According to the Chinese, its tasks will include the tracking of deep-space vehicles. This instrument will cost up to $1 billion. Sources also hint at the use of the Kiribati and Namibia Chinese tracking stations, and even the three Yuanwang (Long View) tracking ships that are used to support human missions.

The third and fourth phases of the lunar exploration programme envisage the use of rovers and other robots. This appears to be a topic of extreme interest to the Chinese, with several papers published every year in refereed journals. No less than five universities and academies are currently working on lunar robotics: Beijing Control Engineering Institute, Tsinghua University, Harbin Industrial University, the National Defence Science and Technology University, and the Chinese Science and Technology University. Preliminary studies concern 30–60-kg vehicles moving on either wheels or legs. Moreover, at the beginning of 2001 a rover prototype was placed on show by Tsinghua University. From a description by associate professor Zhu Ji Hong, this appears very similar to the Sojourner rover of NASA's Mars Pathfinder mission in 1997. It has six wheels powered by independent motors, and a rocker-bogie system similar to that on Sojourner, said to be able to avoid obstacles up to 18 cm high. It is powered by solar panels, has four floodlights at the front, and a three-dimensional camera to take navigational pictures and panoramas to be relayed to Earth. The definitive rover may carry a colour camera, alpha and X-ray spectrometers, an accelerometer, and a robotic arm to collect samples. This type of rover does not appear to be able to survive for more than a single (lunar) day on the Moon's surface. It is also understood that the Russian VNII TransMash – which built the Lunokhods – will provide consultancy on Chinese lunar rovers.

Another study by Tsinghua University has identified the robotic technologies required to collect material for a sample-return mission. A control system prototype for deep-space robotic arms has been implemented using industrial robots and a virtual reality system. According to a leading Chinese robotic scientist, Sun Zenqi, these robots could investigate future landing areas, deploy scientific instruments, collect samples, and take pictures.

The Chinese lunar missions will undoubtedly provide further proof of technical maturity, but so far little has been said about their scientific objectives. Concerning this, Wu Ji, the Deputy Director of the Chinese Academy of Sciences' Centre for Space Science and Applied Research, has recently declared that Chinese lunar probes will aim at problems not addressed by previous missions, stating the importance of doing 'something unique'. Other sources state that the goals of the first probes will include mapping the lunar surface in detail in three dimensions, analysing its chemical composition, and studying the regolith and the interplanetary environment near the Moon.

The first decade of the new millennium may also see another newcomer to the exploration of the Solar System. Since 1980, when it launched its first satellite on an indigenous launcher (the seventh country to be able to do so; or the eighth, including Europe), India has concentrated its efforts on a low-profile programme of Earth-observation, meteorological and telecommunication satellites, in addition to some small scientific and astronomical spacecraft (the Rohini satellites carried astronomical detectors). Finally, in 1992 the Indian Space Research Organisation (ISRO) announced that it would ask for funds for a programme of deep-space probes to be carried by the proposed Geostationary Satellite Launch Vehicle (GSLV). Although missions to Venus and Mars were said to be under study, in the end India's attention turned to the least known of the inner planets: Mercury. Although the deep-space exploration programme was not funded, work began on the GSLV – a three-stage launcher consisting of a solid-fuel first stage, a liquid-fuel second stage, a cryogenic third stage, and four liquid-fuel boosters. For the first test launches, the controversial decision was taken to use Russian technology for the cryogenic stage, thereby carrying into space, for the first time, the engine developed for the N-1F Soviet manned lunar launcher. GSLV is based on the Polar Satellite Launch Vehicle (PSLV), which is still in service and has a good record of success. The new launch vehicle made its debut in 2001.

Meanwhile, after more than five years of oblivion, word of the Indian deep-space missions emerged in 1999, when a former President of the space agency boasted that the GSLV might provide the basis for a launcher to carry humans to the lunar surface. The following year, the preliminary plan for a lunar probe was presented by ISRO to the Indian government for approval. If the PSLV were to be fitted with an upper stage, it would be able to send a 530-kg spacecraft on a fly-by trajectory, or to put a 350–400 kg (plus fuel) spacecraft in a 100-km-high orbit. Alternatively, the GSLV could launch an 850–950 kg spacecraft on a fly-by trajectory, or a 600-kg orbiter with an expected mission duration of two years. At the moment, a soft-lander is considered too complex.

The development of a spacecraft based on the Indian Resource Satellite (IRS) bus, and carrying a stereoscopic camera with a ground resolution of 5 metres, a multispectral camera, two X-ray spectrometers and a gamma-ray spectrometer, will unfold in two phases. The first phase, up to 2004, will analyse the technical feasibility and scientific interest of such a mission. During the second phase – lasting at least three years – the Chandrayan Pratham (first journey to the Moon) probe will be built at a cost of 7–8.5 billion rupees (140 million and 170 million Euros), including 1.2 billion rupees (2.4 million Euros) for the launcher. After several studies and extensive simulation, go-ahead for the mission was given by prime minister Atal Bihari Vajpayee himself during his Indian Independence Day address on 15 August 2003. The 525-kg spacecraft will be boosted into geostationary transfer orbit by a PSLV and will use its own engine to attain lunar orbit.

To provide an impression of the Indian space industry's capabilities, it is worth remembering that the first stage of both GSLV and PSLV is the fourth most powerful solid-fuel engine in the world, and is exceeded only by the boosters of the Space Shuttle, Titan IV, and Ariane 5.

All of the Indian launches take place from Shriharikota island, 100 km from Madras, at a latitude of 13° N.

It has recently been suggested by Indian space officials that the second decade of the new millennium might see a second race to the Moon – but this time between India and China.

Besides being the two most populous countries in the world, China and India are both nuclear powers, and since 1962 have been mutually hostile concerning the ownership of a part of Kashmir. Meanwhile, a scenario well exercised during the Cold War is being repeated. Chinese scientists are accusing their Indian colleagues of 'reinventing the wheel', of understating the costs of a lunar mission, and of putting the priorities of their space plan in second place. As during the 1980s, when Soviet scientists were very critical of the US Space Shuttle, in order to criticise their own shuttle programme, these Chinese critics may in fact be addressing their own lunar programme. In their favour, the main Indian motivation appears to be the will of the ruling Hindu nationalist party, and of the Indian President, Abdul Kalam – the former director of the national nuclear weapons programme – to reach the Moon before the Chinese. India has no real planetary scientists, and ISRO's Physical Research Laboratory is attempting to set up a small planetary research group to design the payloads for the automatic probes. India's lunar programme differs from other lunar projects currently under study or recently proposed, because it will probably rely, to some degree, on international cooperation. The Canadian Space Agency hopes to contribute a 10-kg, $7-million scientific payload, ESA may provide a far-infrared sensor, and Germany and Israel might also contribute. The satellite-tracking facility in Bangalore will probably be upgraded with larger dishes to track lunar missions.

Other countries have succeeded in sending small satellites into orbit using indigenous launchers, and others may do so soon – including Israel, which launched its first satellite in 1988 on a modified Jericho 2 ballistic missile; Brazil, whose Vehiculo Lancador de Satelites has failed twice and recently suffered a terrible explosion on the pad during preparations for launch; South Korea and North Korea, the latter with a failed first space launch in August 1998; and Iran and Pakistan (using North Korean technology); but none of these countries presently have either the technology or the political will to aim at the Moon. Other countries have abandoned indigenous space launcher projects. Having led Europe in the development of large liquid-fuelled rockets in the late 1950s, Britain developed its own three-stage Black Arrow satellite launcher in the 1960s, then cancelled it after placing its first 70-kg satellite (Prospero) into orbit in 1971. Egypt developed, with German help, its Al Ared (Pioneer) in the early 1960s. South Africa abandoned its RSA-3 after cancelling a nuclear weapons programme. Argentina cancelled a project based on the controversial Condor 2 missile. Iraq – having had its missile power obliterated after the 1991 war – abandoned its Al Abid (Worshipper) missile, consisting of a cluster of four or five Scuds as the first stage, a single uprated Scud as the second stage, and an SA-2 surface-to-air missile solid-fuel booster as the third stage, and of course the situation in that country was transformed in 2003.

7.7   SOLAR SAILS TO THE MOON

Mention must also be made of solar sails, or solar propulsion. These spacecraft use the radiation pressure of solar light as a propulsive force, and thus need to have as large a collecting area as possible, together with a reasonably low mass. Solar sails usually consist of a small satellite, and a sail made of mylar a few micrometres thick and with an area of several thousand square metres. The main technological problem of a solar sail is that of deploying it in zero gravity and at the same time maintaining its shape.

Although the principles of solar propulsion were known at the beginning of the twentieth century, and several studies were carried out during the first thirty-five years of astronautical history, the first experiment in solar sail deployment was on 4 February 1993, when a 20-metre-diameter circular sail was deployed from the Russian cargo spacecraft Progress M-15. The experiment, called Znamya (Banner), was intended to demonstrate the use of a sail as a mirror rather than for generating thrust, and was successful, but a second trial some years later failed when the sail deployed incorrectly. The first American experiment was funded by the Planetary Society (a space advocacy organisation) and built by the Russian Babakin space centre. The test was to deploy two wings of a sail during the sub-orbital flight of a submarine-launched ballistic missile in July 2001, but the third stage of the rocket failed to separate from the payload, and the experiment failed.

Despite these problems, however, studies have been carried out into many lunar exploration projects incorporating solar sails, including a race, sponsored by the government of the French region of Midi-Pyrenees in the 1980s, to be won by the first sail to reach the Moon. Several countries showed their interest in entering the race, but due to lack of development funds it was soon forgotten – as was a similar race to Mars, to celebrate the five-hundredth anniversary of Columbus's first voyage to the Americas.

Another project involving a solar sail designed to visit the Moon was the Soviet Regatta – a 575-kg spacecraft with a 20-m-diameter sail. The mission would explore the terrestrial magnetosphere, and was to fly by the Moon while on a very elliptical orbit. Other solar propulsion missions are being studied for the near future. Cosmos 1, produced by the Planetary Society and the Babakin space centre, is scheduled to enter low Earth orbit during 2004. Another is the $2-million Solar Blade Heliogyro prototype developed by Carnegie–Mellon University, and funded by the Pentagon and NASA. This will be launched as a secondary item on a Delta or Ariane, and after the primary payload has been deployed it will be jettisoned and will split in two parts. One part will be a tiny camera-equipped nanosatellite, and the other will be the heliogyro. With a mass of less than 5 kg, it will deploy four 20-metre-long, 1-metre-wide wings made of aluminised kapton, 8 μm thick. After deployment – which will be imaged by both satellites – the heliogyro will be gently pushed by the solar radiation pressure, and will be spin-stabilised by varying the pitch angle of the wings, so that it will look like a windmill. It will then move away from Earth on a spiral trajectory, pass the Moon, and finally become lost in solar orbit.

One of the mylar blades of the Planetary Society's solar sail undergoing deployment tests at the Russian Babakin space centre. Note the Soviet-era portrait of Georgi Babakin, and slogans hanging on the walls. (Images copyright Louis Friedman/ Planetary Society.)

Europe, too, is studying a technology demonstration mission to send a solar sail into lunar orbit. The project – called Odissée – uses a square, 125-m-wide sail with a mass of 100 kg. This probe may be equipped with a high-resolution camera to take pictures of the sail deployment and of the lunar surface. Finally, plans for a solar sail mission to the Moon may already be under way in China.

7.8 THE NEXT STEPS

Between 31 May and 3 June 1994, in the small Swiss town of Beatenberg, an ESA congress was held to define the steps to be taken to establish a common lunar base in the twenty-first century. The world's chief space agencies were represented, and there were many delegates from the United States, Russia, Japan, and several other countries. The recommendation, which has become the official ESA approach, envisages four phases, with increasing funding and international cooperation. In the first phase – exemplified by national programmes such as Clementine, Lunar

Prospector, Lunar-A and SMART-1 – the single agencies will launch small and cheap missions aimed at improving our knowledge of lunar topography, mineralogy and gravimetry. This first phase will also aid in improving the technology required for the next steps. In the second phase a permanent robotic presence will be established, leading to the installation of the first complex instruments, including a radio telescope prototype. In this second phase – more so than in the first – virtual reality and telepresence technologies will be employed. These techniques will not be restricted to mission engineers and scientists, but will also be used for entertainment and education in order to stimulate public support. In the third phase, the first robotic 'factories' will be established to begin producing oxygen from lunar rocks. Larger astronomical observatories will also be built. The fourth and last phase will see the return of humans, and the first permanent lunar base.

The Beatenberg congress identified three principal scientific benefits of such a programme. The first benefit would be an improved knowledge of the Moon and of the Earth–Moon system. Although the Apollo and Luna missions and, more recently, Clementine and Lunar Prospector, have greatly improved our knowledge of the geology and morphology of the Moon, many questions remain unanswered. The Moon has, furthermore, preserved traces of events dating from the origin of the Solar System, and this will help in studying the evolution of the Sun and the Earth. A recent study suggests that every 100 km^2 of lunar surface should have collected about 20 tonnes of terrestrial rocks launched by asteroidal impacts, and these rocks may preserve traces of the origin of life. A second, more controversial benefit would derive from studies into the effects of radiation on biology, in preparation for the human colonisation of the Solar System – if and when it happens. The third, more fascinating, benefit would derive from astronomical studies that may be carried out from the lunar surface. A network of large optical telescopes could be built using interferometry to directly observe planets orbiting other stars, and to obtain spectra to study the composition of their atmospheres and thereby indirectly detect signs of life like ours. Few people remember, however, that the Surveyors, Apollo, and most of all Lunokhod 2, noticed that the lunar sky background, when the Sun is above the horizon, is fifteen times brighter than the darkest terrestrial night.

Very interesting radio astronomy studies could be carried out if an observatory were to be built on the far-side, for which the Moon would act as a shield against terrestrial electromagnetic pollution. Antennae could be built inside small craters to search for signs of intelligent life. The most interesting proposal, however, is for the construction of a very-low-frequency radio telescope on the far-side, as this is the only part of the electromagnetic spectrum that cannot be observed either from the ground, due to the ionosphere, or from Earth orbit, due to the size of the required antenna.

An American project named ALLFA (Astronomical Lunar Low-Frequency Array), presented in 1992, called for the launch of a lander carrying a central station, a rover equipped with a six-degree-of-freedom robotic arm, and up to forty small autonomous receiving stations. The spacecraft would land on the far-side near the equator, and would release the rover, which would deploy the forty stations along an ellipse with a 25-km-long major axis lying on the lunar meridian, and a 17-km-long minor axis. Each station would include a non-moving solar panel for power supply,

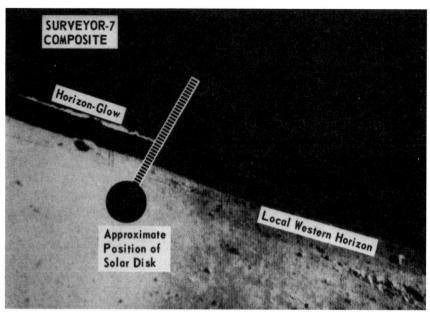

SURVEYOR-7
COMPOSITE

Horizon-Glow

Local Western Horizon

Approximate
Position of
Solar Disk

The worst enemy of lunar optical astronomy may prove to be a thin layer of suspended dust. Surveyor 7 took this image one hour after sunset.

a package of sodium–sulphur batteries to provide power and heat during the night, and a steerable 30-cm-diameter parabolic antenna. Using its own 30-metre-tall antenna, the central station would collect data from the autonomous stations and relay it to Earth through a repeater satellite orbiting close to the L2 Lagrangian point. The project called for the launch of all the components, including the repeater satellite, on a single Titan IV–Centaur – the most powerful (and most expensive) American launcher. A similar proposal being studied at ESA envisages the use of some three hundred autonomous stations, each slightly larger than a lap-top computer, deployed along a spiral inside the crater Tsiolkovsky as a Very-Low-Frequency Array designed to observe the sky in a spectrum of frequencies between 500 kHz and 16 MHz. As with ALLFA, a central station would relay the data from the autonomous stations via a satellite in orbit around the L2 point. These arrays would observe supernova remnants, neutron star magnetospheres, the distribution of low-energy electrons in interstellar space, quasars and Seyfert galaxies, and aurorae and lightning in the atmospheres of the largest planets in the Solar System.

There is no doubt that the opening of another 'window' on the electromagnetic spectrum will lead to the discovery of new and unpredictable phenomena. This happened during the twentieth century – first with radio astronomy, and then with satellites having infrared, ultraviolet, X-ray and gamma-ray detectors.* Other

---

* The first known X-ray source outside the Solar System was discovered on 18 June 1962 by an Aerobee rocket intended to study the fluorescence induced on the lunar regolith by solar X-rays. For this discovery and the ensuing studies, Riccardo Giacconi was awarded the Nobel prize for physics in 2002.

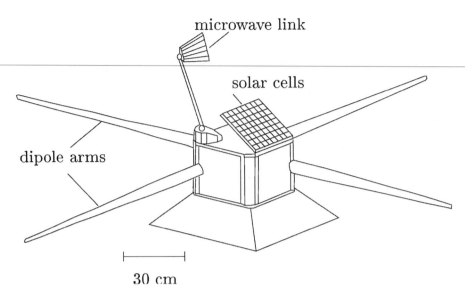

A line drawing of one of the three hundred receiving stations of the proposed ESA lunar far-side Very-Low-Frequency Array. (Courtesy Graham Woan, from 'A very low frequency radio telescope on the far-side of the Moon', in Proceedings of CRIS98, *Measuring the Size of Things in the Universe: HBT Interferometry and Heavy Ion Physics.*)

proposals include radio telescopes that would be constructed by astronauts. However, the operation of such a sensitive astronomical instrument might easily be disturbed by the presence of humans on the Moon.

Still in the future, some propose to use the Moon as a source for raw materials which are becoming much rarer on Earth, but it is debatable whether this would be economically viable. Nevertheless, the Moon may prove to be the closest and cheapest source of helium 3 – a helium isotope thought to be the ideal fuel for nuclear fusion reactors, which are widely expected to begin producing power from the mid-twenty-first century.

The project to return humans to the Moon and to build a first outpost will be so expensive that international collaboration will be necessary. In 1994, the cost of a lunar base was estimated at $150 billion – although this is ten times as much as NASA's annual budget, it is only one-third of the annual budget of the Pentagon.

But before seriously considering a return of humans to the Moon, the cost of the transportation of materials in space will need to be reduced to a tenth of the present cost – and this may well happen within the next two decades.

Appendix A

Soviet launcher nomenclature

The nomenclature of Soviet space launchers is extremely complex. During the design phase, the rockets were assigned alphanumeric codes in the style of military hardware. ICBMs were 8Knn, where the first n indicated the design bureau (Korolyov was 5, 7 and 9, and Chelomei 8), and the second n indicated the progressive project number. After the ICBM entered service it received an R (Raketa – Rocket) designator followed by a progressive digit indicating the service entry order. Space launchers could be assigned an 8Ann code, but this rule was not enforced. Beginning in the 1960s, space launchers were assigned the code 11Knn or11Ann.

Space rockets were often given a (non-official) name that usually indicated the most significant payload; names that became well known – Proton, Molniya, Vostok, Soyuz and so on. One of the few launchers which was never given a name was the lunar N-1, for which the names Lenin and Communism are said to have been considered. For internal design bureau use, the rocket could be assigned a non-official designation such as UR-500, N-1, GR-1, and so on.

Service ICBMs were then displayed on Moscow's Red Square (usually during the victory parade of 9 May). Once the new missile had been observed, the NATO information services designated it with an SS (Surface-to-Surface) code, followed by a progressive number. To aid in identification, the rocket also received a nickname (supposedly derogatory, although this was not the rule) that for surface-to-surface missiles had to start with an 'S' (for example, SS-1b, Scud; SS-6, Sapwood; SS-18, Satan; SS-20 Sabre).

Space launchers identified in spy satellites pictures received an SL (Space Launcher) code followed by a progressive number. In the case of the N-1, which never entered service, an 'X' (eXperimental) is interposed between SL and the number (15). Space launchers were not given a nickname.

To make matters more complicated, a second identification system was quite often used in the West. This Sheldon system (after its creator, Charles Sheldon). identified the launcher family with a letter ('A', for example, indicated the R-7 and its modifications). A number identified the version, and another letter sometimes identified a sub-version; an 'e', for example indicated the incorporation of an upper stage capable of achieving 'escape velocity'.

The following example is an attempt at clarification. The 8K71 entered service as an ICBM as R-7, and its many space versions include 8K72 and 8A91, respectively launchers of the first lunar probes and of Object D/Sputnik 3. Of these two, only 8K72 was given a name: Vostok. On discovering the existence of the R-7 ICBM, NATO called it SS-6, Sapwood. The aforementioned space versions were called SL-3 and SL-2, and the Sheldon names are A-1 and A, respectively.

The nomenclature of Soviet lunar rockets

Code	Name	Other names	NATO	Sheldon
8K72	Vostok	–	SL-3	A-1
8K78	Molniya	–	SL-6	A-2-e
8K82K	Proton	UR-500K	SL-12	D-1-e
11A51	–	N-1	SL-X-15	G-1
15A18M	Dnepr	–	SL-24	–

Appendix B

Glossary

B1 DEFINITIONS

Aphelion The point of maximum distance from the Sun of a solar orbit. Its opposite is *perihelion*.

Apogee The point of maximum distance from the Earth of an Earth orbit. Its opposite is *perigee*. For simplicity, in this book it is also used for lunar orbits.

Apsides The points of maximum and minimum distance from the central body of an elliptical orbit.

Booster An auxiliary rocket used to boost the lift-off-thrust of a launcher.

Bus A structural part common to several satellites; *and* electric power distribution service.

CCD Charge-coupled device; an electronic chip that has replaced vidicons for photography and television.

Cosmic velocities Three characteristic velocities of spaceflight: *first cosmic velocity*, the minimum velocity required to place a satellite in a low Earth orbit (about 8 km/s); *second cosmic velocity* (also called *escape speed*), the velocity required to permanently leave the terrestrial sphere of attraction (about 11 km/s, beginning from the ground); and *third cosmic velocity*, the velocity required to permanently leave the Solar System.

Cryogenic fuels Fuels that can be stored in their liquid state under atmospheric pressure at very low temperature (oxygen, for example, is a liquid below a temperature of $-183°$ C).

Direct ascent A trajectory on which the deep-space probe is launched directly from the Earth's surface to another celestial body without entering an Earth parking orbit.

Direct descent A trajectory on which a probe lands on a celestial body directly from its transfer orbit, without first entering orbit around that body.

Ecliptic The plane of the Earth's orbit.

Fly-by A high-relative-speed and short-duration close encounter between a space-craft and a celestial body.

Hypergolic propellants A fuel and an oxidiser which ignite spontaneously on contact, without requiring an ignition system. Typical hypergolics are hydrazine (N_2H_6) and nitrogen tetroxide (N_2O_4).

ICBM (InterContinental Ballistic Missile) A militarily strategic and usually nuclear-tipped missile with a range of 6,400 km or more. Many space launchers were born as ICBMs; for example, the Russian Semiorka, the American Atlas and the Chinese Long March 2.

Lander A spacecraft designed to land on another celestial body.

Launch window A time interval during which it is possible to launch a spacecraft to ensure that it reaches its target on the assigned trajectory.

Orbit The trajectory on which a celestial body or spacecraft is travelling with respect to its central body. There are three possible cases. *Elliptical orbit*: a closed orbit in which the body passes from minimum distance to maximum distance from its central body every semi-period. This is the orbit of natural and artificial satellites around planets, and of planets around the Sun. *Parabolic orbit*: an open orbit in which the body passes through minimum distance from its central body and reaches infinity at zero velocity in infinite time. This is a pure abstraction, but the orbits of many comets around the Sun can be described as such. *Hyperbolic orbit*: an open orbit in which the body passes through minimum distance from its central body and reaches infinity at non-zero speed. This adequately describes the trajectory of spacecraft with respect to planets during fly-by manoeuvres.

Orbiter A spacecraft designed to orbit a celestial body.

Parking orbit A low Earth orbit used by deep space probes before heading to their targets, to counteract the constraints of launch windows and to eliminate launcher trajectory errors. Its opposite is *direct ascent*.

Perigee The minimum distance point from the Earth of an Earth orbit. Its opposite is *apogee*. For simplicity, in this book it is also used for lunar orbits.

Perihelion The minimum distance point from the Sun of a solar orbit. Its opposite is *aphelion*.

Rendezvous An encounter between two spacecraft or celestial bodies, at low relative speed.

Retro-rocket A rocket with a thrust directed opposite to the motion of a spacecraft, in order to brake it.

Rover A mobile vehicle used for exploring the surface of another celestial body.

Solar flare A solar chromospheric eruption – a powerful source of high-energy particles.

Solid propellant A rocket in which the fuel and oxidiser are mixed to form a putty-like material that can be poured into a motor to solidify.

Space probe A spacecraft designed to explore other celestial bodies from a short range.

Spectrometer An instrument for measuring the radiation in a particular part of the electromagnetic spectrum; for example, ultraviolet, infrared, gamma-ray, and so on.

Spin stabilisation A spacecraft stabilisation system in which the attitude is maintained by spinning the spacecraft around one of its main inertial axes.

Telemetry The transmission, via a radio system or a laser, of engineering data and measures acquired by a satellite.

Three-axis stabilisation A spacecraft stabilisation system in which the axes of the spacecraft are retained in a fixed attitude with respect to the stars.

Ullage rockets Small rockets – usually solid-fuelled – producing a small acceleration sufficient to force fluids in tanks towards the pump intakes to start larger liquid-fuel engines in weightlessness.

Vernier engines Small attitude control engines mounted in clusters and firing through a common point of the spacecraft. Attitude control is obtained by simple differential throttling of the engines.

Vidicon A television system based on resistance changes of certain materials when exposed to light. This system is now obsolete, and has been replaced by the *CCD*.

B2   ACRONYMS

ALSEP	Apollo Lunar Surface Experiment Package
ARIA	Apollo Range Instrumentation Aircraft
ARPA	Advanced Research Project Agency
BMDO	Ballistic Missile Defense Organisation
CCD	Charge-Coupled Device
CIA	Central Intelligence Agency
CNES	Centre National d'Etudes Spatiales [the French National Centre for Space Studies]
CSM	Command and Service Module
EASEP	Early Apollo Surface Experiment Package
ESA	European Space Agency
HGS	Hughes Global Services
ICBM	InterContinental Ballistic Missile
IMP	Interplanetary Monitoring Platform
ISAS	Institute for Space and Astronautical Sciences [Japanese space agency]
ISS	International Space Station
JPL	Jet Propulsion Laboratory [Caltech laboratory working under contract with NASA]
KH	KeyHole [a generic name for American imaging spy satellites]
KORD	KOntrol Roboti Dvigatyelyei [Engine Working Control System]
KREEP	K [potassium], Rare Earth Elements, Phosphorus
KTDU	Korrektiruyushaya Tormoznaya Dvigatyelnaya Ustanovka [Trajectory Correction and Braking Engine]
LK	Lunnii Korabl [Lunar Module]
LM	Lunar Module
LOI	Lunar Orbit Insertion
LOK	Lunnii Orbitalnii Korabl [Lunar Orbital Spaceship]
LRL	Lunar Receiving Laboratory
LRV	Lunar Rover Vehicle

LTTV	Lunar Landing Training Vehicle
MET	Modular Equipment Transporter
MUSES	MU [rocket] Space Engineering Satellite
N-1	Nossitel 1 [(Soviet) First Launcher]
NASA	National Aeronautics and Space Administration
NASDA	National Space Development Agency [Japanese space agency]
P&F	Particles and Fields [satellite]
PC + 2	PeriCynthion plus Two [NASA's way of specifying two hours after closest approach to the Moon]
PrOP	Pribori Ochenki Prokhodimosti [Cross-Country Evaluation Instrument]
RAE	Radio Astronomy Explorer
RD	Reaktinvnyi Dvigatyel [Rocket Engine]
RIFMA	Röntgen Isotopic Fluorescent Method of Analysis
SEI	Space Exploration Initiative
SIM	Scientific Instrument Module
SMSS	Soil Mechanics Surface Sampler
SNAP	System for Nuclear Auxiliary Power
SS	Surface-to-Surface [missile]
STS	Space Transportation System [Space Shuttle]
TL	Télémetrie Laser [Laser Telemetry]; Terre–Lune [Earth–Moon]
TLI	TransLunar Injection
UDMH	Unsymmetrical Di-Methyl Hydrazine
USAF	United States Air Force
V2	Vergeltungswaffe 2 [Vengeance Weapon 2]
VAB	Vehicle Assembly Building
VLF	Very Low Frequency
WSB	Weak Stability Boundary [orbits]

Appendix C

Chronology

C1 LUNAR EXPLORATION

1946	Jan 10	US Army successfully bounces radio waves off the Moon for the first time
1956	Jan 30	The Supreme Soviet approves the Korolyov memorandum: the Soviet Union will launch a satellite and probes to the Moon
1957	Oct 4	The Soviet Union launches the first artificial satellite
1958	Feb 1	The United States launch its first artificial satellite: Explorer 1
	Mar 27	President Eisenhower approves the military lunar project
	Aug 17	The first lunar probe, the American Pioneer 0, fails
	Sep 23	The first Soviet lunar probe fails
	Oct 1	NASA is founded
1959	Jan 2	Luna 1 is the first probe to fly by the Moon
	Sep 13	Luna 2 is the first probe to hit the Moon
	Oct 7	Luna 3 takes the first pictures of the lunar far-side
1961	May 25	President Kennedy initiates the Apollo programme, to land a man on the Moon before 1970
1962	Jan 26	The first American second-generation lunar probe Ranger 3 is launched, but its upper stage fails
	Apr 2	Luna 4 is launched to soft land on the Moon, but fails
1964	Jul 31	Ranger 7 hits the Moon after taking 4,316 pictures; the first completely successful American lunar probe
	Aug 3	The Soviet Union approves its own lunar manned project, three years after America.
1965	Mar 24	Ranger 9 hits the Moon inside the crater Alphonsus
	Jul 20	Zond 3 completes the first reconnaissance of the far-side
1966	Feb 3	Luna 9 is the first successful lunar soft lander
	Apr 3	Luna 10 is the first lunar artificial satellite
	Jun 2	Surveyor 1 is the first American probe to land on the Moon

	Aug 14	Lunar Orbiter 1 is the first American lunar artificial satellite
1967	Jan 27	Astronauts Grissom, White and Chaffee die in the fire in their Apollo capsule
	Apr 24	Cosmonaut Komarov dies during re-entry on Soyuz 1, a prototype of a lunar piloted spacecraft
	May 8	Lunar Orbiter 4 is the first lunar polar orbiter
	Nov 6	The first Saturn V rocket is successfully launched
1968	Jan 10	Surveyor 7 lands near the crater Tycho
	Sep 21	Zond 5 re-enters over the Indian Ocean, after circumnavigating the Moon
	Oct 11	First manned Apollo flight, Apollo 7
	Nov 17	Zond 6 crashes after circumnavigating the Moon
	Dec 24	Apollo 8 enters lunar orbit carrying three astronauts
1969	Feb 21	The Soviet N-1 launcher explodes during its first test flight
	Jul 20	Neil Armstrong and Edwin Aldrin, of Apollo 11, land on the Moon
	Jul 26	The first returned lunar sample container is opened
	Aug 14	Zond 7 circumnavigates the Moon and lands without incident
	Nov 19	Apollo 12 lands near Surveyor 3
1970	Apr 14	Apollo 13 is aborted because of an oxygen tank explosion
	Sep 24	Luna 16 automatically returns 101 g of lunar soil to Earth
	Oct 27	Zond 8 lands after the last lunar circumnavigation mission
	Nov 17	Luna 17 deploys the rover Lunokhod 1 on the Moon
1971	Feb 5	Apollo 14 lands in the Fra Mauro region
	Jul 30	Apollo 15 lands in the lunar Apennines
1972	Feb 25	Luna 20 returns 30 g of soil to Earth
	Apr 21	Apollo 16 lands in the Descartes highlands
	Nov 23	The fourth N-1 launch fails – the end of the Soviet dream to reach the Moon
	Dec 11	Apollo 17 – the last manned lunar mission – lands on the Moon, carrying the first geologist astronaut, Harrison Schmitt
1973	Jan 16	Lunokhod 2 lands inside the crater Lemonnier
	Jun 15	RAE-2 is the first radio astronomy observatory to orbit the Moon
1976	Aug 23	Luna 24 returns 170 g of soil to Earth
1977	Sep 30	All of the instruments deployed on the Moon by Apollo missions are shut off
1983	Dec 22	ICE carries out the last of five fly-bys before heading towards comet Giacobini–Zinner
1990	Jan 24	Japan launches two lunar probes, Hiten and Hagoromo
	Dec 8	Galileo observes the Moon for the first time
1992	Dec 8	Galileo observes the Moon for the second time
1993	Apr 10	Hiten hits the Moon – the first probe for seventeen years to do so
1994	Feb 19	The American military probe Clementine enters lunar orbit
1998	Jan 11	Lunar Prospector enters lunar polar orbit
	May 13	HGS-1 carries out the first 'commercial' fly-by of the Moon
1999	Jul 31	Lunar Prospector impacts on the Moon near the south pole

2003   Mar 3   The Chinese National Space Agency announces the Chang'e lunar exploration programme
        Aug 15   Indian Prime Minister announces the official approval of the Indian lunar exploration programme
        Sept 27 SMART-1, the first European lunar probe is launched

C2   MOON-SHOTS

Name	Launch date	Launcher	Country
Pioneer 0*	1958 Aug 17	Thor–Able	USA
E-1 n.1*	1958 Sep 23	8K72 Vostok	USSR
Pioneer 1*	1958 Oct 11	Thor–Able	USA
E-1 n.2*	1958 Oct 11	8K72 Vostok	USSR
Pioneer 2*	1958 Nov 8	Thor–Able	USA
E-1 n.3*	1958 Dec 4	8K72 Vostok	USSR
Pioneer 3*	1958 Dec 6	Juno II	USA
Luna 1 Mechta*	1959 Jan 2	8K72 Vostok	USSR
Pioneer 4*	1959 Mar 3	Juno II	USA
E-1A n.5*	1959 Jun 18	8K72 Vostok	USSR
Luna 2	1959 Sep 12	8K72 Vostok	USSR
Luna 3	1959 Oct 4	8K72 Vostok	USSR
Pioneer P-3*	1959 Nov 26	Atlas–Able	USA
E-3 n.1*	1960 Apr 15	8K72 Vostok	USSR
E-3 n.2*	1960 Apr 19	8K72 Vostok	USSR
Pioneer P-30*	1960 Sep 25	Atlas–Able	USA
Pioneer P-31*	1960 Dec 15	Atlas–Able	USA
Ranger 3*	1962 Jan 26	Atlas–Agena B	USA
Ranger 4*	1962 Apr 23	Atlas–Agena B	USA
Ranger 5*	1962 Oct 18	Atlas–Agena B	USA
E-6 n.1*	1963 Jan 4	8K78 Molniya	USSR
E-6 n.2*	1963 Feb 3	8K78 Molniya	USSR
Luna 4*	1963 Apr 2	8K78 Molniya	USSR
3MV-1 n.1*	1963 Nov 11	8K78 Molniya	USSR
Ranger 6*	1964 Jan 30	Atlas–Agena B	USA
E-6 n.4*	1964 Mar 21	8K78 Molniya	USSR
E-6 n.5*	1964 Apr 20	8K78 Molniya	USSR
Ranger 7	1964 Jul 28	Atlas–Agena B	USA
Ranger 8	1965 Feb 17	Atlas–Agena B	USA
Ranger 9	1965 Mar 21	Atlas–Agena B	USA
E-6 n.8*	1965 Apr 10	8K78 Molniya	USSR
Luna 5*	1965 May 9	8K78 Molniya	USSR
Luna 6*	1965 Jun 8	8K78 Molniya	USSR
Zond 3	1965 Jul 18	8K78 Molniya	USSR

Name	Launch date	Launcher	Country
Luna 7*	1965 Oct 4	8K78 Molniya	USSR
Luna 8*	1965 Dec 3	8K78 Molniya	USSR
Luna 9	1966 Jan 31	8K78M Molniya	USSR
E-6S n.204*	1966 Mar 1	8K78M Molniya	USSR
Luna 10	1966 Mar 31	8K78M Molniya	USSR
Surveyor 1	1966 May 30	Atlas–Centaur D	USA
Explorer 33*	1966 Jul 1	Thor–Delta E1	USA
Lunar Orbiter 1	1966 Aug 10	Atlas–Agena D	USA
Luna 11	1966 Aug 24	8K78M Molniya	USSR
Surveyor 2*	1966 Sep 20	Atlas–Centaur D	USA
Luna 12	1966 Oct 22	8K78M Molniya	USSR
Lunar Orbiter 2	1966 Nov 6	Atlas–Agena D	USA
Luna 13	1966 Dec 21	8K78M Molniya	USSR
Lunar Orbiter 3	1967 Feb 5	Atlas–Agena D	USA
Surveyor 3	1967 Apr 17	Atlas–Centaur D	USA
Lunar Orbiter 4	1967 May 4	Atlas–Agena D	USA
Surveyor 4*	1967 Jul 14	Atlas–Centaur D	USA
Explorer 35	1967 Jul 19	Thor–Delta E1	USA
Lunar Orbiter 5	1967 Aug 1	Atlas–Agena D	USA
Surveyor 5	1967 Sep 8	Atlas–Centaur D	USA
L-1 n.4L*	1967 Sep 27	8K82K Proton K/D	USSR
Surveyor 6	1967 Nov 7	Atlas–Centaur D	USA
L-1 n.5L*	1967 Nov 22	8K82K Proton K/D	USSR
Surveyor 7	1968 Jan 7	Atlas–Centaur D	USA
E-6LS n.112*	1968 Feb 7	8K78M Molniya	USSR
Luna 14	1968 Apr 7	8K78M Molniya	USSR
L-1 n.7L*	1968 Apr 22	8K82K Proton K/D	USSR
Zond 5	1968 Sep 14	8K82K Proton K/D	USSR
Zond 6*	1968 Nov 10	8K82K Proton K/D	USSR
Apollo 8	1968 Dec 21	Saturn V	USA
L-1 n.13L*	1969 Jan 20	8K82K Proton K/D	USSR
E-8 n.201*	1969 Feb 19	8K82K Proton K/D	USSR
L-1S n.3*	1969 Feb 21	11A52 N1	USSR
Apollo 10	1969 May 18	Saturn V	USA
E-8-5 n.402*	1969 Jun 13	8K82K Proton K/D	USSR
L-1S n.5*	1969 Jul 3	11A52 N1	USSR
Luna 15*	1969 Jul 13	8K82K Proton K/D	USSR
Apollo 11	1969 Jul 16	Saturn V	USA
Zond 7	1969 Aug 7	8K82K Proton K/D	USSR
E-8-5 n.403*	1969 Sep 23	8K82K Proton K/D	USSR
E-8-5 n.404*	1969 Oct 22	8K82K Proton K/D	USSR
Apollo 12	1969 Nov 14	Saturn V	USA
E-8-5 n.405*	1970 Feb 6	8K82K Proton K/D	USSR

Name	Launch date	Launcher	Country
Apollo 13*	1970 Apr 11	Saturn V	USA
Luna 16	1970 Sep 12	8K82K Proton K/D	USSR
Zond 8	1970 Oct 20	8K82K Proton K/D	USSR
Luna 17	1970 Nov 10	8K82K Proton K/D	USSR
Apollo 14	1971 Jan 31	Saturn V	USA
Apollo 15	1971 Jul 26	Saturn V	USA
Luna 18*	1971 Sep 2	8K82K Proton K/D	USSR
Luna 19	1971 Sep 28	8K82K Proton K/D	USSR
Luna 20	1972 Feb 14	8K82K Proton K/D	USSR
Apollo 16	1972 Apr 16	Saturn V	USA
7K-LOK n.6A*	1972 Nov 23	11A52 N1	USSR
Apollo 17	1972 Dec 7	Saturn V	USA
Luna 21	1973 Jan 8	8K82K Proton K/D	USSR
Explorer 49	1973 Jun 10	Delta 1913	USA
Luna 22	1974 May 29	8K82K Proton K/D	USSR
Luna 23*	1974 Oct 28	8K82K Proton K/D	USSR
E-8-5M n.412*	1975 Oct 16	8K82K Proton K/D	USSR
Luna 24	1976 Aug 9	8K82K Proton K/D	USSR
ICE	1978 Aug 12	Delta 2914	USA
Galileo	1989 Oct 18	OV-104 Atlantis + IUS	USA
Muses-A	1990 Jan 24	Mu-3S-II	Japan
Geotail	1992 Jul 24	Delta 6925	Japan/USA
DSPSE Clementine	1994 Jan 25	Titan II SLV	USA
Wind	1994 Nov 1	Delta 7925	USA
HGS-1	1997 Dec 24	8K82K Proton K/DM3	Commercial
Lunar Prospector	1998 Jan 7	Athena 2	USA
Nozomi	1998 Jul 3	M-V	Japan
WMAP	2001 Jun 30	Delta 7425-10	USA
SMART-1	2003 Sep 27	Ariane 5G	(ESA)

* Failed missions

C3   OBJECT E: SOVIET LUNAR PROBES

Name	Characteristics	Launcher	Launched as
E-1	Lunar impact, 170-kg mass	8K72	Luna 1 and three failures
E-1A	Improved E-1	8K72	Luna 2 and one failure
E-2	Far-side reconnaissance, Yenissei-1 camera	8K73	Cancelled
E-2A	Improved E-2, Yenissei-2 camera	8K72	Luna 3

Name	Characteristics	Launcher	Launched as
E-2F	Improved E-2	8K73?	Cancelled
E-3	1. Far-side reconnaissance	8K73	Cancelled
	2. Improved E-2A	8K72	Two failures
E-4	Lunar atom bomb	8K73	Cancelled
E-5	E-1-based orbiter	8K72	Cancelled
E-6	Lander. Cruise stage and	8K78	Eleven failures,
	100-kg mini-lander	8K78M	including Luna 4–8
E-6A	150-kg lander E-6	8K78	Cancelled
E-6LF	Photographic orbiter	8K78M	Luna 11 and Luna 12
E-6LS	N1-L3 support orbiter	8K78M	Luna 14
E-6M	Improved E-6	8K78M	Luna 9 and Luna 13
E-6S	Simplified orbiter	8K78M	Luna 10 and one failure
E-7	Photographic orbiter	8K78	Cancelled
E-8	Lander and rover	8K82K	Luna 17, Luna 21 and one failure
E-8LS	Heavy orbiter	8K82K	Luna 19, Luna 22
E-8-5	Lander, robot arm and sample return capsule	8K82K	Luna 16, Luna 20 and six failures. including Luna 15 and Luna 18
E-8-5M	E-8-5 with rail mounted drill	8K82K	Luna 24 and two failures, including Luna 23
E-8/M-69	E-8-based Mars orbiter	8K82K	Cancelled
E-8/Zvezda	Deep (10 m) drilling lander in support of Zvezda base	8K82K(?)	Cancelled
E-8 Beacon	Manned landing E-8 mounted beacon	8K82K(?)	Cancelled

C4 SATURN LAUNCHERS

Serial numbers: SA-1–SA-10 – Saturn 1; SA-2NN – Saturn IB; SA-5NN – Saturn V

Date		Name	Launcher	CSM	LM	Notes
1961	Oct 27		SA-1	–	–	Inert stages 2 and 3
1962	Apr 25		SA-2	–	–	Inert stages 2 and 3
	Nov 16		SA-3	–	–	Inert stages 2 and 3
1963	Mar 28		SA-4	–	–	Inert stages 2 and 3
1964	Jan 29		SA-5	–	–	S-IV stage test
	May 28		SA-6	BP-13	–	Boilerplate test
	Sep 18		SA-7	BP-15	–	LES test
1965	Feb 16	Pegasus 1	SA-9	BP-16	–	Boilerplate test
	May 25	Pegasus 2	SA-8	BP-26	–	Boilerplate test
	Jul 30	Pegasus 3	SA-10	BP-9A	–	Boilerplate test

Date		Name	Launcher	CSM	LM	Notes
1966	Feb 26	'Apollo 1'	SA-201	009	–	First Saturn IB
	Jul 5	'Apollo 2'	SA-203	–	–	S-IVB stage test
	Aug 25	'Apollo 3'	SA-202	011	–	
–			SA-204	012	–	Fire	CSM test
1967	Nov 9	Apollo 4	SA-501	017	LTA-10	First Saturn V
1968	Jan 22	Apollo 5	SA-204	–	1	LM test
	Apr 4	Apollo 6	SA-502	020	LTA-2R	Non-nominal test
	Oct 11	Apollo 7	SA-205	101	–	First manned flight
	Dec 21	Apollo 8	SA-503	103	LTA-B	Lunar flight
1969	Mar 3	Apollo 9	SA-504	104	3	Complete test
	May 18	Apollo 10	SA-505	106	4	Lunar test
	Jul 16	Apollo 11	SA-506	107	5	Landing
	Nov 14	Apollo 12	SA-507	108	6	Lightning strike
1970	Apr 11	Apollo 13	SA-508	109	7	Aborted mission
1971	Jan 31	Apollo 14	SA-509	110	8	To Fra Mauro
	Jul 26	Apollo 15	SA-510	112	10	First J mission
1972	Apr 16	Apollo 16	SA-511	113	11	To Descartes
	Dec 7	Apollo 17	SA-512	114	12	Last lunar flight
1973	May 14	Skylab 1	SA-513	–	–	Skylab launch
	May 25	Skylab 2	SA-206	116	–	
	Jul 28	Skylab 3	SA-207	117	–	
	Nov 16	Skylab 4	SA-208	118	–	
1975	Jul 15	ASTP	SA-210	111	DM-2	Last Saturn

C5   APOLLO CREWS

A crew consists of Commander, CSM pilot and LM pilot

Mission	CSM	LM	Crew	Back-up
AS–204	–	–	Grissom	Schirra
			White	Eisele
			Chaffee	Cunningham
Apollo 7	–	–	Schirra	Stafford
			Eisele	Young
			Cunningham	Cernan
Apollo 8	–	–	Borman	Armstrong
			Lovell	Aldrin
			Anders	Haise
Apollo 9	Gumdrop	Spider	McDivitt	Conrad
			Scott	Gordon
			Schweickart	Bean

Mission	CSM	LM	Crew	Back-up
Apollo 10	Charlie Brown	Snoopy	Stafford	Cooper
			Young	Eisele
			Cernan	Mitchell
Apollo 11	Columbia	Eagle	Armstrong	Lovell
			Collins	Anders
			Aldrin	Haise
Apollo 12	Yankee Clipper	Intrepid	Conrad	Scott
			Gordon	Worden
			Bean	Irwin
Apollo 13	Odyssey	Aquarius	Lovell	Young
			Mattingly*	Swigert
			Haise	Duke
Apollo 14	Kitty Hawk	Antares	Shepard	Cernan
			Roosa	Evans
			Mitchell	Engle
Apollo 15	Endeavour	Falcon	Scott	Gordon
			Worden	Brand
			Irwin	Schmitt
Apollo 16	Casper	Orion	Young	Haise
			Mattingly	Roosa
			Duke	Mitchell
Apollo 17	America	Challenger	Cernan	Haise
			Evans	Worden
			Schmitt	Irwin

* Mattingly was replaced by Swigert a few days prior to launch.

C6   APOLLO LUNAR MISSIONS

Mission	Landing date	Total stay	EVA duration	Samples mass (kg)
Apollo 11	1969 Jul 20	21h 36m	2h 31m	21.0
Apollo 12	1969 Nov 19	31h 31m	3h 56m	16.7
			3h 49m	17.6
			7h 45m	34.3
Apollo 14	1971 Feb 5	33h 30m	4h 47m	20.5
			4h 34m	22.3
			9h 22m	42.8

Mission	Landing date	Total stay	EVA duration	Samples mass (kg)
Apollo 15	1971 Jul 30	66h 55m	0h 33m	–
			6h 32m	14.5
			7h 12m	34.9
			4h 50m	27.3
			19h 08m	76.7
Apollo 16	1972 Apr 21	71h 02m	7h 11m	29.9
			7h 23m	29.0
			5h 40m	35.4
			20h 14m	94.3
Apollo 17	1972 Dec 11	74h 59m	7h 12m	14.3
			7h 37m	34.1
			7h 15m	62.0
			22h 04m	110.4

C7   LUNOKHOD MISSIONS

Mission	Lunar day	Dates	Metres travelled
Lunokhod 1	1	1970 Nov 17–24	197
	2	1970 Dec 8–23	1,522
	3	1971 Jan 7–21	1,936
	4	1971 Feb 7–20	1,573
	5	1971 Mar 7–20	2,004
	6	1971 Apr 6–20	1,029
	7	1971 May 6–20	197
	8	1971 Jun 4–11	1,560
	9	1971 Jul 3–17	219
	10	1971 Aug 2–16	215
	11	1971 Sep 8–15	88
Lunokhod 2	1	1973 Jan 16–24	1,148
	2	1973 Feb 7–22	9,919
	3	1973 Mar 9–23	~16,533
	4	1973 Apr 8–23	~8,600
	5	1973 May 7–9	~800

C8   OBJECTS ON THE MOON

Date		Name	Latitude (º)	Longitude (º)	Notes
1959	Sep 13	Luna 2	~31 N	~1 W	Impact
	Sep 13	Stage-E I1-7B	30 N?	1 W?	Impact
1962	Apr 26	Ranger 4	15.5 S	130.7 W	Impact
1964	Feb 2	Ranger 6	9.39 N	21.51 E	Impact
	Jul 31	Ranger 7	10.63 S	20.60 W	Impact
1965	Feb 20	Ranger 8	2.7 N	24.63 E	Impact
	Mar 24	Ranger 9	12.83 S	2.37 W	Impact
	May 9	Luna 5	31 S?	8 W?	Impact
	Oct 7	Luna 7	9 N	49 W	Impact
	Oct 7	Auxiliary modules (2)	9 N	49 W	Impact
	Dec 6	Luna 8	9.13 N	63.3 W	Impact
	Dec 6	Auxiliary modules (2)	9.13 N	63.3 W	Impact
1966	Feb 5	Luna 9	7.13 N	64.37 W	Soft
	Feb 5	Cruise stage	7.13 N	64.37 W	Impact
	Feb 5	Auxiliary modules (2)	7.13 N	64.37 W	Impact
	Jun 2	Surveyor 1	2.46 S	43.23 W	Soft
	Jun 2	Retro-rocket	2.46 S	43.23 W	Impact
	Sep 23	Surveyor 2	5.5 N	12 W	Impact
	Oct 29	Lunar Orbiter 1	6.7 N	162 E	Impact
	Dec 24	Luna 13	18.9 N	62.05 W	Soft
	Dec 24	Cruise stage	18.9 N	62.05 W	Impact
	Dec 24	Auxiliary modules (2)	18.9 N	62.05 W	Impact
1967	Apr 20	Surveyor 3	2.97 S	23.34 W	Soft
	Apr 20	Retro-rocket	2.97 S	23.34 W	Impact
	Jul 17	Surveyor 4	0.4 N	1.33 W	Impact?
	Sep 11	Surveyor 5	1.42 N	23.2 E	Soft
	Sep 11	Retro-rocket	1.42 N	23.2 E	Impact
	Oct 9	Lunar Orbiter 3	14.6 N	91.7 W	Impact
	Oct 11	Lunar Orbiter 2	4 S	98 E	Impact
	Nov 10	Surveyor 6	0.51 N	1.39 W	Soft
	Nov 10	Retro-rocket	0.51 N	1.39 W	Impact
1968	Jan 10	Surveyor 7	40.86 S	11.47 W	Soft
	Jan 10	Retro-rocket	40.86 S	11.47 W	Impact
	Jan 31	Lunar Orbiter 5	0 S	70 W	Impact
1969	Jul 20	Eagle	0.674 N	23.472 E	Soft
	Jul 20	EASEP	0.673 N	23.472 E	
	Jul 21	Luna 15	17 N?	60 E?	Impact?
	Nov 19	Intrepid	3.013 S	23.419 W	Soft
	Nov 19	ALSEP	3.010 S	23.424 W	
	Nov 20	Intrepid AS	5.5 S	23.4 W	Impact

Date		Name	Latitude (º)	Longitude (º)	Notes
1970	Apr 15	S-IVB 508	2.4 S	27.9 W	Impact
	Sep 20	Luna 16	0.68 S	56.30 E	Soft
	Nov 17	Luna 17 Lunokhod 1	38.28 N	35 W	Soft
1971	Feb 4	S-IVB 509	8.0 S	26.6 W	Impact
	Feb 5	Antares	3.645 S	17.471 W	Soft
	Feb 5	ALSEP	3.644 S	17.477 W	
	Feb 7	Antares AS	3.42 S	19.67 W	Impact
	Jul 29	S-IVB 510	0.99 S	11.89 W	Impact
	Jul 30	Falcon	26.132 N	3.634 E	Soft
	Jul 30	ALSEP	26.134 N	3.629 E	Plus LRV
	Aug 4	Falcon AS	26.36 N	0.25 E	Impact
	Sep 11	Luna 18	3.57 N	56.5 E	Impact?
1972	Feb 21	Luna 20	3.53 N	56.55 E	Soft
	Apr 19	S-IVB 511	1.83 N	23.30 W	Impact
	Apr 21	Orion	8.973 S	15.498 E	Soft
	Apr 21	ALSEP	8.975 S	15.496 E	Plus LRV
	May 29	P&F	10.16 N	111.9 E	Impact
	Dec 10	S-IVB 512	4.33 S	12.37 W	Impact
	Dec 11	Challenger	20.188 N	30.774 E	Soft
	Dec 11	ALSEP	20.189 N	30.767 E	Plus LRV
	Dec 15	Challenger AS	19.96 N	30.50 E	Impact
1973	Jan 15	Luna 21 Lunokhod 2	25.85 N	30.45 E	Soft
1974	Nov 6	Luna 23	13 N	62 E	Soft?
1976	Aug 18	Luna 24	12.75 N	62.2 E	Soft
1993	Apr 10	Hiten	34.0 S	55.3 E	Impact
1999	Jul 31	Lunar Prospector	87.7 S	42 E	Impact

C9   OBJECTS IN LUNAR ORBIT

Date		Name	Periastron (km) (V = variable)	Apastron (km)	In orbit until
1966	Apr 3	Luna 10	350	1,017	
	Apr 3	Luna 10 cruise stage	350	1,017	
	Aug 14	Lunar Orbiter 1	V	V	1966 Oct 29
	Aug 27	Luna 11	160	1,193	
	Oct 25	Luna 12	V	V	
	Nov 10	Lunar Orbiter 2	V	V	1967 Oct 11
1967	Feb 6	Lunar Orbiter 3	V	V	1967 Oct 9
	May 8	Lunar Orbiter 4	V	V	1967 October?

Date		Name	Periastron (km)	Apastron (km)	In orbit until
			(V = variable)		
1967	Jul 21	Explorer 35	800	7,692	
	Jul 21	Explorer 35 retro-rocket	800	7,692	
	Aug 5	Lunar Orbiter 5	V	V	1968 Jan 31
1968	Apr 10	Luna 14	160	870	
	Dec 12	Apollo 8	V	V	1968 Dec 25
1969	May 21	Charlie Brown	V	V	1969 May 24
	May 22	Snoopy	V	V	
	May 22	Snoopy AS	V	V	1969 May 24
	Jul 17	Luna 15	V	V	1969 Jul 21
	Jul 17	Auxiliary tanks (2)	16	110	
	Jul 19	Columbia	V	V	1969 Jul 22
	Jul 20	Eagle	V	V	1969 Jul 20
	Jul 21	Eagle AS	V	V	
	Nov 18	Yankee Clipper	V	V	1969 Nov 21
	Nov 19	Intrepid	V	V	1969 Nov 19
	Nov 20	Intrepid AS	V	V	1969 Nov 20
1970	Sep 17	Luna 16	V	V	1970 Sep 20
	Sep 20	Auxiliary tanks (2)	15	105	
	Nov 15	Luna 17	V	V	1970 Nov 17
	Nov 7	Auxiliary tanks (2)	19	85	
1971	Feb 4	Kitty Hawk	V	V	1971 Feb 7
	Feb 5	Antares	V	V	1971 Feb 5
	Feb 7	Antares AS	V	V	1971 Feb 7
	Jul 29	Endeavour	V	V	1971 Aug 4
	Jul 30	Falcon	V	V	1971 Jul 30
	Aug 2	Falcon AS	V	V	1971 Aug 4
	Aug 4	P&F	100	139	
	Sep 6	Luna 18	V	V	1971 Sep 11
	Sep 11	Auxiliary tanks (2)	~15	~100	
	Oct 3	Luna 19	V	V	
1972	Feb 18	Luna 20	V	V	1972 Feb 21
	Feb 19	Auxiliary tanks (2)	21	100	
	Apr 19	Casper	V	V	1972 Apr 25
	Apr 20	Orion	V	V	1972 Apr 21
	Apr 24	Orion AS	V	V	
	Apr 24	Mass spectrometer	100?	100?	
	Apr 24	P&F	96	122	1973 May 29
	Dec 10	America	V	V	1972 Dec 16
	Dec 11	Challenger	V	V	1972 Dec 11
	Dec 14	Challenger AS	V	V	1972 Dec 15

Date		Name	Periastron (km) (V = variable)	Apastron (km)	In orbit until
1973	Jan 12	Luna 21	V	V	1973 Jan 15
	Jan 16	Auxiliary tanks (2)	16	~100	
	Jun 15	Explorer 49	1,053	1,063	
	Jun 15	Explorer 49 retro-rocket	1,123	1,334	
1974	Jun 2	Luna 22	V	V	
	Nov 1	Luna 23	V	V	1974 Nov 6
	Nov 5	Auxiliary tanks (2)	17	105	
1976	Aug 13	Luna 24	V	V	1976 Aug 18
	Aug 18	Auxiliary tanks (2)	12	120	
1990	Mar 18	Hagoromo	7,400	20,000	
1992	Feb 15	Hiten	422	49,400	1993 Apr 10
1994	Feb 19	DSPSE Clementine	V	V	1994 May 5
1998	Jan 11	Lunar Prospector	V	V	1999 Jul 31

(Many of the objects that are supposedly still orbiting the Moon have probably decayed.)

C10 OBJECTS WHICH FLEW BY THE MOON

Date		Name	Periastron (km)	Notes
1959	Jan 4	Luna 1 Mechta	6,400	
	Jan 4	Stage-E B1-6	6,400 ?	Luna 1 launcher
	Mar 4	Pioneer 4	59,500	
	Mar 4	Baby Sergeant AM-14	59,500 ?	Pioneer 4 launcher
	Mar 4	Masses (2)	59,500 ?	De-spin system
	Oct 6	Luna 3	7,900	
	Oct 6	Stage-E I1-8	7,900 ?	Luna 3 launcher
1960	Jan 24	Luna 3	50,500 ?	
1962	Jan 28	Ranger 3	36,785	
	Jan 28	Agena B 6003	32,000 ?	Ranger 3 launcher
	Apr 26	Agena B 6004	–	Ranger 4 launcher
	Oct 21	Ranger 5	724	
	Oct 21	Agena B 6005	–	Ranger 5 launcher
1963	Apr 6	Luna 4	8,336	
	Apr 6	Stage L G103-11	8,500 ?	Luna 4 launcher
1964	Feb 2	Agena B 6008	–	Ranger 6 launcher
	Jul 31	Agena B 6009	–	Ranger 7 launcher
1965	Feb 20	Agena B 6006	–	Ranger 8 launcher
	Mar 24	Agena B 6007	–	Ranger 9 launcher

Date		Name	Periastron (km)	Notes
1965	May 12	Stage L U103-30	–	Luna 5 launcher
	Jun 11	Luna 6	160,000	
	Jun 11	Stage L U103-31	–	Luna 6 launcher
	Jul 20	Zond 3	9,220	
	Jul 20	Stage L	9,220 ?	Zond 3 launcher
	Oct 7	Stage L U103-27	–	Luna 7 launcher
	Dec 6	Stage L U103-28	–	Luna 8 launcher
1966	Feb 3	Stage L U103-32	–	Luna 9 launcher
	Apr 3	Stage L N103-42	–	Luna 10 launcher
	Jun 2	Centaur C-10	400 ?	Surveyor 1 launcher
	Aug 14	Agena D	–	Lunar Orbiter 1 launcher
	Aug 27	Stage L N103-43	–	Luna 11 launcher
	Sep 23	Centaur C-7	–	Surveyor 2 launcher
	Oct 25	Stage L N103-44	–	Luna 12 launcher
	Dec 24	Stage L N103-45	–	Luna 13 launcher
	Nov 10	Agena D	–	Lunar Orbiter 2 launcher
1967	Feb 6	Agena D	–	Lunar Orbiter 3 launcher
	Apr 20	Centaur C-12	–	Surveyor 3 launcher
	May 8	Agena D	–	Lunar Orbiter 4 launcher
	Jul 17	Centaur C-11	–	Surveyor 4 launcher
	Jul 21	FW-4D	–	Explorer 35 launcher
	Jul 21	Masses (2)	–	De-spin system
	Aug 5	Agena D	–	Lunar Orbiter 5 launcher
	Sep 11	Centaur C-13	–	Surveyor 5 launcher
	Nov 10	Centaur C-14	–	Surveyor 6 launcher
1968	Jan 10	Centaur C-15	–	Surveyor 7 launcher
	Apr 10	Stage L Ya716-58	–	Luna 14 launcher
	Sep 18	Zond 5	1,950	
	Sep 18	Stage D 234-01	–	Zond 5 launcher
	Nov 14	Zond 6	2,420	
	Nov 14	Stage D 235-01	–	Zond 6 launcher
	Dec 24	S-IVB 503N/LTA-B	1,263	Apollo 8 launcher
	Dec 24	LM adapters (4)	–	
1969	May 21	S-IVB 505N	3,111	Apollo 10 launcher
	May 21	LM adapters (4)	–	
	Jul 17	Stage D 242-01	–	Luna 15 launcher
	Jul 19	S-IVB 506	3,380	Apollo 11 launcher
	Jul 19	LM adapters (4)	–	
	Aug 11	Zond 7	1,985	
	Aug 11	Stage D 243-01	–	Zond 7 launcher
	Nov 19	S-IVB 507	5,708	Apollo 12 launcher
	Nov 19	LM adapters (4)	–	

Date		Name	Periastron (km)	Notes
1970	Apr 14	LM adapters (4)	–	
	Apr 14	Apollo 13	219	
	Sep 17	Stage D 248-01	–	Luna 16 launcher
	Oct 24	Zond 8	1,110	
	Oct 24	Stage D 250-01	–	Zond 8 launcher
	Nov 15	Stage D 251-01	–	Luna 17 launcher
1971	Feb 4	LM adapters (4)	–	
	Jul 29	SIM Door	–	
	Jul 29	LM adapters (4)	–	
	Sep 6	Stage D 256-01	–	Luna 18 launcher
	Oct 3	Stage D 257-01	–	Luna 19 launcher
1972	Feb 18	Stage D 258-01	–	Luna 20 launcher
	Apr 19	SIM Door	–	
	Apr 19	LM adapters (4)	–	
	Dec 10	SIM Door	–	
	Dec 10	LM adapters (4)	–	
1973	Jan 12	Stage D 259-01	–	Luna 21 launcher
	Jun 15	Third stage Delta 1913	–	Explorer 49 launcher
	Jun 15	Masses (2)	–	De-spin system
1974	Jun 2	Stage D 282-02	–	Luna 22 launcher
	Nov 2	Stage D 285-01	–	Luna 23 launcher
1976	Aug 13	Stage D 288-02	–	Luna 24 launcher
1983	Mar 30	ICE	19,570	
	Apr 23	ICE	21,137	
	Sep 27	ICE	22,790	
	Oct 21	ICE	17,440	
	Dec 22	ICE	120	
1990	Mar 18	Hiten	16,472	
	Jul 10	Hiten	76,050	
	Aug 4	Hiten	–	
	Sep 7	Hiten	–	
	Oct 2	Hiten	–	
1991	Jan 2	Hiten	–	
	Jan 28	Hiten	–	
	Mar 3	Hiten	13,300	
	Apr 26	Hiten	–	
	Oct 2	Hiten	–	
1992	Sep 8	Geotail	13,648	
	Oct 14	Geotail	32,637	
	Nov 8	Geotail	23,773	
1993	Jan 10	Geotail	24,320	
	Feb 5	Geotail	22,497	

Date		Name	Periastron (km)	Notes
1993	Apr 8	Geotail	23,459	
	May 2	Geotail	24,187	
	Aug 6	Geotail	32,286	
	Sep 26	Geotail	34,076	
1994	Jan 1	Geotail	32,674	
	Feb 21	Geotail	30,431	
	Jun 28	Geotail	59,089	
	Jul 20	DSPSE Clementine	–	
	Aug 16	Geotail	74,299	
	Oct 25	Geotail	22,445	
	Dec 27	Wind	11,834	
1995	Jul 30	Wind	23,609	
	Aug 23	Wind	45,744	
	Sep 20	Wind	46,370	
	Nov 26	Wind	46,370	
1996	Jan 16	Wind	34,223	
	Mar 24	Wind	34,919	
	May 13	Wind	36,030	
	Aug 17	Wind	28,762	
	Oct 7	Wind	41,860	
	Nov 13	Wind	45,000	
	Dec 6	Wind	44,049	
1997	Jan 4	Wind	–	
	Jun 8	Wind	27,962	
	Jul 29	Wind	44,194	
	Sep 5	Wind	45,731	
	Oct 25	Wind	36,260	
1998	Nov 11	Star 37	–	Lunar Prospector launcher
	May 13	HGS-1	6,248	
	Jun 6	HGS-1	34,300	
	Jun 29	Wind	27,137	
	Aug 19	Wind	33,305	
	Sep 24	Nozomi	5,000	
	Oct 23	Wind	22,183	
	Nov 17	Wind	15,054	
	Dec 18	Nozomi	2,809	
1999	Apr 1	Wind	2,354	
	Apr 15	Wind	20,927	
	May 12	Wind	3,554	
	Aug 15	Wind	29,365	
	Oct 4	Wind	25,777	
	Dec 12	Wind	68,492	

Date		Name	Periastron (km)	Notes
2000	Jan 20	Wind	6,736	
	Feb 2	Wind	3,697	
	Apr 7	Wind	26,973	
	May 29	Wind	41,011	
	Aug 19	Wind	7,600	
2001	Jul 31	MAP	5,200	
	Sep 14	Wind	16,477	
	Dec 5	Wind	13,121	
2002	Jun 15	J002E3	161,000	Apollo 12 launcher?
	Jul 18	Wind	12,383	
	Nov 3	Wind	80,834	
	Nov 30	Wind	29,868	

C11 DEEP-SPACE PROBES WHICH OBSERVED THE MOON

Date		Name	Distance (km)
1973	Nov 3	Mariner 10	110,000
1977	Sep 18	Voyager 1	11,660,000
1990	Dec 8	Galileo	350,000
1992	Dec 8	Galileo	110,000
1996	Feb 21	NEAR	2,900,000
1998	Jan 23	NEAR	390,000
1999	Aug 18	Cassini	377,000
2001	Jan 16	Stardust	108,000
	Apr 19	Mars Odyssey	3,500,000
2003	May 5	Mars Global Surveyor	139,000,000
	July 3	Mars Express	8,000,000

C12 SPACE COUNTRIES

Date		Country	Launcher	Satellite
1957	Oct 4	Soviet Union	8K71PS	Sputnik
1958	Feb 1	United States	Juno-I	Explorer 1
1965	Nov 26	France	Diamant A	Asterix
1970	Feb 11	Japan	Lambda 4S	Oshumi
	Apr 24	China	Chang Zheng 1	Dong Fang Hong
1971	Oct 28	United Kingdom	Black Arrow	Prospero
1979	Dec 24	(Europe)	Ariane 1	CAT 01
1980	Jul 18	India	SLV	Rohini RS-1
1988	Sep 19	Israel	Shavit	Offeq-1

Appendix D

Bibliography

* Available on-line at www.hq.nasa.gov/office/pao/History/on-line.html
† Available on-line at NASA Astrophysics Data System (or European mirror at cdsads.u-strasbg.fr)

CHAPTER 1

Bajcár, R., Vis Novcová, F., Photographs of the Artificial Comet Launched by the Cosmic Rocket, *Bulletin of the Czechoslovak Astronomical Institute*, **11**, 118 (in Russian).†

Barkhouse, L., *et al.*, Operation Dominic I 1962, Washington, Defense Nuclear Agency.

Bay, Z., Reflection of Microwaves from the Moon, *Hungarica Acta Physica*, **1**, 1947, 1–22.

Bille, M., Lishock, E., NOTSNIK, The Secret Satellite, AIAA Paper 2002-0314.

Butrica, A. J. (*editor*), Beyond the Ionosphere: Fifty Years of Satellite Communication, Washington, NASA.*

Capaccioli, M., Stendardo, E., Ernesto Capocci, un astronomo nel crepuscolo dei Borbone, *l'Astronomia*, March 2002, 44–51 (in Italian).

Chikmachev, V. I., Shevchenko, V. V., South Pole–Aitken Basin on the First Images of the Lunar Far-side. Paper presented at the International Jubilee Symposium 'The Scientific Results of Space Research of the Moon', Sternberg State Astronomical Institute, Moscow, 2000.

Ciancone, M. L., Motta Rubagotti, D., Luigi Gussalli – Italian Spaceflight Visionary, Paper IAC-02-IAA.2.P.01.

Clarke, A. C., Extra-Terrestrial Relays, *Wireless World*, October 1945, 305–308.

Curtis, H. D., Voyages to the Moon, PASP, 32, 145–150.†

Davidson, K., *Sagan: A Life*, New York, John Wiley & Sons.

Ducrocq, A., *Le Fabuleux Pari sur la Lune*, Paris, R. Laffont (in French).

Fisher, A. C. Jr, Marden, L., Reaching for the Moon, *National Geographic*, February 1959, 157–171.

Gavaghan, H., *Something New Under the Sun*, New York, Copernicus.

Goddard, R. H., A Method of Reaching Extreme Altitudes, *Smithsonian Miscellaneous Collections*, 71, No. 2, 1919.

Heath, T. L., *Greek Astronomy*, New York, Dover Publications.

Heppenheimer, T. A., Gorin, P., Match Race, *Air & Space Smithsonian*, February/March 1996.

LePage, A., The Great Moon Race, In the Beginning..., *EJASA*, B,No. 10, May 1992.

LePage, A., Recent Soviet Lunar and Planetary Program Revelations, *EJASA*, 4 No. 10, May 1993.

Ley, W., *Rockets: the Future of Travel Beyond the Stratosphere*, New York, The Viking Press.

Link, F., Sur les phénomènes d'impact des projectiles lunaires, *Bulletin of the Astronomical Institute of Czechoslovakia*, 11, 113–118. (in French)[†]

Luczak, W., Poland's Atomic Adventure, *Air International*, July 1996, 18–21.

McLaughlin, W. I., Walter Hohmann's Roads in Space, *Journal of Space Mission Architecture*, No.2, 1–14.

Michaels, J. E.,Wachman, M., Petty, A., Lunik III Trajectory Predictions, in Proceedings of the Sixth Annual Meeting of the American Astronautical Society, 244–251.

Mofenson, J., Radar Echoes from the Moon, *Electronics*, April 1946.

Mudgway, D. J., *Uplink–Downlink: a History of the Deep Space Network, 1957–1997*, Washington, NASA.*

Neugebauer, M., Pioneers of space science: a career in the solar wind, *Journal of Geophysical Research*, 102, No. A12, 26,887–26,894.

Nininger H. H., Thoughts on Exploring the Moon, *Popular Astronomy*, December 1945, 513–514.

Poteat, S. E., *Engineering in the CIA: the Bent of Tau Beta Pi, Fall 1999*, 22–27.

Powell, J. W., Thor–Able and Atlas–Able, *Journal of the British Interplanetary Society*, 37, 219–225.

Rapporto Ufficiale sul volo dello 'Orbitnik' e le foto del volto ignoto della Luna., Milan, Esse (in Italian).

Reiffel, L., Sagan Breached Security by Revealing US Work on a Lunar Bomb Project, *Nature*, 405, 4 May 2000, 13.

Reiffel, L. (*editor*), A Study of Lunar Research Flights, 1, Kirtland AFB, Air Force Special Weapons Center TR-59-39.

Russo, L., *La Rivoluzione Dimenticata*, Milan, Feltrinelli (in Italian). English translation, *The Forgotten Revolution*, Berlin, Springer.

Satterfield, P. H., Akens, D. S., *Army Ordnance Satellite Program*, Huntsville, Redstone Arsenal.

Scott Berg, A., *Lindbergh*, New York, Putnam.

Siddiqi, A. A., First to the Moon, *Journal of the British Interplanetary Society*, 51, 231–238.

Stuhlinger, E., Mesmer, G., *Space Science and Engineering*, New York, McGraw-Hill.

Tsander, F. A., *Problems of Flight by Jet Propulsion*, Jerusalem, Israel Program for Scientific Translations.

Varfolomeyev, T., Soviet Rocketry that Conquered Space, Parts 1–3, *Spaceflight*, **37** (August 1995), 260–263; **38** (February 1996), 49–52; **38** (June 1996),206–208.

Yefimov, V., Kak Bili Poluchyeni Pyerviye Fotografyi Obratnoi Storoni Luni, Novosti Kosmonavtiki, August 2000 (In Russian).

Scientists Wonder if Shot Nears Moon, *New York Times*, 5 November 1957.

First US Lunar Probe Fails After Promising Launch, *Aviation Week*, 25 August 1958, 20–23.

Army Plans Two Moon Shots Next Month, *Aviation Week*, 17 November 1958, 28–30.

Radiation Belt Explored by Army's Pioneer III Probe, *Aviation Week*, 15 December 1958, 28–31.

NASA Reports Data on Pioneer IV Orbit, *Aviation Week*, 16 March 1959, 32.

Photographing the Moon, *Flight*, 6 November 1959, 493.

Attempt to Launch Lunar-Orbiting Payload Fails, *Aviation Week*, 7 December 1959, 52–53.

Atlas–Able IV Instrumentation Detailed, *Aviation Week*, 14 December 1959, 59–63.

Pioneer VI Designed for Moon Orbit, *Aviation Week*, 12 September 1960, 56–59.

'50s Plan Weighed Nuclear Moon Blast, *Albuquerque Journal*, 19 August 2000.

CHAPTER 2

Alter, D., A Suspected Partial Obscuration of the Floor of Alphonsus, *PASP*, **69**, No. 407, 158–161.[†]

Behannon, K. W. *et al.*, Observations of the Lunar Environment by Explorer 35, *Astronomical Journal*, **73**, 5.[†]

Burke, J. D., Past US Studies and Developments for Planetary Rovers, in *Missions, Technologies et Conception des Vehicules Mobiles Planetaires*, Toulouse, Cépaduès.

Burnham, D. L., Mobile Explorers and Beasts of Burden: a History of NASA's Prospector and Lunar Logistic Vehicle Projects, *Journal of the British Interplanetary Society*, **48**, 213–228.

Byers, B., *Destination Moon: a History of the Lunar Orbiter Program*, Washington, NASA.*

Cargill Hall, R., *Lunar Impact: a History of Project Ranger*, Washington, NASA.*

Corliss, W. R., *Space Probes and Planetary Exploration*, Princeton, Van Nostrand.

Cortright, E. M., *Exploring Space with a Camera*, Washington, NASA.*

Doel, R. E., The Lunar Volcanism Controversy, *Sky & Telescope*, October 1996, 26–30.

Gatland, K. W., *Spacecraft and Boosters*, London, Iliffe books.

Gillis, J. J., *et al.*, Digitized Lunar Orbiter IV Images, A Preliminary Step to Recording the Global Set of Lunar Orbiter Images in Browker and Hughes' paper presented at the Lunar and Planetary Science Conference XXX, Houston, March 1999.[†]

Hage, G. H., Boyer, W. J., Lunar Orbiter I: an Obedient Robot. Paper presented at the Seventeenth International Astronautical Congress, Madrid, 1966.

Howard, D., *Astronautics Year 1965*, London, Pergamon.

Johnson, R. W., The Lunar Surface According to Luna IX and Surveyor I, *Astronautica Acta*, **12**, 370–383.

LePage, A., The Great Moon Race: The Commitment, *EJASA*, **4** No. 1 (August 1992); The Long Road to Success, **4** No. 2 (September 1992); The Red Moon, **4** No. 12 (July 1993); The Tide Turns, **5** No. 1(August 1993); The Final Lap, **5** No. 4, November 1993.

Liepack, O., The Ranger Project, IAF Paper IAA-98-IAA.2.3.08.

Parks, R. J., Surveyor I Design and Performance. Paper presented at the Seventeenth International Astronautical Congress, Madrid, 1966.

Robotto, A., *Missilistica e Astronautica*, Turin, UTET (in Italian).

Scherer, L. R., Nelson, C. H., The Preliminary Results from Lunar Orbiter I. Paper presented at the Seventeenth International Astronautical Congress, Madrid, 1966.

Sheehan, W., Dobbins, T., The TLP Myth: a Brief for the Prosecution, *Sky & Telescope*, September 1999, 118–123.

Shoemaker, E. M., Batson, R. M., Larson, K. B., An Appreciation of the Luna 9 Pictures, *Astronautics & Aeronautics*, May 1966, 40–50.

Siddiqi, A., Hendrickx, B., Varfolomeyev, T., The Tough Road Travelled: a New Look at the Second Generation Luna Probes, *Journal of the British Interplanetary Society*, **53**, 319–356.

Varfolomeyev, T., Soviet Rocketry that Conquered Space, Parts 4–6, *Spaceflight*, **40** (January 1998), 28–30; **40** (March 1998) 85–88; **40** (May 1998), 181–184.

Vitkus, G., Lucas, J. W., Saari, J. M., Lunar Surface Thermal Characteristics during Eclipse from Surveyors III, V and after Sunset from Surveyor V, AIAA Paper 68-747.

Wilson, K. T., Rangers 3–5, America's First Lunar Landing Attempt, *Journal of the British Interplanetary Society*, 36, 265–274.

Wolff, E. A., *Spacecraft Technology*, Washington, Spartan Books.

NASA Kills Vega, Adopts USAF Agena, *Aviation Week*, 21 December 1959, 18–19.

Vega Study Shows Early NASA Problems, *Aviation Week*, 27 June 1960, 62–68.

Hughes Unveils Lunar Landing Mockup, *Aviation Week*, 20 February 1961, 70–71.

European 10-year Space Program Proposed, *Aviation Week*, 3 July 1961, 30–31.

Roving Lunar Surface Vehicles Studied, *Aviation Week & Space Technology*, 2 October 1961, 52–69.

Reduction of Ranger Components Studied, *Aviation Week & Space Technology*, 16 October 1961, 91–97.

Europe's Space Programme, *Flight*, 2 November 1961, 678–680.

Lunik 4 Believed to Have Failed in Mission, *Aviation Week & Space Technology*, 14 April 1963, 38.

Lunar IMP Experiments to Be Selected, *Aviation Week & Space Technology*, 18 May 1964, 71–74.

Lunar Orbiter Nears Final Configuration, *Aviation Week & Space Technology*, 1 February 1965, 42–49.

Luna 5 Hits Moon, *Flight International*, 10 May 1965, 806.

Engine Firing Moves Surveyor Instrument, *Aviation Week & Space Technology*, 2 October 1967, 18–19.

Soil Test Indicates Clear Vision for Astronauts in Lunar Landing, *Aviation Week & Space Technology*, 27 November 1967, 23.

Surveyor Receives Earth Laser Beams, *Aviation Week & Space Technology*, 29 January 1968, 27.

Explorer 35 Gathers New Data on Moon, *Aviation Week & Space Technology*, 15 April 1968, 30.

CHAPTER 3

The Apollo Spacecraft Chronology, Washington, NASA.*

Benson, C. D., Faherty, W. B., Moonport , *A History of Apollo Launch Facilities and Operations*, Washington, NASA.*

Beschloss, M. R., *The Crisis Years*, New York, Harper & Row.

Bourque, J., Shooting the Moon, *Air & Space*, April/May 2002, 54–61.

Brooks, C. G., Grimwood, J. M., Swenson, L. S. Jr, *Chariots for Apollo*, Washington, NASA.*

Chaikin, A., *A Man on the Moon*, London, Penguin.

Compton, W. D., *Where No Man Has Gone Before*, Washington, NASA.*

Day, D. A., Mapping the Dark Side of the World, Part 2, *Spaceflight*, **40**, August 1998, 303–310.

Farquhar, R. W., Lunar Communications with Libration-Point Satellites, *Journal of Spacecraft and Rockets*, **4**, 1383–1384.

Gatland, K. W., *Manned Spacecraft*, New York, Macmillan.

Gurzadyan, G. A., *Theory of Interplanetary Flights*, Amsterdam, Gordon and Breach.

Harland, D. M. *Exploring the Moon – The Apollo Expeditions*, Chichester, Springer–Praxis.

Lovell, J. and Kluger, J., *Lost Moon: the Perilous Voyage of Apollo 13*, Boston, Houghton Mifflin.

Mason, B., The Lunar Rocks, *Scientific American*, **225**, No. 4, 48–62.

Masursky, H., Colton, G. W., El-Baz, F. (*editors*), *Apollo Over the Moon*, Washington, NASA.*

Mintz Testa, B., A Swirl of Moondust, *Astronomy*, October 1994, 28–35.

Orloff, R. W., *Apollo by the Numbers*, Washington, NASA.*

Pearcy, A., *Flying the Frontier*, Shrewsbury, Airlife.

Runcorn, S. K., The Moon's Ancient Magnetism, *Scientific American*, **257**, No. 6, 60–68.

Stern, A., Where the Lunar Winds Blow Free, *Astronomy*, November 1993, 36–41.

Sutton, G. P., *Rocket Propulsion Elements*, New York, John Wiley.

Taylor Jeffrey, G., The Scientific Legacy of Apollo, *Scientific American*, **271**, No. 1, 26–33.

Taylor, M. J. H. (editor), *Jane's Encyclopedia of Aviation*, New York, Portland House.

Withers, P., Meteor Storm Evidence Against the Recent Formation of Lunar Crater Giordano Bruno. Paper presented at the Lunar and Planetary Science Conference XXXII, Houston, March 2001.[†]

Wood, J. A., The Moon, *Scientific American*, **233**, No. 3, 92–105.

Young, H., Silcock, B., Dunn, P., *Journey to Tranquillity: the History of Man's Assault on the Moon*, London, Cape.

Kennedy's Space Boomerang, *Aviation Week & Space Technology*, 20 September 1963, 21.

A Deep-Space UFO Mystery Solved, *Sky & Telescope*, December 2002, 26.

CHAPTER 4

Glembozky, J. L., Grozdova, T.J., Parfenov, G. P., Experiments with Drosophila on Board the Spacecraft Zond 5. In *Life Sciences and Space Research VII*, North-Holland publishing.

Hendrickx, B., The Kamanin Diaries 1964–1971, *Journal of the British Interplanetary Society*, **51**, 413–440; **53**, 384–428; **55**, 312–360.

Lebedev, D. A., The N1-L3 Programme, *Spaceflight*, 34, September 1992, 288–290.

LePage, A., The Great Moon Race, the Soviet Story Part 1, *EJASA*, **2**, No. 5 (December 1990); the Soviet Story Part 2, **2**, No. 6 (January 1991); The Finish Line, **5**, No. 12, July 1994.

Mishin, V., *Pourquoi Nous ne sommes pas allés sur la Lune*, Toulouse, Cépaduès (in French).

Pirard, T., The Cosmonauts Missed the Moon!, *Spaceflight*, **35**, December 1993, 410–414.

Van Den Abeelen, L., The Persistent Dream – Soviet Plans for Manned Lunar Missions, *Journal of the British Interplanetary Society*, **52**, 123–126.

Varfolomeyev, T., Soviet Rocketry that Conquered Space, Parts 9/11, *Spaceflight*, **41**, May 1999, 207–210; **42** (October 2000), 432–436.

Vick, C. P., CIA/CIO Declassifies N1-L3 Details, *Spaceflight*, **38**, January 1998, 28–29.

Soviets Admit Zond 6 Manned Capability, *Aviation Week & Space Technology*, 8 December 1968, 18–19.

Zond Biological Program Completed, *Aviation Week & Space Technology*, 6 September 1971, 21.

Russian Space Disaster Revealed, *Flight International*, 20 March 1995, 28.

CHAPTER 5

Alexander, J. K., Stone, R. G., A Satellite System for Radio Astronomical Measurements at Low Frequencies. Proceedings of the International Astronomical Union Symposium no.23, Astronomical Observations from Space Vehicles, Liege, August 1964.[†]

Alexander, J. K., *et al.*, Scientific Instrumentation of the Radio-Astronomy-Explorer-2 Satellite, *Astronomy & Astrophysics*, **40**, 365–371.[†]

Bogatchev, A., *et al.*, Wheel Propulsive Devices for Mobile Robots, Designs and Characteristics. Paper presented at the 1999 Field and Service Robotics Conference, Pittsburgh, August 1999.

Burnham, D. L., Back to the Moon with Robots?, *Spaceflight*, **35**, February 1993, 54–57.

Cheng, A. F., Lunar Scout Two Spacecraft Gravity Experiment. Paper presented at Lunar and Planetary Science Conference XXIV, Houston, March 1993.[†]

Farquhar, R. W., The Use of Earth-Return Trajectories for Missions to Comets, *Acta Astronautica*, **44**, 607–623.

Florensky, C. P., *et al.*, The Floor of Crater Le Monier: a Study of Lunokhod 2 Data.[†]

Kemurdjian, A. L. *et al.*, Soviet Developments of Planet Rovers in Period of 1964–1990. In *Missions, Technologies et Conception des Vehicules Mobiles Planetaires*, Toulouse, Cépaduès.

Kemurdjian, A. L., From the Moon Rover to the Mars Rover, *Planetary Report*, **10**, No. 4, July/August 1990, 4–11.

Kupriyanov, V., Perviye Shagi 'Lunokhoda', *Novosti Kosmonavtiki*, November 2000 (in Russian).

Lantratov, K., 25-Let Lunokhodu-1, *Novosti Kosmonavtiki*, No. 23, 79–83; No. 24, 70–79 (in Russian).

Meyer, C., Opportunity for Early Science Return by the Artemis Program. Paper presented at the Lunar and Planetary Science Conference XXIV, Houston, March 1993.[†]

Morrison, D. A., Hoffman, S. J., Lunar Science Strategy, Exploring the Moon with Humans and Machines. In *Missions, Technologies et Conception des Vehicules Mobiles Planetaires*, Toulouse, Cépaduès.

Nein, M. E., Hilchey, J. D., The Lunar Ultraviolet Telescope Experiment (LUTE), Enabling Technology for an Early Lunar Surface Payload, *Journal of the British Interplanetary Society*, **48**, 93–98.

Novaco, J. C., Brown, L. W., Nonthermal Galactic Emission Below 10 MegaHertz, *Astronomical Journal*, **221**, 114–123.[†]

Oberg, J., Soviet Lunar Exploration, Past and Future. In *Lunar Bases and Space Activities in the 21st Century*, Houston, LPI.[†]

Perminov, V. G., *The Difficult Road to Mars*, Washington, NASA.

Portree, D. S. F., *Humans to Mars: Fifty Years of Mission Planning, 1950–2000*, Washington, NASA.

Shevchenko, V. V., Clementine and Lunokhod 2, Iron Content in the Lunar Crater

Le Monnier. Paper presented at Lunar and Planetary Science Conference XXX, Houston, March 1999.[†]

Soviet Space Events in 1972, C.I.A. Scientific and Technical Intelligence.

Specimens of Space Technology, Earth-Based Demonstrators of Planetary Rovers, Running Mock-Ups 1963–1990, Saint Petersburg, Joint-Stock Company Russian Mobile Vehicles Engineering Institute.

Sturms, F. M. Jr., Lunar Geoscience Observer Mission Overview. Paper presented at Lunar and Planetary Science Conference XVIII, Houston, March 1987.[†]

Surkov, Yu. A., Scientific Instrument Making in Space Exploration, *Uspekhi-Fizicheskikh Nauk*, **170**, No. 9, 946–947.

Vostrikov, D., 'Lunokhod' – 30 Let, *Novosti Kosmonavtiki*, November 2000 (in Russian).

Soviet Luna 20 Spacecraft Used Improved Core Drilling Apparatus, *Aviation Week & Space Technology*, 6 March 1972, 15.

Luna 20 Sample Return System Detailed, *Aviation Week & Space Technology*, 20 March 1972, 20–21.

Countdown to co-operation, *Flight International*, December 5, 1987, 30–33.

Bush Sets US Sights on Moon, *Flight International*, 29 July 1989, 6.

High Level Support for US Lunar Base, *Flight International*, 29 July 1989, 16.

US Plans for Moon and Mars, *Flight International*, 19 August 1989, 28–29.

NASA Outlines Manned Mars Mission Plans, *Flight International*, 29 November 1989, 18.

NASA Moon and Mars Mission Plans Criticized by NRC, *Flight International*, 14 March 14 1990, 26.

CHAPTER 6

Andotz, F. J., *Lunar Prospector Mission Handbook*, Sunnyvale, Lockheed Martin Missiles & Space Co.

Anselmo, L., A Feasibility Assessment of Small Italian Lunar Missions, *Journal of the British Interplanetary Society*, **53**, 259–265.

Barbieri, C., *et al.*, LUNAM2000 (Lunar Atmospheric Mission). Paper presented at the EGS XXVI General Assembly, Nice, March 2001.

Berezhnoi, A. A., *et al.*, Radio Emission of the Moon before and after the Lunar Prospector Impact. ArXiv Pre-print astro-ph/0202039.

Brown, R. H., *et al.*, The VIMS/Cassini Observation of the Moon. Paper presented at the Lunar and Planetary Science Conference XXXI, Houston, March 2000.[†]

Burnham, D. L., Hiten Mission Ends, *Spaceflight*, **35**, August 1993, 278–279.

Burnham, R., The Moon-Miner's Daughter, *Astronomy*, February 1994, 34–39.

Chicarro, A., Racca, G., Coradini, M., MORO: a European Moon-Orbiting Observatory for Global Lunar Characterisation, *ESA Journal*, **18**, No. 3, 183–195.

Dunham, D. W., *et al.*, The First Confirmed Videorecordings of Lunar Meteor

Impacts. Paper presented at the Lunar and Planetary Science Conference XXXI, Houston, March 2000.[†]

Feldman, W. C., *et al.*, Fluxes of Fast and Epithermal Neutrons from Lunar Prospector: Evidence for Water Ice at the Lunar Poles, *Science*, **281**, 1496–1500.

Feldman, W. C., *et al.*, Major Compositional Units of the Moon, Lunar Prospector Thermal and Fast Neutrons, *Science*, **281**, 1489–1493.

Foust, J. A., NASA's New Moon, *Sky & Telescope*, September 1998, 48–52.

Goldman, S. J., Clementine Maps the Moon, *Sky & Telescope*, August 1994, 20–24.

Goldstein, D. B., *et al.*, Impacting Lunar Prospector in a Cold Trap to Detect Water Ice, *Geophysical Research Letters*, 26, No. 12 , 1653–1656.

Harland, D. M., *Jupiter Odyssey: the Story of NASA's Galileo Mission*, Chichester, Springer–Praxis.

Iannotta, B.., Slingshot for a Stranded Satellite, *Aerospace America*, September 1998, 38–41.

Kassing, D., Novara, M., LEDA: a First Step in ESA's Lunar Exploration Initiative, *ESA Bulletin*, No. 82, 16–26.

Kawaguchi, J. *et al.*, On the Operation of Lunar and Interplanetary Spacecraft at ISAS, *Acta Astronautica*, **37**, 141–151.

Kelly Beatty, J., The Long Road to Jupiter, *Sky & Telescope*, April 1993, 18–21.

Kelly Beatty, J., New Measures of the Moon, *Sky & Telescope*, July 1995, 32–33.

Kirsch, E., *et al.*, Comparison of Lunar and Terrestrial Ion Measurements Obtained by the Wind and Geotail S/C Outside and Inside of the Earth's Magnetosphere, *Advances in Space Research*, **20**, 4–5, 845–849.

Konopliv, A. S., *et al.*, Improved Gravity Field of the Moon from Lunar Prospector, *Science*, **281**, 1476–1480.

Lawrence, D. J., *et al.*, Global Elemental Maps of the Moon: the Lunar Prospector Gamma-Ray Spectrometer, *Science*, **281**, 1484–1489.

Lawson, S. L., *et al.*, Preliminary Results from the Lunar Prospector Alpha Particle Spectrometer. Paper presented at the Lunar and Planetary Science Conference XXXII, Houston, March 2001.[†]

Lin, R. P., *et al.*, Lunar Surface Magnetic Fields and their Interactions with the Solar Wind, Results from Lunar Prospector, *Science*, **281**, 1480–1484.

Middour, J. W., *et al.*, Clementine Trajectory Analysis. In *Mécanique Spatiale [Spaceflight Dynamics]*, Toulouse, Cépaduès.

Nozette, S., Shoemaker, E. M., Clementine Goes Exploring, *Sky & Telescope*, April 1994, 38–39.

Nozette, S., *et al.*, The Clementine Bistatic Radar Experiment, *Science*, **274**, 1495–1500.

O'Neil, W. J., Galileo Completing VEEGA: a Mid-Term Report, IAF Paper 92-0560.

O'Neil, W. J., Project Galileo Mission Status, IAF Paper 91-468.

Oglivie, K. W., Desch, M. D., The Wind Spacecraft and its Early Scientific Results, *Advances in Space Research*, **20**, 4–5, 559–568.

Oglivie, K. W., *et al.*, Observations of the Lunar Plasma Wake from the Wind Spacecraft on December 27, 1994, *Geophysics Research Letters*, **23** , No. 10, 1255–1258.

Okels, W., *EuroMoon 2000*, Noordwijk, ESA.

Racca, G. D., *et al.*, Smallsat Version of the European Moon Orbiting Observatory (MORO), *Acta Astronautica*, **39**, 121–131.

Regeon, P. A., *et al.*, The Clementine Lunar Orbiter. Paper presented at the 20th Gifu ISTS and 11th IAS, No. 96-e-40.

Roberts, C., Richon, K, Newman, L., Lunar Orbit Design and Orbit Maneuver Computation for the Clementine Mission. In *Mécanique Spatiale [Spaceflight Dynamics]*, Toulouse, Cépaduès.

Robertson, D. F., To Boldly Go..., *Astronomy*, December 1994, 34–41.

Rodriguez, P., *et al.*, Lunar Radar Cross Section at Low Frequency. Paper presented at the Workshop on Moon Beyond 2002, Taos, September 2002.

Scott Hubbard, G., *et al.*, Lunar Prospector, First Results and Lessons Learned, *Acta Astronautica*, **50**, No. 1, 39–47.

Sharer, P., Franz, H., Folta, D., Wind Trajectory Design and Control. In *Mécanique Spatiale [Spaceflight Dynamics]*, Toulouse, Cépaduès.

Shiomi, K., *et al.*, Observations of the Moon's Albedo with the Extreme Ultraviolet Scanner on the Mars Orbiter Nozomi. Paper presented at the Lunar and Planetary Science Conference XXXII, Houston, March 2001.[†]

Spudis, P. D., *The Once and Future Moon*, Washington, Smithsonian Institution Press.

Stacy, S. J. S., Campbell, D. B., Ford, P. G., Arecibo Radar Mapping of the Lunar Poles: a Search for Ice Deposits, *Science*, **276**, 1527–1530.

Uesugi, K., *et al.*, An Adaptive Orbit Design for Multiple Lunar Swingby Missions. In *Mécanique Spatiale [Spaceflight Dynamics]*, Toulouse, Cépaduès.

Uesugi, K., *et al.*, Japanese First Double Lunar Swingby Mission 'Hiten', *Acta Astronautica*, **25**, 347–355.

US Proposes Lunar Mission With Soviets, *Aviation Week & Space Technology*, 28 October 1985, 28.

Call for Lunar Water Probe, *Flight International*, 3 July 1989, 20.

Japan Set for Lunar Orbiter, *Flight International*, 20 December 1989, 23.

Japanese Satellite in Lunar Orbit, *Flight International*, 28 March 1990, 13.

Daring Clementine, *Flight International*, 20 October 1993, 49–50.

Planetary Pathfinder, *Flight International*, 11 May 1994, 37–38.

Computer Error Foils Clementine Mission, *Flight International*, 18 May 1994, 23.

Icy Irony, *Flight International*, 15 January 1997, 42.

AsiaSat 3 Drifts in Space after Failure of Proton Upper Stage, *Flight International*, 7 January 1998, 18.

Ice Find Lead to Renewed Interest in Lunar Probe, *Flight International*, 18 March 1998, 25.

Hughes Attempts Lunar Fly-by Rescue Plan for AsiaSat, *Flight International*, 6 May 1998, 5.

Lunar Assist Eyed for Satcom, *Aviation Week & Space Technology*, 28 August 2000, 49.

International Astronomical Unit *Circular* No. 7589, 25 February 2001.

CHAPTER 7

Ball, A. J., Zarnecki, J. C., Re-use of BepiColombo Technology for a Lunar Landing around 2010? Paper presented at New Views of the Moon, Europe.

Bhandari, N., Joseph, G., Agrawal, P. C., High Resolution Chemical Mapping of the Lunar Surface Using a Lunar Polar Orbiter. Paper presented at New Views of the Moon, Europe.

Boyle, A., Rocks to Riches, *New Scientist*, 16 February 2002, 28–32.

Burns, J. O., *et al.*, Observatories on the Moon, *Scientific American*, **262**, No. 3, 42–49.

Coué, P., *Cosmonautes de Chine*, Paris, L'Harmattan (in French).

Coué, P., Le Programme Lunaire Chinois, *Espace Magazine*, No. 1, pp. 22–25 (in French).

Douglas, J. N., Smith, H. J., A Very Low Frequency Radio Astronomy Observatory on the Moon. In Filippenko, A. V. (*editor*), *Robotic Telescopes in the 1990s*, ASP Conference Series, **34**.[†]

Drean, R. J., *et al.*, Engineering Design of an Unmanned Lunar Radio Observatory. In Filippenko, A. V. (*editor*), *Robotic Telescopes in the 1990s*, ASP Conference Series, **34**.[†]

Duke, M. B., Challenges for Sample Return from the Lunar South Pole–Aitken Basin. Paper presented at the Lunar and Planetary Science Conference XXXIV, Houston, March 2003.

Edery, A., Schiff, C., The Double Lunar Swingby of the MMS Mission. Paper presented at the Sixteenth International Symposium on Space Flight Dynamics, Pasadena, December 2001.

Friedman, L., *Starsailing, Solar Sails and Interstellar Travel*, New York, John Wiley & Sons.

Galimov, E. M., *et al.*, The Russian Lunar Exploration Project, *Solar System Research*, **33**, 327–337.

Harvey, B., *The Chinese Space Programme*, Chichester, Wiley–Praxis.

Harvey, B., Chinese Space Review, Recent Developments, 1998–2000, *Journal of the British Interplanetary Society* 54, 119–126.

Harvey, B., Project 863, *Journal of the British Interplanetary Society* 55, 222–225.

Iwata, T., *et al.*, Global Lunar Gravity Mapping Using SELENE Sub-Satellites. Paper presented at the Workshop on Moon Beyond 2002, Taos, September 2002.

Iwata, T., *et al.*, System Design of SELENE Relay Satellite for Selenodesy. Paper presented at the Lunar and Planetary Science Conference XXXI, Houston, March 2000.[†]

Kaneko, Y, *et al.*, The SELENE Project and the following Lunar Mission, *Acta Astronautica*, **47**, 467–473.

Kruep, J. M., Blase, W. P., Olliver, V., A Proposed Microlander for Low-Cost Lunar Missions. Paper presented at the Lunar and Planetary Science Conference XXX, Houston, March 1999.[†]

Landecker, P. B., *et al.*, Telerobotically Deployed Lunar Farside VLF Observatory. In Filippenko, A. V. (*editor*), *Robotic Telescopes in the 1990s*, ASP Conference Series, **34**.[†]

Lawler, A., The New Race to the Moon, *Science*, 300, 724–727.

Leipold, M., Kassing, D., Eiden, M., Herbeck L., Solar Sails for Space Exploration: the Development and Demonstration of Critical Technologies in Partnership, *ESA Bulletin*, No. 98, 102–107.

Li Teng, Yang Wei, LunarNet: an Innovative Project for Lunar Mission. In Proceedings of the Eight International Space Conference of Pacific-Basin Societies, 698–701.

Li Teng, Xu Min, Yang Wei, Global Scheme of Lunar–Earth Information Network. In Proceeding of the First International Conference on Astronautics and Aeronautics, 477–480.

LM-2C User's Manual, Beijing, CALT (ed. 1999).

Lucey, P. G., Polar Night: a Mission to the Lunar Poles. Paper presented at the Workshop on Moon Beyond 2002, Taos, September 2002.

Malakhov, Yu. I, ISTC Projects in the Field of Space and Aviation, paper presented at the 6th Workshop on Advanced Technology in Russia/CIS, Tokio, January 1998.

Manifold, J. D., Robotic Sample Return Missions: Steppingstones to the Establishment of a Lunar Infrastructure. Paper presented at the Commercial Lunar Base Development Symposium, League City, July 1999.

Manifold, J. D., Norris, D. A., Commercial Lunar Geological Exploration, A Synergistic Model of Science and Private Enterprise. Paper presented at the Lunar and Planetary Science Conference XXX, Houston, March 1999.[†]

Meyer, C., An Outline of the Jules Verne Discovery Mission to Explore the Lunar Mantle via South Pole–Aitken Basin. Paper presented at the Lunar And Planetary Science Conference XXVI, Houston, March 1995.[†]

Nagae, Y., Takizawa, Y., Sasaki, S., The System Concept of SELENE, *Acta Astronautica*, **45**, 197–205.

Nakajima, T., *et al.*, Lunar Penetrator Program, Lunar-A, *Acta Astronautica*, **39**, 111–119.

Nakajima, T., *et al.*, An Updated Lunar-A Mission Scenario and its Sesmic Penetrator Deveropment [*sic*]. Paper presented at the 16th International Symposium on Space Flight Dynamics, Pasadena, December 2001.

New Frontiers in the Solar System: an Integrated Exploration Strategy, Washington, National Research Council.

Normile, D., Ding Yimin, Science Emerges from Shadows of China's Space Program, *Science*, **296**, 1788–1791.

Peterson, C., Return to the Moon: a Lunar Giant Basin Sample Return Mission. Paper presented at the 4th IAA International Conference on Low Cost Planetary Missions.

Pirard, T., Europe Prepares for Lunar Exploration, *Spaceflight*, **41**, February 1999, 48.

Pirard, T., Return to the Moon, *Spaceflight*, **42**, November 2000, 470–471.

Racca, G., Foing, B., and the SMART-1 Project Team, The SMART-1 Mission: a Solar Powered Visit to the Moon, *ESA Bulletin*, No. 113, 14–27.

Racca, G. D., Whitcomb, G. P., Foing, B. H., The Smart-1 Mission, *ESA Bulletin*, No. 95, 72–81.

Russell, C. T., *et al.*, The Diana Discovery Mission: a Solar Electric Propulsion Mission to the Moon and a Comet. Paper presented at the Lunar and Planetary Science Conference XXVI, Houston, March 1995.[†]

Saccoccia, G., Gonzalez del Amo, J., Estublier, D., Electric Propulsion: a Key Technology for Space Missions in the New Millennium, *ESA Bulletin*, No. 101, 62–71.

Schulze, R., Aurora: a European Roadmap for Solar System Exploration, *On Station*, No. 8, 4–5.

Shkuratov, Yu. G., *et al.*, A Prospective Lunar Orbiter Mission, Objectives and Scientific Payload. Paper presented at New Views of the Moon, Europe.

Simon, M., International Space Enterprises, *Spaceflight*, **36**, July 1994, 237–240.

Sweeting, M. *et al.*, Low Cost Planetary Exploration, Surrey Lunar Minisatellite and Interplanetary Platform Missions, *Acta Astronautica*, **48**, No. 5–12, 669–680.

Tanaka, S., *et al.*, In Situ Lunar Heat Flow Experiment Using the Lunar-A Penetrator. Paper presented at the Lunar and Planetary Science Conference XXXII, Houston, March 2001.[†]

Tanaka, S., *et al.*, The SELENE Project, Mission and Science. Paper presented at New Views of the Moon, Europe.

Tompkins, P., Stroupe, A., IceBreaker: an Exploration of the Lunar South Pole. In Proceedings of the 14th SSI Conference on Space Manufacturing, May 1999.

Whittaker, W., *et al.*, Sun-Sychronous Planetary Exploration. Paper AIAA-2000-5300.

Woan, G., A Very Low Frequency Radio Telescope on the Far Side of the Moon. In *Measuring the Size of Things in the Universe*, HBT Interferometry and Heavy Ion Physics, Proceedings of CRIS98.

Wu Xiaotao, *et al.*, Design and Implementation of a Telerobotic System with Large Time Delay, Proceedings of the International Symposium on Artificial Intelligence, Robotics and Automation in Space (i-SAIRAS'97) 321–324.

Zhu Yilin, Development of Chinese Satellites under Prof. Tsien, *Journal of the British Interplanetary Society*, **52**, 185–188.

1993 launch mooted for international Regatta, *Flight International*, 22 May 1991, 19.

Cosmic Klieg Light, *Sky & Telescope*, June 1993, 13–14.

China Aims for Moon, *Flight International*, 25 January 1995, 24.

Moon Mission to Lead Russian Space Reforms, *Aviation Week & Space Technology*, 19 May 1997, 64–65.

Orbital Aspirations, *Flight International*, 22 April 1998, 60.

Mechanical Fault Delays Japan's Moon Probe Launch, *Flight International,* 30 June 1999, 27.

Sweden Wins Contract to Develop Smart for ESA, *Flight International*, 15 December 1999, 29.

NASA Reveals Future-X 'Pathfinder' Studies, *Flight International*, 27 June 27 2000, 26.

Return to the Moon, *Spaceflight*, **42**, February 2000, 55–57.

US Attacks SSTL Technology Sale, *Flight International*, 23 January 2001, 28.

Indian Research Body Seeks Go-Ahead for Moon Mission, *Flight International*, 18 September 2001, 39.

Lunar Allure, *Flight International*, 19 February 2002, 56–57.

India Fires up New GSLV Powerplant, *Flight International*, 24 September 2002, 29.

Kosmotras to Launch Commercial Moon Flights, *Flight International*, 10 December 2002, 26.

GENERAL

Ball, A. J., Lorenz, R. D., Penetrometry of Extraterrestrial Surfaces: an Historical Overview. Paper presented at International Workshop on Penetrometry in the Solar System, Graz, 18–20 October 1999.

Biesbroek, R., Janin, G., Ways to the Moon?, *ESA Bulletin*, No. 103, 92–99.

Burrows, W. E., *This New Ocean: the Story of the First Space Age*, New York, The Modern Library.

Cadogan, P., *The Moon , Our Sister Planet*, Cambridge University Press.

Clark, P. S., Launch Profiles Used by the Four-Stage Proton-K, *Journal of the British Interplanetary Society*, 53, 197–214.

Davies, J. K., *Satellite Astronomy*, Chichester, Ellis Horwood.

Day, D. A., Logsdon, J. M., Latell, B., *Eye in the Sky: the Story of the Corona Spy Satellites*, Washington, Smithsonian Institution Press.

Grichkov, S., *Guide des Lanceurs Spatiaux*, Guilford, Tessier & Ashpool (in French).

Hall, R., Shayler, D. J., *The Rocket Men*, Chichester, Springer–Praxis.

Harvey, B., *The New Russian Space Programme*, Chichester, Wiley–Praxis.

Harford, J., *Korolev*, New York, John Wiley & Sons.

Isakowitz, S. J., Hopkins, J. P. Jr, Hopkins, J. B., *International Reference Guide to Space Launch Systems* (third edition, 1999), Reston, AIAA.

Koppes, C. R., *JPL and the American Space Program*, Yale University Press.

Kotenikov, M. A., Savich, N. A., Yakovlev, O. I., Spacecraft Radiophysical Investigations of the Sun and Planets. In Kotelnikov, V. A. (*editor*), *Problems of Modern Radio Engineering and Electronics*, Moscow, Nauka.

Lardier, C., *L'Astronautique Soviétique*, Paris, Armand Colin (in French).

Pedlow, G. W., Welzenbach, D. E., *The CIA and the U-2 Program, 1954–1974*, Washington, CIA.

Rükl, Antonin, *Atlas de la Lune*, Paris, Gründ, (in French).

Semenov, Yu. P. (*editor*), *Rakyetno-Kosmiceskaya Korporatsiya 'Energhiya' imieni S. P. Karaliova 1946–1996*, Moscow, RKK Energiya, (in Russian).

Shayler, D. J., *Disasters and Accidents in Manned Spaceflight*, Chichester, Springer–Praxis.

Sheehan, W. P., Dobbins, T. A., *Epic Moon: a History of Lunar Exploration in the Age of the Telescope*, Richmond, Willmann-Bell.

Siddiqi, A. A., *Challenge to Apollo*, Washington, NASA.

Siddiqi, A. A., *Deep Space Chronicle*, Washington, NASA.*

Space Handbook, *Astronautics and its Applications*, Washington, United States Government Printing Office.*

Surkov, Yu. A., *Exploration of Terrestrial Planets from Spacecraft*, (second edition), Chichester, Wiley–Praxis.

Turnill, R. (*editor*), *Jane's Spaceflight Directory 1984*, London, Jane's Publishing.

Wertz, J. R. (*editor*), *Spacecraft Attitude Determination and Control*, Dordrecht, Reidel.

Wilson, A., *The Eagle has Wings*, London, British Interplanetary Society.

Wilson, A., *Solar System Log*, London, Jane's Publishing.

Wotzlaw, S., Käsmann, F. C. W., Nagel, M., Proton: Development of a Russian Launch Vehicle, *Journal of the British Interplanetary Society*, **51**, 3–18.

MAGAZINES

l'Astronomia (Italian)
Ciel et Espace (French)
Flight International
Scientific American
Sky & Telescope

Aviation Week and Space Technology
ESA Bulletin
Letectví + Kosmonautika (Czech)
le Stelle (Italian)
Spaceflight

INTERNET SITES

Aerospace China	www.space.cetin.net.cn/docs/ht97-b.htm
Babakin Space Centre	russianscientists.com/babakin/babakin.php3
Encyclopedia Astronautica	www.astronautix.com
ESA	www.esa.int
Federation of American Scientists	www.fas.org
Go Taikonaut!	www.geocities.com/CapeCanaveral/Launchpad/1921
Indian Space Page	www.bharat-rakshak.com/SPACE/
Jonathan's Space Report	hea-www.harvard.edu/QEDT/jcm/space
JPL	www.jpl.nasa.gov
Lunacorp	www.lunacorp.com
Lunar Prospector	lunar.arc.nasa.gov/
NASA NSSDC	nssdc.gsfc.nasa.gov
National Air and Space Museum	www.nasm.org
NEAR Mission	sd-www.jhuapl.edu/NEAR/
Novosti Kosmonavtiki	www.novosti-kosmonavtiki.ru
Nozomi	www.planet-b.isas.ac.jp/index-e.html
RSC Energiya	www.energia.ru
Solar Blade Heliogyro	www.frc.ri.cmu.edu/projects/blade/solarblade.html
Space Daily	www.spacedaily.com

Space.Com   www.space.com
Spaceflight Now   www.spaceflightnow.com
Sven's Space Place   www.svengrahn.se
Swedish Space Corp.   www.ssc.se
The FP Space Archives   www.friends-partners.org/pipermail/fpspace
The Planetary Society   planetary.org
Transorbital   www.transorbital.net

SOFTWARE

Lunar phases and other astronomical computations were carried out using Project Pluto's *Guide 5.0* software.

CHINESE

Chen Zong-hai, Lunar Probe Path Planning Using Case-Based Learning Algorithm, *Aeronautical Computer Technique*, **30**, No.2.

Gu Lixiang, Liu Zhusheng, Research on Phasing Loops Earth to Moon Transfer Orbit, *Missiles and Space Vehicles*, No. 3, 2002.

Hao Yingming, *et al.*, Direct Adaptive Control of a Lunar Robot Position with High Precision Using Fuzzy Neural Network, *High Technology Letters*, **11**, No.8, 89–92.

Huang, C., Hu, X. G., Li, X., Lunar-landing trajectory designing under certain constraints, *Acta Astronomica Sinica*, **42**, no.2, 161–172.

Jia Shijin, She Mingshang, The Theory of Changing Satellite Orbit Inclination Using Moon Gravitation, *Chinese Space Science and Technology*, **21**, No. 5.

Li Dong *et al.*, A Tentative Idea about Lunar Exploration, *Missiles and Space Vehicles*, 2002, No. 5, pp. 20–28.

Li Jun, Sun Demin, The Vision System and Autonomous Navigation System for the Lunar Rover, *Aerospace Control*, **18**, No. 2.

Li Jun, Sun Demin Path Planning Using Case-Based Learning and its Neural Network Implementation of the Lunar Vehicle's Self-Autonomous Navigation. Proceedings of the 3rd World Congress on Intelligent Control and Automation, 1182–1186.

Li Litao, *et al.*, Collectivity Scheme Design for Modern Small Lunar Explorer, *High Technology Letters*, **11**, No.5, 2001, 80–84.

Liu Fang-Hu, A Five-Wheeled Lunar Robot and its Characteristics Analysis, *Journal of Machine Design*, **18**, No. 5, 15–18.

Liu Fang-hu, *et al.*, Kinematic Modeling of a Five-Wheel Articulated Lunar Robot, *Robot*, **23**, No.6, 481–485.

Liu Fang-hu, *et al.*, Research Status and Development Trends towards Planetary Exploration Robots, *Robot*, **24**, No.3, 268–275.

Liu Fang-hu, *et al.*, Intelligent Fuzzy Control for the Five-Wheel Articulated Lunar Rover (FWALR), *Journal of Shanghai Jiaotong University*, **36**, No. 3, 297–301.

Liu Lin, Wang Jia-song, An Analytic Solution of the Orbital Variation of Lunar

Satellites, Acta Astronomica Sinica, 39, No.1, 1998. English translation published in, *Chinese Astronomy and Astrophysics*, **22**, No. 3, 1998.

Liu Lin, To Guide a probe to the Moon with Light Pressure, *Acta Astronomica Sinica*, **42**, No.1. English translation published in *Chinese Astronomy and Astrophysics*, **25**, 343–348.

Liu Lin, Liu Yingchun, Precise Orbit Determination for Lunar Satellite, *Chinese Journal of Space Science*, **22**, No. 3, 249–255. English translation published in *Acta Astronautica*, **51**, No. 1–9, 501–506.

Long Lehao, LM-3A Launch Vehicle Series, *Missiles and Space Vehicles*, No.4, 1999.

Long Lehao, National Lunar Launcher Configuration Study, *International Space*, No. 2, 2003.

Luo Xun-ji, Sun Zeng-qi, Research on the Lunar Rover Simulation Platform, *Journal of System Simulation*, **14**, No.9 1235–1238.

Ma Kemao, Chen Lijia, Wang Zicai, Practical Design of Control Law for Flight Vehicle Soft Landing, *Missiles and Space Vehicles*, No.2, 2001.

Ping Jinsong, Y. Kono, N. Kawano, How spin of a stabilized S/C affects two-way Doppler tracking, *Journal of Beijing Normal University (Natural Science)*, **36**, No. 4.

Ping J., *et al.*, SELENE Mission: Mathematical Model for SST Doppler Measurements, *Progress in Astronomy*, 2001.

Ruan Xiaogang, A Nonlinear Neurocontrol Scheme for Lunar Soft Landing, *Journal of Astronautics*, **9**, No.1, 1998.

Su Y., The Prospect of FAST in Deep Space Exploration, *Acta Astronomica Sinica*, **42**, No.1.

Tan Zhengming, *et al.*, Hard Landing Impact of Planet Probe, *Missiles and Space Vehicles*, No.4, 1999.

Wang Dayi, *et al.*, Neuro-Optimal Guidance Law for Lunar Soft Landing, *Systems Engineering and Electronics*, **21**, No.12.

Wang Dayi, Li Tieshou, Ma Xingrui, Numerical Solution of TPBVP in Optimal Lunar Soft Landing, *Aerospace Control*, **18**, No.3.

Wang Dayi, *et al*, Guidance Control for Lunar Gravity-Turn Descent, *Chinese Space Science and Technology*, **20**, No. 5.

Wang Dayi, *et al.*, Explicit Guidance Control for Lunar Soft Landing, *High Technology Letters*, **10**, No. 7.

Wang Dayi, *et al.*, A Sub-Optimal Fuel Guidance Law for Lunar Soft Landing, *Journal of Astronautics*, No. 4, 2000.

Wang Jie, Cui Nai-gang, Liu Dun, Preliminary Study on Minimum-Fuel Lunar Probe Trajectories, *Flight Dynamics*, **18**, No.2.

Wang Jie, Cui Naigang, Liu Dun, Study of Lunar Soft Landing by the Method of Establishment of the Lunar Perpendicular, *Missiles and Space Vehicles*, No. 4, 2000.

Wang Jie, *et al.*, Fuel Consumption Estimation of Limited-Thrust Lunar Probe, *Missiles and Space Vehicles*, No. 6, 2000.

Wang Jie, Cui Naigang, Liu Dun, On Constant-Amplitude Low Thrust Lunar Probe Trajectory, *Acta Aeronautica et Astronautica Sinica*, No. 1, 2001.

Wang Wei, *et al.*, Study on the Critical Technology for a Lunar Rover, *Robot*, **23**, No. 3.

Wang Wei, *et al.*, GPS Navigation of the Lunar Probe in the Close Earth Orbit Phase, *Journal of National University of Defence Technology*, **23**, No. 2.

Wang Wei, *et al.*, Navigation for the Lunar Probe based on Ground Tracking Sites, *Journal of National University of Defence Technology*, **23**, No. 6, 33–37.

Wang Wei, Liang Bin, Simulation Study on a Lunar Robot Locomotion System Based on the Virtual Prototyping Technique, *High Technology Letters*, **12**, No. 2, 84–87.

Xi Xiao-Ning, Zeng Guo-Qiang, Zhu Wen-Yao, Window Selection for the Lunar Probe Launched from the Earth, *Acta Astronomica Sinica*, **41**, No. 4.

Xu Rui, *et al.*, The Active Nutation Control of the Small Lunar Explorer, *High Technology Letters*, **11**, No. 3.

Yan Hui, Wu Hongxin, Lunar Trajectories and Tracking Arcs, *Journal of Astronautics*, **9**, No. 4.

Yu Menglun, Chinese Lunar Probe Launcher Research, *Aerospace China*, No. 11, 2002.

Zeng Guo-Qiang, Xi Xiao-Ning, Ren Xuan, A Study on the Optimal Low-Thrust Orbit Maneuver of Lunar Satellite, *Acta Astronomica Sinica*, **41**, No. 3.

Zeng Guo-Qiang, Xi Xiao-Ning, Ren Xuan, An Algebraic Method for Fast Design of Lunar Satellite Transfer Trajectory, *Journal of the National University of Defence Technology*, **22**, No.2.

Zeng Guoqiang, Xi Xiaonin, Ren Xuan, A Study of Lunar Swing-by Technique, *Journal of Astronautics*, No. 4, 2000.

Zhang Lao, *et al.*, Chinese Deep Space Probe Technology Development and Forecast, *International Space*, No. 2, 2003.

Zhang Wei, *et al.*, Moon Rabbit, A Concept for a Small Lunar Probe, *Aerospace China*, No. 9, 2002.

Zhang Zhen Min, Cui Hu Tao, Yang Di, High-Precision Pointing Control for Small Lunar Explorer, *High Technology Letters*, **12**, No. 4, 80–82.

Zhang Zhen Min, Li Litao, Yang Di, Trajectory Design of Lunar Polar Probe, *Chinese Space Science and Technology*, **22**, No.3.

Zhu Renzhang, Yu Nanjia, Yu Menglun, Studies of Earth–Moon Transfer Trajectory with Gravitational Capture, *Journal of Astronautics*, No. 4, 2000.

Zhuang Jun, Qiu Ping, Sun Zongqi, Distributed Telerobotic System with Large Time Delay, *Journal of Tsinghua University*, **40**, No.1.

Deep space communication and tracking problems, international solutions, present situation and our country's response, *Journal of Flight Vehicle Observation and Control*, **19**, No. 3.

India aims at the Moon, *International Space*, 2001, No. 12, 7–9.

Lunar transfer orbit approximated model and influence estimate of the approximation, *Journal of Flight Vehicle Observation and Control*, **20**, No. 1, 55–62.

Orbital Design of Vertical Hitting Moon Probe, *Chinese Space Science and Technology*, **18**, No. 2.

The Whole World Starts the Return to the Moon, *International Space*, No. 3, 2002, 7–11.

Index